PROPERTY OF
NOVO NORDISK
ENTOTECH, INC.

ADVANCES IN PROTEIN CHEMISTRY

Volume 43

PROPERTY OF
NOVO NORDISK
ENTOTECH, INC.

ADVANCES IN PROTEIN CHEMISTRY

EDITED BY

C. B. ANFINSEN
Department of Biology
The Johns Hopkins University
Baltimore, Maryland

JOHN T. EDSALL
Department of Biochemistry
and Molecular Biology
Harvard University
Cambridge, Massachusetts

FREDERIC M. RICHARDS
Department of Molecular Biophysics
and Biochemistry
Yale University
New Haven, Connecticut

DAVID S. EISENBERG
Department of Chemistry
and Biochemistry
University of California, Los Angeles
Los Angeles, California

VOLUME 43

ACADEMIC PRESS, INC.
Harcourt Brace Jovanovich, Publishers
San Diego   New York   Boston
London   Sydney   Tokyo   Toronto

This book is printed on acid-free paper. ∞

Copyright © 1992 by ACADEMIC PRESS, INC.
All Rights Reserved.
No part of this publication may be reproduced or transmitted in any form or by any means, electronic or mechanical, including photocopy, recording, or any information storage and retrieval system, without permission in writing from the publisher.

Academic Press, Inc.
1250 Sixth Avenue, San Diego, California 92101-4311

United Kingdom Edition published by
Academic Press Limited
24–28 Oval Road, London NW1 7DX

Library of Congress Catalog Number: 44-8853

International Standard Book Number: 0-12-034243-X

PRINTED IN THE UNITED STATES OF AMERICA
92 93 94 95 96 97   MM   9 8 7 6 5 4 3 2 1

CONTENTS

Biochemical, Structural, and Molecular Genetic Aspects of Halophilism

HENRYK EISENBERG, MOSHE MEVARECH, AND GIUSEPPE ZACCAI

I.	Introduction	1
II.	Purification of Halophilic Enzymes	5
III.	Biochemical Aspects	12
IV.	Structure	25
V.	Molecular Genetic Aspects	43
	References	56

Structure and Stability of Bovine Casein Micelles

C. HOLT

I.	Caseins	63
II.	Physicochemical Properties of Caseins	85
III.	Structure of the Native Bovine Casein Micelles	105
IV.	Stability of Casein Micelles	133
	References	143

Proton Nuclear Magnetic Resonance Studies on Hemoglobin: Cooperative Interactions and Partially Ligated Intermediates

CHIEN HO

I.	Introduction	154
II.	Experimental Procedures	166
III.	Resonance Assignments: Subunit Interfaces and Heme Environments	192
IV.	^{1}H NMR Investigations of Ligand Binding to α and β Chains of Hemoglobin	214
V.	^{1}H NMR Investigations of Partially Oxygenated Species of Hemoglobin: Evidence for Nonconcerted Structural Changes during the Oxygenation Process	240

VI.	¹NMR Investigations of Structures and Properties of Symmetric Valency Hybrid Hemoglobins: Models for Doubly Ligated Species	252
VII.	¹H NMR Investigations of Structures and Properties of Asymmetric Valency Hybrid Hemoglobins: Models for Singly and Doubly Ligated Species	261
VIII.	Influence of Salt Bridges on Tertiary and Quarternary Structures of Hemoglobin	273
IX.	Other Evidence for Existence of Ligation Intermediates of Hemoglobin	280
X.	X-Ray Crystallographic Investigations of Structural Characteristics of Partially Ligated Species of Hemoglobin	285
XI.	Possible Pathways for Heme–Heme Communication	291
XII.	Concluding Remarks	296
	References	303

Thermodynamics of Structural Stability and Cooperative Folding Behavior in Proteins

Kenneth P. Murphy and Ernesto Freire

I.	Introduction	313
II.	Folding/Unfolding Partition Function	314
III.	Estimation of Forces Required to Specify Partition Functions	316
IV.	Calculation of Folding/Unfolding Thermodynamics from Protein Structure	335
V.	Thermodynamic Stability and Cooperative Interactions	340
VI.	Cooperativity of Two-Domain Proteins	347
VII.	Cooperativity of Single-Domain Proteins	351
VIII.	Energetics of Molten Globule Intermediates	355
IX.	Concluding Remarks	357
	References	358

AUTHOR INDEX	363
SUBJECT INDEX	381

BIOCHEMICAL, STRUCTURAL, AND MOLECULAR GENETIC ASPECTS OF HALOPHILISM

By HENRYK EISENBERG,* MOSHE MEVARECH,† and GIUSEPPE ZACCAI‡

*Structural Biology Department, Weizmann Institute of Science,
Rehovot 76100, Israel
†Department of Molecular Microbiology and Biotechnology,
George S. Wise Faculty of Life Sciences,
Tel Aviv University, Tel Aviv 69978, Israel
‡Institut Laue Langevin (CNRS URA 1333), 156X, 38042 Grenoble, France

I.	Introduction	1
II.	Purification of Halophilic Enzymes	5
	A. Ammonium Sulfate-Mediated Chromatography	7
	B. Affinity Chromatography	11
	C. Other Methods	11
III.	Biochemical Aspects	12
	A. Some Metabolic Aspects	12
	B. Enzymology of Extremely Halophilic Archaebacteria	14
	C. Effect of Salt Concentration on Structural Integrity of hMDH	15
	D. hDHFR	20
IV.	Structure	25
	A. Ribosomal Subunits	26
	B. Surface Layers	29
	C. Purple Membrane	30
	D. Ferredoxin from *Haloarcula marismortui*	31
	E. Methods for Solution Studies	31
	F. Thermodynamics of Protein–Salt–Water Interactions and Structural Models	35
	G. Halophilic Malate Dehydrogenase	36
	H. Model for Halophilic Protein Stabilization	39
	I. Halophilic EF-Tu and Other Halophilic Proteins under Study	41
	J. Conclusions	42
V.	Molecular Genetic Aspects	43
	A. Genome Organization	43
	B. Genetic Tools	46
	C. Isolation of Genes	49
	D. Transcript Organization and Structure	50
	E. What Can We Learn about Halophilic Enzymes?	53
	References	56

I. INTRODUCTION

The complicated process termed "life" is associated with a very delicate balance in the interactions among the different components of the living cell. Many of the components are macromolecules that have an extremely nonlinear response to external stimuli. Small variations in the physicochemical state of the cell could, therefore, have enormous effects on the life process. The sum total of the physicochemical conditions enabling normal functioning of the cell is termed "physiological conditions." In general, living organisms are adapted to function in a rather limited set of physiological conditions: chemical and ionic composition of the medium, pH, temperature, and pressure. Significant deviations from physiological conditions will lead to disaggregation of complex structures, denaturation of protein and DNA molecules, and consequently to cell death. In this context, it is worthwhile mentioning that although in many cases the physiological conditions as determined *in vitro* resemble the living processes, in many other cases the conditions for stability and effectiveness of biological components and interactions as determined in laboratory experiments do not necessarily apply to *in vivo* conditions for phenomena occurring in the crowded living cell (Minton, 1983; Richey *et al.*, 1987).

It is now believed that on our 4.5-billion-year-old planet, unicellular microorganisms originated about 3.5 billion years ago. Microorganisms dominated life on Earth for a considerable length of time, and higher organisms appeared only much later. In this period of time the earth cooled considerably, and oxygen appeared in the atmosphere about 2 billion years ago. Ancient microorganisms therefore had to adapt to an evolving habitat. They survive to date mostly in niches characterized by unusual environmental circumstances. A striking example of an unusual habitat, close to the topic of this article, is the Dead Sea, which was recognized only in the late 1930s [Wilkansky (Volcani), 1936]. This unique body of water, rich in ancient human history and site of biblical events, has many names, such as the Sea of Asphalt, indicating extensive organic deposits, in Hebrew, it is named the Sea of Salt. It is now known to be a dynamically evolving stratified ecological niche, hosting microbial and algal populations. An interesting aspect of the Dead Sea is that, in contrast to other well-known saltwater bodies, such as the Great Salt Lake in Utah, it is extremely rich in magnesium salts (Nissenbaum, 1975). Early studies of the Halobacteriaceae (cf. Section V,A), microorganisms that are stable and active only in extremely high concentrations of salt, have been reviewed (Larsen, 1986; Kushner, 1985).

Physiological conditions for most living organisms are very similar, but it is well documented that many organisms are adapted to grow under extreme conditions of salt concentration, pH, temperature, and pressure. It is possible to divide these organisms into two groups according to their mode of adaptation. One group of organisms developed various mechanisms to preserve a benign inner environment in extreme surroundings, for instance, bacteria that are capable of surviving in extreme pH environments by activating powerful proton pumps to maintain a close to neutral intracellular pH (Edwards, 1990). In the other group, the entire biochemical machinery is adapted to function in the particular extreme conditions. When adaptation to hypersaline conditions is considered, there are fundamental differences between the extremely halophilic archaebacteria and the other halophilic organisms regarding their mode of adaptation (see recent reviews edited by Rodriguez-Valera, 1988, and references therein). The halophilic eubacteria and eukaryotes accumulate mostly organic neutral compatible solutes and exclude most of the inorganic salts. On the other hand, the halophilic archaebacteria balance the external high salt concentration by accumulating within the cell inorganic ions at concentrations that exceed that of the medium. Therefore, all the cellular components of the halophilic archaebacteria have to be adapted to function at the extremely high intracellular salt concentration. A short explanation regarding the phylogenetic definition of the archaebacteria follows.

The original proposal by C. R. Woese and colleagues that living organisms should be classified into three different kingdoms was based on an extensive comparison of sequences of oligonucleotides derived from rRNA molecules of many organisms (Woese and Fox, 1977). This original phylogenetic distance analysis of oligonucleotides was extended to longer 16S rRNA sequences as modern, rapid DNA sequencing techniques became available, and similar results were obtained. Three unique molecular features, shared by all the members of the archaebacteria, help to distinguish these microorganisms from the eubacteria. (1) Archaebacterial lipids are made up of isopranyl glycerol ethers rather than the fatty acid ester-linked glycerol lipids that predominate in eubacteria and eukaryotes (Langworthy and Pond, 1986). (2) The DNA-dependent RNA polymerases of all archaebacteria are more complex than their eubacterial counterparts and their structure and sequence resemble more that of RNA polymerase II of eukaryotes (Zillig et al., 1988). (3) Archaebacteria lack the typical eubacterial peptidoglycan cell wall and instead have an S layer composed mainly of glycoproteins (Koenig and Stetter, 1986, and references therein).

Basic interest in the study of halobacteria thus relates also to a better understanding of evolutionary relationships extending to the dawn of life in a world quite unlike our present-day environment. If we consider the current processes of increasing atmospheric CO_2

levels, increasing salinity, and cooling or heating of the seas, and also the possibility of extended human journeys into space, then we must accept that adaptation of life to changing environments constitutes an important riddle deserving close examination. To understand adaptation it is of course essential to understand both the physiological and physical bases of life on a broad level, as well as the physicochemical characteristics of interactions of cell components—proteins, nucleic acids, sugars, lipids, and so forth—at a molecular and cellular level. Site-specific mutagenesis coupled to sequence analysis and macromolecular structure is becoming a powerful tool to modify the structure and properties of macromolecular components along a path of adaptive change. An important bonus in the study of modified and adapted systems is a better understanding of factors important in the function and regulation of components and systems comprising the bulk of organisms in our present-day environment.

A practical consequence of the study of proteins from organisms adapted to extreme environments consists in their utilization in biotechnological applications. Experimental dimensions are added that may make processes feasible, cheaper, or more reliable in high salt concentration or at a higher temperature, for instance. These developments are still at an early stage and will receive much stimulation with the development of suitable vectors and cloning systems in extreme environments. On a more practical level, commercial production of β-carotene and glycerol has been achieved from *Dunaliella* algae grown in open-air saline ponds (Ben-Amotz and Avron, 1990). A recent unusually successful achievement is the use of the heat-resistant *Taq* polymerase from *Thermus aquaticus* in the polymerase chain reaction (PCR), which has revolutionized processes based on the quick multiplication of minute amounts of DNA (Mullis and Faloona, 1987).

Work on extremely halophilic archaebacteria in our and other laboratories, to be described in the following review, went through a number of stages. Early work on these microorganisms was mainly aimed at developing enrichment procedures and physiological studies. Most of the biochemical studies were performed on impure enzymatic preparations. Even when some enzymes could be purified to homogeneity, the yields were too low to enable physical characterization of the proteins. The development of efficient purification protocols by which large amounts of halobacterial proteins were fractionated at high salt concentration, thus avoiding losses due to inactivation, enabled detailed biochemical and biophysical characterization of several enzymes. Studies in the ultracentrifuge and application of thermodynamic considerations developed for multicomponent systems led to interesting results relating to salt and water binding (Pundak and Eisenberg, 1981; Pundak *et al.*, 1981). These observations were considerably extended by the use of X-ray and neutron scattering, and a model could be obtained for halophilic protein stabilization (Zaccai *et al.*, 1989; Zaccai and Eisenberg, 1990). Recent developments in molecular genetics enabled the isolation of a number of halobacterial genes, allowing quick determination of the amino acid sequences of the corresponding coded proteins. The development of transformation protocols for halobacteria (Cline and Doolittle, 1987) and the construction of halobacterial shuttle vectors (Lam and Doolittle, 1989; Holmes and Dyall-Smith, 1990) opened the way to exploitation of the methodology of site-specific mutagenesis as an extremely powerful tool in the elucidation of the relationship between the structure of the halobacterial proteins and their adaptation to function at extremely high salt concentration.

The purpose of this review article is to familiarize the reader with recent developments in the molecular characterization of halobacterial proteins, starting with the methodology of their purification. Then we describe the biochemical and biophysical structural analyses of some enzymatic systems for which extensive knowledge has been accumulated. Finally, very recent developments in the field of the molecular genetics of halobacteria are discussed. For previous descriptions of the subject, the reader is referred to recent review articles already mentioned. Other reviews dealing with more specific issues will be mentioned throughout this article.

II. Purification of Halophilic Enzymes

Halophilic enzymes are very unstable in low salt concentrations. Because some of the important fractionation methods in protein chemistry, such as electrophoresis or ion-exchange chromatography, cannot be applied at high salt concentrations, the available fractionation methods are rather limited. This basic difficulty is the main reason why the number of halophilic enzymes studied in pure form is very small.

The existing purification procedures fall into two groups: the nonhalophilic approach and the halophilic approach. In the first, at certain stages in the purification procedure, the salt concentration is reduced and techniques that are suitable to low salt concentrations are applied. Inactivation in these conditions can be overcome partially either by protecting the native enzyme with its substrate or cofactors or by reactivating the enzyme at a later stage by exposure to high salt concentration.

The first reported procedure for the purification of a halophilic enzyme was that of malate dehydrogenase, described by Holmes and Halvorson (1965). In this procedure, the salt concentration was reduced at the very beginning of the procedure and methods such as ion-exchange chromatography on DEAE-cellulose and electrophoresis were applied. The enzyme was recovered after reactivation by dialysis against 25% (w/v) NaCl. The yield was very poor (~0.5%). In the purification of the halophilic enzyme isocitrate dehydrogenase (Hubbard and Miller, 1969), an inactivation step was performed prior to ammonium sulfate fractionation in order to improve selectivity. In this case as well, the recovery was very low (2.7%), mainly due to the inactivation step. On the other hand, in the purification of dihydrolipoamide dehydrogenase (Danson *et al.*, 1986), although several purification steps were used at the low salt concentration at which the enzyme was inactivated, this inactivation could be completely reversed by increasing NaCl concentration to 2 M. In several cases large losses in activity were prevented by using a protective agent. In the purification of NADH dehydrogenase (Hochstein and Dalton, 1973), NADH was used at a concentration of 0.1 mM to protect the activity of the enzyme when exposed to 0.35 M NaCl. This protection enabled the use of ion-exchange chromatography without major loss of activity.

The enzyme DNA-dependent RNA polymerase isolated from halobacteria presents an interesting example by being active *in vitro* only at salt concentrations below 0.4 M and by lacking the ability to initiate transcription at the specific transcription initiation sites. It was purified from *Halobacterium halobium* (Madon and Zillig, 1983) and from *Halococcus morrhuae* (Madon *et al.*, 1983) using purification protocols that include polymer partitioning methods and heparin-cellulose and DEAE-cellulose chromatographies. No special attempts were made to keep the enzyme at high salt concentration throughout the purification procedure. The enzyme was stabilized, though, by adding glycerol to the various buffers to a final concentration of 40%. The fact that the purified enzyme is active *in vitro* only at salt concentrations much lower than that existing in the cell and the fact that it lacks specificity might indicate that some essential factors were lost during the purification.

According to the halophilic approach, all the purification steps are performed in high salt concentrations. The advantage of this approach is the high level of recovery achieved in each step. This approach was used to purify the halobacterial enzyme ornithine carbamoyltransferase (Dundas, 1970), but its applicability was for many years limited due to the lack of suitable fractionation methods for multimolar salt solutions. Subsequently, several new methods were introduced that enabled the purification of many halophilic proteins at high salt concentration. The rest of this section will be devoted to a review of these methods.

A. Ammonium Sulfate-Mediated Chromatography

Differential salting-out by ammonium sulfate salt was used for many years as a means for differential precipitation and crude fractionation of soluble proteins. This salting-out electrolyte facilitates hydrophobic interactions by reducing the solubility of amino acid side chains in salt solution. The solubility of different proteins depends, therefore, on the distribution and exposure of the various amino acids on the surface. However, because the solubility of proteins in ammonium sulfate does not vary much from one protein to another, the applicability of this method has been limited in most cases to the first steps of the purification process. Halobacterial proteins are in general more soluble in ammonium sulfate than are nonhalophilic proteins, and therefore when halobacterial proteins are to be fractionated this limitation is pronounced even more. In the early 1970s several modifications to the ammonium sulfate fractionation method were introduced. These modifications were based on the fact that high concentrations of ammonium sulfate cause the adsorption of proteins to solid surfaces or gels. Among the useful supports were Celite (King, 1972), DEAE-cellulose (Mayhew and Howell, 1971), and alkylaminoagaroses (Rimerman and Hatfield, 1973). The technique involved adsorption of the unfractionated proteins to the matrix at high ammonium sulfate concentration followed by separation of the proteins by applying a decreasing concentration gradient of the same salt. The binding of the proteins to alkylaminoagarose was interpreted to be due to the facilitation by the ammonium sulfate of "hydrophobic interactions" between the aliphatic side chains covalently bound to the matrix and the protein surface. It was surprising, therefore, to discover that unsubstituted agarose could also adsorb large quantities of proteins in the presence of concentrated ammonium sulfate solutions (von der Haar, 1976; Mevarech et al., 1976). The explanation of this phenomenon given by von der Haar was that the solvation sphere on the gel surface differs from the solvation in solution and therefore the precipitation of proteins on this surface occurs at lower salt concentration than is needed for precipitation out of solution. This explanation assumes that the proteins are precipitated at the solvation layer of the gel. Another explanation is that the proteins accumulate in the solvation layer of the hydrophilic polysaccharide matrix, in a way analogous to partition of proteins between two phases of hydrophilic polymers dissolved in salt solution (Albertsson, 1970).

Because ammonium sulfate-mediated chromatography requires high salt concentrations and halobacterial proteins are very stable in ammonium sulfate concentrations higher than 1 M, this method is very useful for the purification of halobacterial enzymes. Detailed analyses of the adsorption properties of the halobacterial enzymes malate dehydrogenase (hMDH) and glutamate dehydrogenase (hGDH) on various supports were performed by Mevarech et al. (1976). A crude protein extract of sonicated *Haloarcula marismortui* was dialyzed against 2.5 M ammonium sulfate and applied on columns prepared from various materials. A decreasing concentration gradient of ammonium sulfate was then applied, and fractions were collected and assayed for enzymatic activity. On Sepharose 4B this procedure enabled the separation of several enzymes, with purification factors ranging between four- and sixfold. The ammonium sulfate concentrations at which the enzymes hGDH and hMDH were eluted from various solid supports are compared in Table I together with the ammonium sulfate concentration at the midpoint of the solubility curve. From this comparison it is clear that the elution from Celite is governed mainly by the solubility properties of the two halophilic enzymes in ammonium sulfate. As for the other solid supports, the ammonium sulfate concentrations at which the two enzymes are eluted are much lower than those at which they are precipitated, although the order of elution is related to that of decreasing solubility. It is worth noting that the elution concentration depends on the charge of the matrix. This observation is very surprising—it is totally unexpected that at concentrations of salt as high as 2 M, ionic interactions between the matrix and the proteins are effective, and this might be related to hydration-mediated interactions such as the ones postulated to contribute to the stabilization of the proteins (Zaccai et al., 1989).

To summarize this set of observations, halophilic proteins adsorb to polysaccharide matrices at high ammonium sulfate concentration. When the matrix is charged the adsorption of proteins having the same charge on the matrix is reduced whereas the adsorption of proteins having opposite charge is greatly facilitated. These observations enabled the development of several powerful purification

TABLE I
Elution of Two Halophilic Enzymes by Decreasing Concentration Gradients of Ammonium Sulfate[a]

Column	GDH (M)	MDH (M)
Sepharose 4B	1.44	1.70
CM-cellulose	1.84	2.06
HMD-agarose	1.18	
DEAE-cellulose	ne[b]	ne[b]
Celite	3.11	3.53
Solubility[c]	3.01	3.32

[a] Ammonium sulfate concentration at which the enzymes eluted from various solid supports. From Mevarech et al. (1976), with permission.

[b] ne, Not eluted: the enzymes did not elute until the concentration was 0.4 M $(NH_4)_2SO_4$ [0.3 M $(NH_4)_2SO_4$ in the case of GDH]. They could, however, be eluted by a NaCl gradient.

[c] Ammonium sulfate concentration at which 50% of the activity was found in the supernatant.

schemes by which halophilic proteins were purified to homogeneity. In order to demonstrate the full potential of the ammonium sulfate-dependent chromatographies, the applicability of these methods to the purification of some halophilic enzymes will be reviewed.

The capacity of Sepharose to adsorb proteins in 2.5 M ammonium sulfate is at least 30 mg/ml gel (Leicht and Pundak, 1981). Even higher amounts of protein can be adsorbed on DEAE-cellulose under the same conditions. It is, therefore, advantageous to use Sepharose or DEAE-cellulose, either batchwise (Zusman et al., 1989) or by loading on a column (Leicht and Pundak, 1981; Mevarech et al., 1977), in the early stages of the purification. In addition to the severalfold purification achieved in this step, it is possible to eliminate most of the cellular debris as well as viscous material, which interferes in later stages. The desorption of the proteins from the Sepharose can be achieved by an ammonium sulfate decreasing concentration gradient. The adsorption of proteins to DEAE-cellulose is governed by both the interaction with the polysaccharide backbone, which is facilitated by the salt, and the interaction with the positive charges of the matrix, which are enhanced as the salt concentration is reduced. Therefore, when decreasing concentration gradients are applied to a DEAE-cellulose column, the proteins will start to move as soon as the interactions with the polysaccharide gel weaken. However, when the salt concentration is reduced too much, the interaction with the positive charges becomes stronger and the proteins are retarded again on the gel. There are two ways to desorb the proteins from the positively charged gel, either by eluting the protein with ammonium sulfate solution at the concentration at which the protein is bound most weakly to the gel, or by using solutions having lower ammonium sulfate concentrations, to which sodium chloride is added to overcome the electrostatic interactions. For every protein there is a different optimum combination of the two interactions, thus the ammonium sulfate concentration at which the protein moves most quickly in the gel is different. Using this principle, it was possible to purify the ferredoxin of *H. marismortui* (Werber and Mevarech, 1978) and the dihydrofolate reductase of *Haloferax volcanii* (Zusman *et al.*, 1989) to homogeneity using only three purification steps. In the first case, after a step of adsorption to Sepharose 4B at 2.3 M ammonium sulfate and elution with a decreasing concentration gradient from 2.3 to 1.3 M, there were two steps at which the protein was adsorbed on DEAE-cellulose at 2.3 M ammonium sulfate and eluted by 1 M ammonium sulfate. In the case of the dihydrofolate reductase, the order of the steps was reversed. In the first two steps the enzyme was adsorbed on a DEAE-cellulose column at 2.5 M ammonium sulfate and eluted with 1.5 M ammonium sulfate. These steps were followed by adsorption to Sepharose 4B at 2.5 M ammonium sulfate and elution with a decreasing concentration gradient of 2.5 to 1 M.

The other principle was used in the large-scale purification of hMDH and hGDH (Leicht and Pundak, 1981). After fractionation on Sepharose 4B, the two enzymes were adsorbed on DEAE-cellulose and eluted with a concentration gradient ranging from 1.3 M ammonium sulfate to 1.3 M ammonium sulfate containing 2 M sodium chloride. In the purification of superoxide dismutase from *Halobacterium cutirubrum*, the enzyme was eluted from DEAE-Sepharose using a NaCl concentration gradient ranging from 0 to 0.8 M in the presence of 0.8 M ammonium sulfate (May and Dennis, 1987).

Although ammonium sulfate-mediated adsorption to Sepharose has been widely used in the purification of many other halophilic enzymes [i.e., halobacterial translation elongation factors by Kessel and Klink (1981); NAD-dependent glutamate dehydrogenase (E.C. 1.4.1.2) by Bonete *et al.* (1986); 2-oxoacid:ferredoxin oxidoreductases by Kerscher and Oesterhelt (1981a)], the systemic exploitation of ionic interactions between the proteins and charged gels at high ammonium sulfate concentrations still lags behind the other purification methods.

B. Affinity Chromatography

The principle of affinity chromatography, by which proteins are purified according to their specific ability to bind immobilized ligands, is particularly suitable to halophilic enzymes. Because halophilic enzymes are active at high salt concentrations, it is reasonable to assume that they will also bind their substrates and cofactors in these high salt concentrations. The enzyme hMDH of *H. marismortui* was purified to homogeneity using 8-(6-aminohexyl)amino-NAD-agarose (Mevarech *et al.*, 1977) and the hGDH of the same organism was purified using 8-(6-aminohexyl)amino-NADP-agarose (Leicht *et al.*, 1978). Leicht (1978) has shown that not only does the ammonium sulfate not interfere with the binding of the enzyme to the ligand, it actually enhances the interactions. This enhancement was shown to be biospecific, suggesting that the strength of the interaction between the enzyme and the immobilized coenzyme is a function of the sulfate concentration.

A very interesting application of affinity chromatography to the purification of halophilic enzymes was reported by Sundquist and Fahey (1988). These authors have purified the enzymes bis-γ-glutamylcysteine reductase and dihydrolipoamide dehydrogenase from *H. halobium* using immobilized metal ion affinity chromatography in high-salt buffers.

C. Other Methods

Two widely used methods that are not affected by high salt concentrations are gel-permeation chromatography and chromatography on hydroxylapatite gels. Columns for gel-permeation chromatography are prepared in either NaCl or KCl, usually in low phosphate concentration buffer. These salts are neutral in the sense that the migration of the proteins in the gel is dictated mostly by their Stokes radii. However, due to the rather low resolution of regular gel-permeation matrices in several cases [i.e., malate dehydrogenase (Mevarech *et al.*, 1977) and the two 2-oxoacid:ferredoxin oxidoreductases (Kerscher and Oesterhelt, 1981a)], the enzyme is recycled three or more times in the column in order to increase the effective length of the columns.

Chromatography in hydroxylapatite gel is particularly suitable for application after the gel-permeation chromatography steps. The adsorption of proteins to the matrix is affected specifically by the presence of phosphate ions. Therefore, eluents of gel permeation chromatography can be readily adsorbed to the hydroxylapatite gel without further treatment, no matter how diluted the enzyme.

III. Biochemical Aspects

A. Some Metabolic Aspects

Studies of the metabolic requirements of the extremely halophilic archaebacteria, as well as of many other aspects of these microorganisms, suffer from the fact that general conclusions were drawn from the examination of a rather limited number of species. For instance, it was believed that most halophilic archaebacteria utilize proteins and amino acids rather than carbohydrates or other nonnitrogenous carbon compounds as their main sources for energy and carbon (Larsen, 1981). This conclusion was probably the result of the fact that the original studies were performed on species isolated from sources rich in nutrients, for instance, salted fish. More recent work shows that many, if not most, halophilic archaebacteria can utilize a variety of nonnitrogenous organic compounds as sole source for energy and carbon (Torreblanca et al., 1986; Javor, 1984). Another example that demonstrates this point is the question of the ability of halobacteria to grow fermentatively. In the description of the genera *Halobacterium* and *Halococcus* in "Bergey's Manual of Determinative Bacteriology" (Gibbons, 1974) it is stated that fermentation was never found in these genera. It seems, however, that most halobacteria can ferment simple carbon compounds (Javor, 1984), and the anaerobic growth of *H. halobium* can be supported by arginine (Hartmann et al., 1980). Descriptions of the available literature concerning the central metabolic pathways of the extremely halophilic archaebacteria were written by Hochstein (1988) and by Danson (1988, 1989). The lipid metabolism of the halophilic archaebacteria was reviewed by Kamekura and Kates (1988) and the glycoprotein biosynthesis was reviewed by Sumper (1987).

Whereas we have no intention to describe in this review the various aspects of halobacterial metabolism, we would like to mention several unique features of their metabolic system. The conversions of the two 2-oxoacids (pyruvate and oxoglutarate) to their corresponding acyl-CoA thioesters are crucial steps in the two pathways described above. In most eukaryotes and aerobic eubacteria these reactions are catalyzed by the 2-oxoacid dehydrogenase multienzyme complexes that use NAD^+ as the final electron acceptor. These complexes are large aggregates (molar mass 3×10^6 to 7×10^6 g/mol) of three different enzymes that catalyze the three partial reactions (Koike and Koike, 1976). Two alternative reactions exist in the strict and facultative anaerobes in order to prevent excessive reduction of the NAD^+ pool. One alternative avoids the oxidation of the pyruvate by converting it to acetyl-CoA and formate, a reaction catalyzed by the enzyme pyruvate formate-lyase (formate acetyltransferase). The other alternative is to use an enzymatic system of the 2-oxoacid:ferredoxin oxidoreductases, in which the reducing power of the pyruvate is transferred to ferredoxin. The latter has a stronger reducing potential than NADH and therefore can be further used for various reactions, for instance, nitrogen fixation, synthesis of other 2-oxoacids, reduction of carbon dioxide to formate, or evolution of hydrogen.

The molar masses of the 2-oxoacid:ferredoxin oxidoreductases are 200,000–300,000 g/mol and they are composed of four subunits of the kind $\alpha_2\beta_2$. It has been shown that halobacteria have only these systems of 2-oxoacid:ferredoxin oxidoreductases. The two enzymes of *H. halobium* (pyruvate and oxoglutarate) were isolated and characterized by Kerscher and Oesterhelt (1981a). These systems proved to be thiamin diphosphate-containing iron–sulfur proteins. The relative stability of the halobacterial enzymes enabled detailed analysis of the various steps of the catalytic cycles (Kerscher and Oesterhelt, 1981b), demonstrating two distinct steps of one-electron transfer reactions.

Whereas the electron acceptors in the anaerobic organisms are the bacterial-type ferredoxins that contain [4Fe–S] clusters as the redox center, in the case of the halobacteria the electrons are transferred to [2Fe–S] ferredoxins. These ferredoxins were isolated from two different halobacteria and their amino acid sequences were determined (Hase *et al.*, 1977, 1980) and shown to be highly homologous to the chloroplast (and cyanobacterial) ferredoxins. The implications of these perplexing findings for the question of the molecular evolution of the system is discussed in detail in Kerscher and Oesterhelt (1982).

The absence in halobacteria of the oxoacid dehydrogenase complexes creates another puzzle. In most known systems, the role of the enzyme lipoamide dehydrogenase is to reoxidize the lipoic acid that is involved in the oxidation of the oxoacids in the oxoacid dehydrogenase complexes. This enzyme was nonetheless found in *H. halobium* and purified to homogeneity by Danson *et al.* (1986). What, then, is its function? It is likely that lipoamide dehydrogenase assumes a different role in halobacteria. Another reducing system unique to the halobacteria was discovered. Newton and Javor (1985) found that halobacteria contain millimolar quantities of the glutathione analog γ-glutamylcysteine and Sundquist and Fahey (1988) purified the corresponding enzyme bis-γ-glutamylcysteine reductase. It is, therefore, possible that these two reductases are part of yet unknown oxidation–reduction system.

Another unique property of at least some of the halobacteria is the ability to grow phototrophically by employing the light-driven proton pump bacteriorhodopsin. The proton gradient that is produced is used directly to generate ATP (Hartmann et al., 1980; Oesterhelt and Kripphal, 1983). Photoassimilation of CO_2 by halobacteria was shown by Danon and Caplan (1977) and Oren (1983). In vivo CO_2 fixation was demonstrated by Javor (1988) and the existence of the enzyme ribulose-bisphosphate carboxylase activity in several halobacteria was shown by Altekar and Rajagopalan (1990).

B. Enzymology of Extremely Halophilic Archaebacteria

As mentioned earlier, the unique feature of extremely halophilic archaebacteria is the existence in the cell of multimolar concentrations of KCl (Christian and Waltho, 1962; Ginzburg et al., 1970; Lanyi and Silverman, 1972). The entire biochemical machinery has therefore adapted to the high intracellular KCl concentration. This aspect of the extreme halophilic archaebacteria has attracted much interest and research and the effect of salt concentration on the activity and stability of halobacterial enzymes has been studied extensively (for reviews of the literature, see Lanyi, 1974; Bayley and Morton, 1978; Kushner, 1985). However, due to the lack of suitable purification procedures enabling the preservation of halophilic enzymes in their native state, only a few studies were performed on homogeneous enzyme preparations. Also, in many of the early studies, measurements of the effect of salt concentration on the activity of the halophilic enzymes were performed under conditions in which the enzymes were unstable and were inactivated during the time of the activity measurements. The most extensive examination of the properties of the halophilic enzymes was undertaken by Lanyi (1974). In the review article by Lanyi, the effect of salt concentration on the activities and stabilities of halophilic enzymes was discussed in length in terms of screening of excess surface charges, of enhancing hydrophobic interactions, and of specific ion–protein interactions. The conclusion of that illuminating review was that, despite the fact that the complexity of the protein structure precludes any single explanation for the salt dependencies observed, some generalities can be stated. At salt concentrations below 0.5 M the electrostatic shielding of charged groups predominates. Though such shielding decreases the stability of charge interactions owing to the high negative charge densities of the halophilic proteins, the overall effect of salt is structure stabilization. At high salt concentrations, hydrophobic interactions predominate and the protein molecules assume more tightly folded conformations. Finally, specific interactions of ions with amino acid side chains should explain specific ion effects. The relative contributions of these effects appear to differ from one protein to the other. Therefore, comparative studies of halophilic proteins and their nonhalophilic counterparts may help to uncover details of structure–function relationships and the mode of adaptation to high salt concentration.

We shall describe two halophilic enzymes in more detail: malate dehydrogenase from *H. marismortui* (*h*MDH) and dihydrofolate reductase from *H. volcanii* (*h*DHFR). *h*MDH was purified to homogeneity (Mevarech *et al.*, 1977; Leicht and Pundak, 1981) and was subjected to extensive enzymatic and physical studies (cf. Sections IV,G and IV,H). This enzyme will be used to demonstrate the methodologies used for the determination of the effect of salt concentration on the stability of the halophilic proteins. *h*DHFR was also purified to homogeneity (Zusman *et al.*, 1989), its amino acid sequence was deduced from the nucleotide sequence of its coding gene, and its catalytic properties were studied (Zusman, 1990). This enzyme will be used to demonstrate an approach to understanding the effect of salt concentration on the catalytic properties of a halophilic enzyme.

C. *Effect of Salt Concentration on Structural Integrity of hMDH*

Under conditions in which substrates are in large excess with respect to enzymes, the initial rate v_0 of NADH oxidation by *h*MDH obeys the relationship

$$v_0 = K[\mathrm{E_a}] \tag{1}$$

in which $[\mathrm{E_a}]$ is the concentration of the active enzyme and K is a proportionality constant containing a combination of elementary rate constants of the various steps in the enzymatic reaction. Because both

K and [E_a] are dependent on salt concentration C_S, we may write in general that

$$(dv_0/dC_S) = K(d[E_a]/dC_S) + [E_a](dK/dC_S) \qquad (2)$$

The first term on the right-hand side of Eq. (2) measures the dependence of the stability of the active enzyme on salt concentration; the second term expresses the dependence of the enzymatic activity (the elementary rate constants) on the salt concentration.

Under conditions in which the enzyme is stable, Eq. (2) reduces to

$$(dv_0/dC_S) = [E_a](dK/dC_S) \qquad (3)$$

Equation (3) reflects the dependence of the elementary kinetic parameters of the enzymatic reaction on salt concentration. Hecht and Jaenicke (1989b) have shown that in the case of hMDH the main kinetic parameter that is affected by salt concentration is the K_m of oxaloacetate. Therefore, under conditions when the oxaloacetate concentration is limited, there is a maximum in the enzymatic activity at 1 M NaCl. When the oxaloacetate concentration is raised above 1.5 mM, the enzymatic activity becomes independent of salt concentration.

In order to explore the effect of salt concentration on the stability of the active enzyme, the enzymatic activity was determined under standard conditions of 4 M NaCl after exposure of the enzyme for various times to various salt concentrations. If the transition of the enzyme between the active form, E_a, and the inactive form is very slow as compared to the time of the assay (as is the case in hMDH for various lower salt concentrations), the distribution of enzyme molecules between the active and inactive states during the assay reflects that distribution at the end of the preincubation period. For this case, Eq. (2) is reduced to

$$dv_0/dC_S = K(d[E_a]/dC_S) \qquad (4)$$

and a change in v_0 reflects a change in the concentration of the active enzyme.

The above assumptions can be applied to the establishment of the kinetics of the inactivation after exposure of the enzyme to low salt concentrations. When there is only one form of active enzyme, E_a, which undergoes an irreversible transition to an inactive form, E_i, the transition kinetics should follow a first-order reaction

$$E_a \rightarrow E_i$$

in which

$$k_i = -d \ln[E_a]/dt \qquad (5)$$

Measurement of the inactivation rates of hMDH were performed under various conditions of salt types, salt concentrations, temperatures, and buffers (Mevarech and Neumann, 1977; Pundak et al., 1981; Zaccai et al., 1986b, 1989; Hecht and Jaenicke, 1989a). It was found that (1) the inactivation process is of first order (which means that only one active form of the enzyme exists), (2) the rate constant for inactivation increases as the salt concentration decreases, (3) the temperature dependence of the rate constants of inactivation depends on type of salt, and (4) the dependence of the rate constants on salt type follows the Hofmeister series (von Hippel and Schleich, 1969), being lower for salting-out salts. The different models for the role of the salts in the stabilization of the hMDH will be discussed in Section IV,G.

The fact that the inactivation reaction follows first-order kinetics does not explain the mechanism of the inactivation process. An insight into this process may be gained by following the reactivation kinetics. Reactivation experiments were conducted in which a fully inactive enzyme was introduced into a concentrated salt solution and the rate of appearance of activity was determined (Mevarech and Neumann, 1977; Hecht and Jaenicke, 1989a). These experiments showed that the reactivation process is bimolecular and follows second-order kinetics, suggesting that the inactivation is the result of dissociation of the active dimeric enzyme into inactive monomeric subunits. This apparent reversibility can also be achieved when the inactivation is caused by the exposure of the enzyme to high concentration of the salting-in salt guanidinium chloride (Hecht and Jaenicke, 1989a). In both cases, however, prolonged exposure of the inactive form of the enzyme to the inactivating conditions reduces the extent to which the inactive enzyme can be reactivated. It appears, therefore, that the inactive monomer, M, undergoes an additional, as yet unexplained process of irreversible denaturation. The existence of this denatured state can be also detected, as we shall see later, by spectral analysis of hMDH at different salt concentrations. The emerging model for the inactivation–reactivation process is

$$\text{D} \underset{\text{High salt}}{\overset{\text{Low salt}}{\rightleftharpoons}} 2\text{M} \rightarrow 2\text{M}^*$$

in which D is an active dimer, M is a form of the inactive monomer that can be reactivated under proper conditions, and M* is an irreversibly denatured monomer.

The model shown gains support from spectral and sedimentation analyses. When the UV absorption spectra and the fluorescence spectra of the native and the low-salt inactive hMDH are compared, considerable differences are observed. On inactivation the absorption maximum is shifted from 280 to 276 nm and the specific absorption decreases (Mevarech et al., 1977). Similarly, the inactivation process is associated with a red shift in the emission spectrum of the inactive enzyme and a considerable reduction in the quantum yield is observed (Mevarech et al., 1977; Hecht and Jaenicke, 1989a). These spectral changes seem to reflect the exposure of the tryptophan residues of the enzyme from an environment of low dielectric constant (in the native enzyme) to the aqueous medium (in the inactive form). The kinetics of the fluorescence spectral changes correspond precisely to the kinetics of inactivation (Mevarech et al., 1977), suggesting that these two phenomena are highly correlated.

Far-UV circular dichroism spectral analyses show that on inactivation at low salt concentration the α-helical fraction is decreased from ~30 to 0% (Pundak et al., 1981; Hecht and Jaenicke, 1989a). The kinetics of the decrease are not simple. A sharp decrease in the α-helix content of the enzyme is followed by a prolonged moderate decrease of the helical structure, resulting, finally, in the complete disruption of all organized structures. It is possible that the initial decrease in the α-helical content corresponds to the transition D → M, which is followed by the slow transition M → M* as discussed above.

The dissociation of hMDH was also followed by sedimentation velocity measurements (Pundak et al., 1981) and neutron scattering (Zaccai et al., 1986b). Samples of hMDH were incubated at various NaCl concentrations at 20°C and aliquots were withdrawn at various times and analyzed by ultracentrifugation. Dissociation of the enzyme was visualized by the appearance of a slower sedimenting boundary (Fig. 1). The relative amounts of the protein in both boundaries could be measured and the dissociation rates could be calculated. These

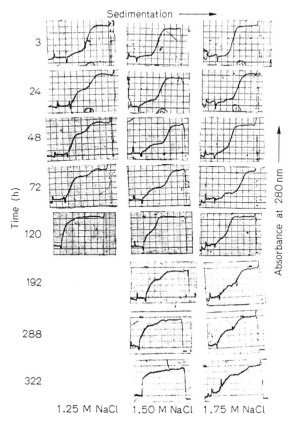

FIG. 1. Sedimentation patterns of hMDH at various times of incubation at the indicated salt concentrations at 20°C. The appearance of an additional sedimentation boundary indicates dissociation of the native enzyme. From Pundak et al. (1981), with permission.

rates were found to correlate very nicely with the rates of inactivation of the enzyme in corresponding salt conditions. In the neutron scattering experiments, the scattering curve was measured at regular time intervals with the protein in 1 M NaCl solution. The dissociation of the enzyme was followed both through its effect on the molecular weight of the scattering particles and their radii of gyration (see Section IV).

D. hDHFR

Dihydrofolate reductase (5,6,7,8-tetrahydrofolate:NADP oxidoreductase, EC 1.5.1.3) (DHFR) catalyzes the NADPH-dependent reduction of 7,8-dihydrofolate (H_2F) to 5,6,7,8-tetrahydrofolate (H_4F). This enzyme is necessary for maintaining intracellular pools of H_4F and its derivatives, which are essential cofactors in the biosynthesis of purines, thymidylate, and several amino acids. For some reason, in *Escherichia coli* the deletion of the gene coding for DHFR is lethal unless the gene coding for the enzyme thymidylate synthase is mutated as well (Howell *et al.*, 1988). DHFR is the target enzyme of a group of antifolate drugs that are widely used as antitumor (methotrexate) and antimicrobial (trimethoprim) agents (Blakley, 1985).

The crystal structures of the *E. coli* DHFR–methotrexate binary complex (Bolin *et al.*, 1982), of the *Lactobacillus casei* (DHFR–NADPH–methotrexate ternary complex (Filman *et al.*, 1982), of the human DHFR–folate binary complex (Oefner *et al.*, 1988), and of the mouse (DHFR–NADPH–trimethoprim tertiary complex (Stammers *et al.*, 1987) have been resolved at a resolution of 2 Å or better. The crystal structures of the mouse DHFR–NADPH–methotrexate (Stammers *et al.*, 1987) and the avian DHFR–phenyltriazine (Volz *et al.*, 1982) complexes were determined at resolutions of 2.5 and 2.9 Å, respectively. Recently, the crystal structure of the *E. coli* DHFR–NADP$^+$ binary and DHFR–NADP$^+$–folate tertiary complexes were resolved at resolutions of 2.4 and 2.5 Å, respectively (Bystroff *et al.*, 1990). DHFR is therefore the first dehydrogenase system for which so many structures of different complexes have been resolved. Despite less than 30% homology between the amino acid sequences of the *E. coli* and the *L. casei* enzymes, the two backbone structures are similar. When the coordinates of 142 α-carbon atoms (out of 159) of *E. coli* DHFR are matched to equivalent carbons of the *L. casei* enzyme, the root-mean-square deviation is only 1.07 Å (Bolin *et al.*, 1982). Not only are the three-dimensional structures of DHFRs from different sources similar, but, as we shall see later, the overall kinetic schemes for *E. coli* (Fierke *et al.*, 1987), *L. casei* (Andrews *et al.*, 1989), and mouse (Thillet *et al.*, 1990) DHFRs have been determined and are also similar. That the structural properties of DHFRs from different sources are very similar, in spite of the considerable differences in their sequences, suggests that in the absence, so far, of structural information for *h*DHFR it is possible to assume, at least as a first approximation, that the α-carbon chain of the halophilic enzyme will not deviate considerably from those of the nonhalophilic ones.

The primary structure of DHFR of the extremely halophilic archaebacterium *H. volcanii* (hDHFR) was determined from the nucleotide sequence of its coding gene. *Haloferax volcanii* is very sensitive to trimethoprim but spontaneously resistant mutants can be found at frequencies of 10^{-10}–10^{-9}. The molecular basis of this resistance is an amplification of the chromosomal region coding for DHFR (Rosenshine *et al.*, 1987). This gene amplification causes an overproduction of the enzyme, which constitutes in the mutants as much as 5% of the total cellular protein. The gene coding for *H. volcanii* DHFR was isolated from the amplified DNA of a trimethoprim-resistant mutant and its nucleotide sequence was determined (Zusman *et al.*, 1989).

The alignment of the amino acid sequence of hDHFR with those of *L. casei* and *E. coli* shows a homology of 23 and 30%, respectively (see Fig. 2). Some amino acid residues are conserved in almost all known DHFRs (Beverley *et al.*, 1986; Volz *et al.*, 1982). Most of these residues are also conserved in hDHFR. The functions of some of the residues that are also conserved in hDHFR were inferred from the crystal structures. For instance, in *E. coli* DHFR, Ala-7, Ser-49, and Leu-54 are involved in binding of dihydrofolate and its analogous inhibitor methotrexate (Bolin *et al.*, 1982). The bond between Gly-

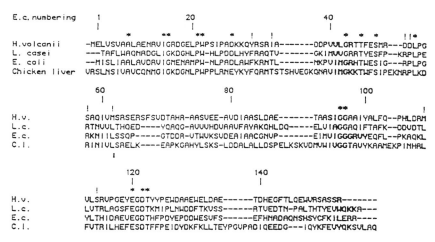

FIG. 2. Alignment of the amino acid sequence of *H. volcanii* DHFR with the amino acid sequences of DHFRs of *E. coli*, *L. casei*, and chicken liver. From Zusman *et al.* (1989), with permission.

95 and Gly-96 has an unusual cis configuration that seems to have an essential role in the conformation of the enzyme. Replacement of Gly-95 by Ala abolishes the activity entirely (Villafranca et al., 1983). Asp-27 (or the corresponding Glu in eukaryotic DHFRs) has an essential role in the catalytic mechanism, as will be discussed later.

There are, however, residues that are conserved in all the DHFRs except for hDHFR. These replacements are Leu-24 to Ile, Phe-31 to Tyr, Thr-35 to Ile, Arg-57 to Ser, Leu-62 to Met, and Thr-113 to Ser. The roles of Phe-31 and Thr-113 were suggested from the crystallographic data and the effects of replacing these residues by others using site-specific mutagenesis were studied (Chen et al., 1987). Phe-31 forms part of the hydrophobic pocket and interacts with the pteroyl moiety through van der Waals contacts such that the edge of the phenyl ring is oriented toward the faces of both the pteridine ring and the p-aminobenzoyl group in an edge-to-face aromatic–aromatic interaction. The replacement in the E. coli DHFR (eDHFR) of Phe-31 by the smaller amino acid Val reduces the binding of dihydrofolate, tetrahydrofolate, and methotrexate. It also destabilizes the protein structure, probably by disrupting the close packing found in the interior of the protein. The replacement, in eDHFR, of Phe-31 by Tyr (the wild-type residue of hDHFR) causes reduction in the binding efficiency of the enzyme to the folates, but in this case this loss may reflect difficulties in solvating as well as accommodating the extra hydroxyl group in the binary complex (Benkovic et al., 1988; Wagner and Benkovic, 1990). In both of these replacements V_{max} increases two- to threefold probably due to acceleration in the release of the tetrahydrofolate from the enzyme.

The hydroxyl of Thr-113 forms a hydrogen bond to the carboxylate of Asp-27 and interacts with the 2-amino group of the pteridine moiety through a hydrogen-bonded water. Its replacement, in eDHFR, by Val decreases the binding of dihydrofolate 25-fold and causes a destabilization of the protein. In hDHFR, this residue is replaced by Ser, which, like Thr, can participate in the hydrogen bonding.

Similar to other halobacterial proteins, hDHFR is highly acidic, having an excess of 15 acidic residues over basic residues (9 mol %). In comparison, in eDHFR and in the L. casei DHFR (lDHFR), there are 5 (3 mol %) and 4 (2.5 mol %) acidic residues in excess, respectively. The negatively charged residues of hDHFR are spread throughout the primary structure, unlike the halobacterial ferredoxins, in which a large fraction of the excess negative charges is concentrated in the extra amino and carboxyl 22 amino acid residues possessed by these molecules, when compared to other 2[Fe–S] ferredoxins (Hase et al., 1977, 1980). An attempt was made to localize the charged amino acids by a simple approach in which the three-dimensional structure of eDHFR was used as a backbone that carried the hDHFR primary sequence. Using this naive approach it was found that all the charges, except that of Asp-29, were located on the surface of the molecule, creating a highly charged sphere (J. Moult and M. Mevarech, 1990, unpublished). Asp-29 is localized, like its eDHFR Asp-27 counterpart, in a hydrophobic pocket that is a part of the catalytic site.

The catalytic properties of eDHFR were determined by measuring the various ligand association and dissociation rate constants and pre-steady-state reaction transients (Fierke et al., 1987). The key features of the kinetic scheme are (1) at low pH the rate-limiting step for the steady-state turnover is the dissociation of the product H_4F and (2) the rate constant for the hydride transfer from NADPH to H_2F is rapid, favorable, and pH dependent, becoming rate limiting at pH values higher than 7. This pH-dependent hydride transfer reaction was thought to be the result of the involvement of a single amino acid residue, Asp-27 (in eDHFR and the corresponding Asp and Glu residues in other DHFRs), in the catalysis. It is thought that the hydride transfer reaction follows a rapid protonation of N-5 (of the C-6–N-5 imine of the pteridine ring) by the protonated Asp (or Glu) (Benkovic et al., 1988). Substitution of Asp-27 by Ser or Asn results in a dramatic decrease in V_{max} and apparently forces the enzyme to utilize protonated H_2F directly as a substrate (Howell et al., 1986). These features of the catalytic reactions are strikingly similar for the E. coli (Fierke et al., 1987) and L. casei (Andrews et al., 1989) as well as for the mouse (Thillet et al., 1990) enzymes. The virtual identity of the kinetic schemes of eDHFR and lDHFR, despite the low amino acid homology and the variations in the active-site structures, shows that the same catalytic surface can be constructed by different combinations of amino acids (Andrews et al., 1989).

It is tempting to try to explain the halophilic features of hDHFR even in the absence of a detailed kinetic scheme for this enzyme, assuming that the main features of the kinetic scheme of the non-halophilic enzymes hold true also for the halophilic enzyme. The salt concentration might have an effect on the rates of binding or dissociation of the various substrates or on the rate of the hydride transfer reaction. Because, as we saw, the hydride transfer reaction is largely dependent on the protonation of Asp-27, it becomes the rate-limiting step at pH values higher than the pK_a of this residue. The effect of salt concentration on the steady-state turnover can be explained, therefore, as the effect of salt concentration on the pK_a of Asp-29 (the hDHFR equivalent of Asp-27). A similar effect of salt concentration on reaction rates was observed by Russel and Fersht (1987) in the enzyme subtilisin. In that system, unshielded surface charge affects the pK_a of His-64, the protonation of which is essential for catalytic activity.

The explanation given above implies that stimulation of enzymatic activity by increase of salt concentration is pH dependent and that it will be less pronounced at lower pH values. The dependence of the steady-state reaction rate on salt concentration and pH as shown in Fig. 3 (Zusman, 1990) exactly reflects this prediction. It seems, therefore, at least as a preliminary explanation, that salt concentration affects the pK_a of Asp-29, which, unlike its counterparts in eDHFR and lDHFR, has a much lower pK_a. The question, then, is why the pK_a of Asp-29 is so different. To answer this question and the validity of the entire model will require elucidation of the detailed kinetic scheme of hDHFR at different salt concentrations, determination of the three-dimensional structure of this enzyme, and modification of the amino acid residues constituting the active site.

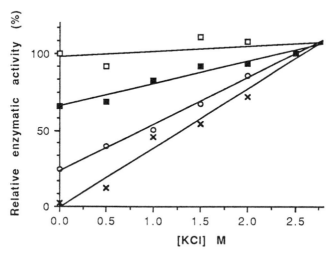

FIG. 3. The relative enzymatic activity of hDHFR as a function of salt concentration and pH; (□) pH 4, (■) pH 5, (○) pH6, and (x) pH 7. From Zusman (1990), with permission.

IV. STRUCTURE

Primary, secondary, tertiary, and quaternary structure are familiar concepts for proteins and refer to the amino acid sequence, local folding arrangement, three-dimensional organization, and subunit interactions of polypeptide chains, respectively. Here, tertiary and quaternary structure shall be considered in the most general way, to include also the small molecules or ions that are essential for the conformational stability of the polypeptide chains. This is especially relevant for halophilic proteins, which have extensive interactions with solvent components (water molecules and salt ions). The known structure of a protein (at any level) always results from experiment, and as such is known only within appropriate error limits.

For tertiary and quaternary structure, the concept of resolution is related to how finely a structure is known. A refined protein structure obtained from X-ray crystallography is usually defined to higher precision than the experimental resolution of the diffraction experiment because a number of chemical bond lengths within each amino acid are given fixed values obtained from highly accurate small molecule crystal structures. Experiments on crystals that diffract to high angles could then yield structures at atomic resolution, i.e., spatial coordinates and a temperature factor (a measure of the mean displacement due to disorder and thermal motion) for all atoms in the protein complex. Experiments on protein complexes in solution, on the other hand, yield structures of very low resolution. Information is obtained on the composition of the solution particle formed by the protein complex and the small molecules and ions associated with it. It establishes if the protein is a monomer or if it has a complex subunit structure—how many molecules of water are associated in its structure, how many of salt or other small molecules. It is also possible to obtain an indication of the spatial configuration or shape of the particle. The experimental approaches for solution studies include ultracentrifugation, quasi-elastic light scattering, and small-angle X-ray and neutron scattering (SAXS and SANS, respectively). With suitable sample preparation, electron microscopy usually provides structural information at a resolution intermediate between that of solution studies and X-ray crystallography. Exciting recent developments, however, show that it could also provide atomic resolution information on ordered arrays (Henderson *et al.*, 1990).

The halophilic proteins and macromolecular complexes on which structural studies are in progress include ferredoxin from *H. marismortui* (X-ray crystallography), ribosomal subunits from *H. marismor-*

tui (electron microscopy, X-ray, and neutron diffraction), bacteriorhodopsin in the purple membrane of *H. halobium* (electron microscopy and diffraction and neutron diffraction), the S layers around eubacteria and archaebacteria, including halophilic bacteria (electron microscopy), and malate dehydrogenase from *H. marismortui* (solution structure).

A. Ribosomal Subunits

Ribosomal subunits from *H. marismortui, Bacillus stearothermophilus,* and *Thermus thermophilus* have been crystallized from different solvents and are being studied by X-ray and neutron diffraction and electron microscopy (Yonath and Wittmann, 1988). The project poses a formidable methodological problem because the ribosome has no symmetry, is conformationally heterogeneous, is most unstable, and contains about 2.5×10^5 atoms. But the possibility of describing a part of the cellular protein factory at an atomic level justifies the necessary enormous efforts in crystallographic development. Already, electron microscopy image reconstruction experiments on two-dimensional arrays show a very interesting low-resolution structure with a space that might host the tRNA molecules and a "tunnel" that might be related to the exit site of the growing polypeptide chain (Yonath and Wittmann, 1989).

Yonath and Wittmann (1988) have described the crystallization of ribosomal subunits from eubacteria, thermophilic bacteria, and halophiles. They observed that crystallized particles retained their biological activity for several months, in contrast with isolated particles in solution that lose activity rapidly. Ribosomes from other than halophilic organisms are inactivated if solvent salt concentrations are increased much beyond the physiological values and the presence of alcohols was required for crystallization. This causes technical difficulties in handling and data collection (because of the volatility of the alcohols) and in the search for isomorphous derivatives, which are essential for the phasing of diffraction data. Halophilic ribosomes, on the other hand, are stable and active in high salt concentrations. They crystallize from salt solutions containing minute amounts of poly(ethylene glycol) and do not have the disadvantages described above (Fig. 4). In fact, the crystals contain the *minimum* salt concentration requirements, which are relatively high—$2.5\ M$ KCl, $0.5\ M$ NH_4Cl, and 10 mM $MgCl_2$ for activity; $0.8\ M$ KCl, $0.5\ M$ NH_4Cl, and 10 mM $MgCl_2$ for stability (Yonath and Wittmann, 1988). A

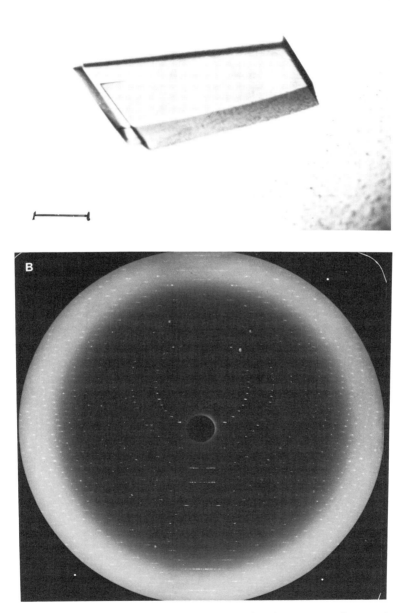

FIG. 4. (a) Crystals (bar = 0.2 mm) and (b) diffraction pattern of *H. marismortui* 50S ribosomal subunits. From Yonath *et al.* (1990), with permission.

TABLE II
Characterized Three-Dimensional Crystals of Ribosomal Particles

Source[a]	Growth form[b]	Cell dimensions (Å); symmetry	Resolution (Å)
70S $T.t$	MPD	524 × 524 × 306; $P4_12_12$	~20
70S $T.t$ + mRNA and tRNA[c]	MPD	524 × 524 × 306; $P4_12_12$	~15
30S $T.t$	MPD	407 × 407 × 170; $P42_12$	7.3
50S $H.m.$	PEG	210 × 300 × 581; $C222_1$	3.0[d]
50S $T.t.$	AS	495 × 495 × 196; $P4_12_12$	8.7
50S $B.st$[e]	A	360 × 680 × 920; $P2_12_12$	~18
50S $B.st.$[e,f]	PEG	308 × 562 × 395; 114°; $C2$	~11

[a]$B.st.$, Bacillus stearothemophilus; $T.t$, Thermus thermophilus; $H.m.$, Haloarcula marismortui. From Yonath et al. (1990), with permission.

[b]Crystals were grown by vapor diffusion in hanging drops from solutions containing methylpentanediol (MPD), poly(ethylene glycol) (PEG), ammonium sulfate (AS), or low-molecular-weight alcohols (A).

[c]A complex including 70S ribosomes, 1.5–2 equivalents of Phe-tRNA^{Phe}, and an oligomer of 35 ± 5 uridines, serving as mRNA.

[d]von Boehlen et al. (1991).

[e]Same form and parameters for crystals of large ribosomal subunits of a mutant (missing protein BL11) of the same source and for modified particles with an undecagold cluster.

[f]Same form and parameters for crystals of a complex of 50S subunits, one tRNA molecule, and a segment (18 to 20-mers) of a nascent polypeptide chain.

summary of three-dimensional crystals of ribosomal particles is given in Table II.

Recently von Boehlen et al. (1991) have obtained an improved crystal form of the large (50S) ribosomal subunit of $H.$ marismortui, which diffracts to 3 Å resolution, by the addition of 1 mM Cd^{2+} to a crystallization medium that contained over 1.9 M of other salts. These improved crystals are isomorphous with the previously reported ones, and, as was the case for the previous crystals, they show no measurable decay after a few days of synchrotron irradiation at cryotemperatures. The new crystals are of adequate mechanical strength. Initial phasing studies by specific and quantitative derivatization with super-dense heavy-atom clusters and by real- and reciprocal-space rotation searches are in progress.

B. Surface Layers

Extreme or moderate halophilic archaebacteria lack a peptidoglycan component in their cell wall. Early electron microscopy studies, however, disclosed a honeycomb of surface morphological units (S layers) comprising six subunits in a hexagonal arrangement (reviewed in Kessel *et al.*, 1988). The cell wall of halobacteria is characterized by a surface glycoprotein that is believed to be required for maintaining the rod shape of the cell (Mescher and Strominger, 1976). The molecular organization of the halobacterial cell surface is essential for survival of these organisms at extreme salt concentrations, but our knowledge concerning this structure is limited to fairly low resolution. Progress in this area has led to the determination of the primary structure of the surface protein of *H. halobium* (Lechner and Sumper, 1987) and *H. volcanii* (Sumper *et al.*, 1990) by cloning and sequencing of the corresponding genes.

Early difficulties in electron microscopy studies of halobacteria relating to the necessity of maintaining the integrity of the regular

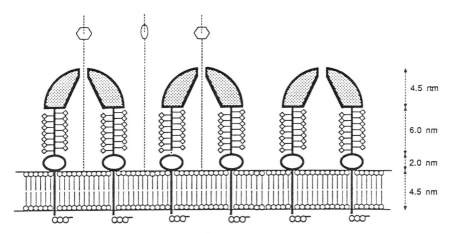

FIG. 5. Schematic diagram summarizing the available structural information on the *Halobacterium* cell envelope from X-ray studies of the envelopes (Blaurock *et al.*, 1976), from the primary structure of the surface glycoprotein (Lechner and Sumper, 1987), and from the three-dimensional structure described by Kessel *et al.* (1988). The three-dimensional structure determined by electron microscopy depicts only the upper dome-shaped region of the structure, which is separated from the cell membrane by the "spacer elements." As indicated by the crystallographic symbols, the section runs from sixfold to sixfold axis via the twofold axis. From Kessel *et al.*, (1988), with permission.

surface arrays by high salt concentrations have been overcome by the finding that morphological integrity can be preserved by the use of divalent cations at concentrations as low as 10 mM CaCl$_2$ alone (Kessel *et al.*, 1988). The surface layers around eubacteria and archaebacteria were subsequently characterized in greater detail by electron microscopy (Baumeister *et al.*, 1989). The cell envelope of halobacteria is made up of the plasma membrane and the S layer, which is a two-dimensional protein lattice, on the outside of the cell (Fig. 5) (Kessel *et al.*, 1988). The lattice is anchored into the plasma membrane by proteins in such a way that the space enclosed may serve as a kind of periplasmic space. A special role of the S layer in halophilic conditions has not yet been established, but in eubacteria it is thought to be involved in maintaining cell shape, molecular sieving, adhesion, and cell–cell interactions.

C. Purple Membrane

The purple membrane of *H. halobium* is made up of a single protein, bacteriorhodopsin, organized on a highly ordered two-dimensional lattice, with lipids between the protein molecules. The structure and unique function (it is a light-driven proton pump) of this retinal-binding membrane protein (the first to be discovered in a noneukaryotic cell) have been reviewed extensively (e.g., Stoeckenius and Bogomolni, 1982; Khorana, 1988). Bacteriorhodopsin was the first membrane protein for which a three-dimensional low-resolution structure became available. The experiments, done by a combination of electron diffraction and image reconstruction electron microscopy, showed the protein to be made up of seven α helices spanning the membrane (Henderson and Unwin, 1975). Neutrons are scattered by nuclei, and isotopes of the same element have different scattering amplitudes. Neutron diffraction experiments on purple membrane samples with specific deuterium labeling have defined certain parts of the structure in greater detail (Popot *et al.*, 1989, and references therein). Recently, the electron microscopy/diffraction study has been extended to much higher resolution, and most of the bacteriorhodopsin structure is now known close to atomic resolution (Henderson *et al.*, 1990).

The natural two-dimensional crystal lattice in the membranes is maintained over a wide temperature range and for different water contents. This allowed a number of neutron diffraction studies of membrane structure as a function of its environment. Neutron diffraction experiments on samples in which H$_2$O was exchanged by

D_2O have shown how the water molecules are distributed in the structure in different conditions (Zaccai, 1987; Papadopoulos et al., 1990). In its halophilic physiological environment, the membrane has a multimolar KCl solution on its cytoplasmic side and a multimolar NaCl solution on the outside of the cell. Most structure and function experiments on purple membranes, however, have been done in low salt concentration conditions. Neutron diffraction experiments have been attempted in high concentrations of KCl and NaCl, but results are not yet available (F. Samatey and G. Zaccai, unpublished data).

D. Ferredoxin from Haloarcula marismortui

Ferredoxin from *H. marismortui* was crystallized from 4 *M* phosphate solution and studied by X-ray diffraction (Sussman et al., 1986). An electron density map of the protein was obtained after phasing of the Bragg reflections by molecular replacement (using the previously resolved structure of ferredoxin from *Spirula platensis*) and anomalous scattering. The map shows disordered regions, however, and a refined high-resolution structure has not yet been published. The primary structure of the halophilic ferredoxin is known (Hase et al., 1977, 1980). It is longer than the algal protein and an alignment of the two sequences shows that the excess is mainly at the N-terminal and C-terminal ends of the halophilic protein, which are also rich in acidic residues. The N-terminal domain is a part of the structure that is not well-defined in the electron density map.

E. Methods for Solution Studies

Despite the limited amount of information obtained when compared with crystallography, solution experiments have certain advantages, including the trivial one that good crystals are not required. The entire particle contributes to the information (in crystallography, disordered parts of the structure are not "seen") and there is a greater freedom in choice of solvent (in crystallography, it is limited to crystallization conditions). Interactions between macromolecules, and between macromolecules and solvent components, can then be studied in solution over a broad range of conditions. Because of these advantages, solution methods have been quite important in increasing our understanding of halophilic proteins.

The theoretical tools for the interpretation of solution experiments are the thermodynamics of multicomponent solutions (Casassa and Eisenberg, 1964; Eisenberg, 1976, 1990) and the theory of small-angle scattering (Guinier and Fournet, 1952; Glatter and Kratky, 1982; Luzzati and Tardieu, 1980). In the ultracentrifuge, particles in solution are exposed to a strong gravitational field, leading to increasing concentration toward the bottom of the tube, for particles that are denser than the solvent, or toward the top, if they are less dense. This gives rise to an osmotic field that tends to oppose the concentration increase. It can be shown that for a macromolecular particle at vanishing concentrations (so that interparticle interactions can be neglected), a concentration distribution will be established at equilibrium, given by

$$d \ln c_2/dr^2 = (\omega^2/2RT)M_2(\partial\rho/\partial c_2)_\mu \qquad (6)$$

where c_2 (g/ml) is the concentration of the macromolecule, r is the distance from the center of rotation, ω is the angular velocity of the centrifuge tube, R and T are the gas constant and absolute temperature, respectively, M_2 (g/mol) is the molar mass of the macromolecule, and $(\partial\rho/\partial c_2)_\mu$ is a buoyancy term, the density increment of the solution at constant chemical potential, μ, of all components in the solution. In most cases of interest, positive values of $(\partial\rho/\partial c_2)_\mu$ lead to sedimentation; negative values lead to flotation and, when $(\partial\rho/\partial c_2)_\mu$ vanishes in a density gradient, localized particle bands are formed (Eisenberg, 1976). The two experimental parameters pertaining to the particle, therefore, are M_2 and $(\partial\rho/\partial c_2)_\mu$, and we shall see in Section IV,B how they contain useful information on the composition of the particle.

In another type of experiment with the ultracentrifuge, the sedimentation coefficient of the particle is obtained. If the centrifugal force greatly exceeds the force due to the osmotic field, the particles will sediment, forming a boundary that will move down from the surface of the solution for positive values of $(\partial\rho/\partial c_2)_\mu$. This movement will be opposed by the buoyancy of the particles and by the frictional force generated by their motion. It can be shown that this leads to a constant velocity (dr/dt) of the boundary and a sedimentation coefficient, s, is defined for the particle by $s = (dr/dt)/\omega^2 r$, where r is the distance to the center of rotation (Schachman, 1959). For vanishing particle concentration,

$$s = [M_2(\partial\rho/\partial c_2)_\mu]/f_p \qquad (7)$$

where f_p is the frictional coefficient per mole of particle.

The Svedberg equation relates s to the molar mass of the particles, M_2, their translational diffusion coefficient, D (related to the frictional force exerted on the particle), and to $(\partial \rho/\partial c_2)_\mu$:

$$(s/D) = (M_2/RT)(\partial \rho/\partial c_2)_\mu \qquad (8)$$

It is easier experimentally to determine s than to perform an equilibrium sedimentation measurement, but, because we are interested in the molar mass and the density increment, D has to be determined independently.

At infinite particle dilution, the translational diffusion coefficient is given by Einstein's equation

$$D = RT/f_p \qquad (9)$$

In this cursory discussion of sedimentation and diffusion experiments, we have tacitly assumed that solutions are monodisperse [i.e., containing only one type of particle with constant M_2 and $(\partial \rho/\partial c_2)_\mu$] and have infinitely sharp sedimentation boundaries. Yet even for a monodisperse solution, initially sharp boundaries will broaden with time because of diffusion, this constituting a classical procedure for determining D in the ultracentrifuge (Schachman, 1959). However, the most reliable modern way to determine D for macromolecules is by quasi-elastic laser light scattering. In these experiments D is obtained from the autocorrelation function observed for the fluctuations in the scattered light field, i.e., the time-averaged quantity relating the amplitude of the field at time t to its amplitude τ seconds later.

X-Ray, neutron, or light radiation scattered elastically in the forward direction (zero-angle scattering for X-rays or neutrons, Rayleigh scattering for light) by macromolecules in solution contains information on molar masses and density increments (but these are now scattering density or refractive index increments rather than mass density increments). This is because the forward scattered wave is simply the sum of all the waves scattered from the particle. For particles that are larger than the wavelength of the radiation, the waves scattered at finite scattering angles contain information on the spatial structure of the particle, because of interference between waves scattered from different points within the same particle. The dimensions of proteins or complexes such as ribosomes are of the order of 100

Å, much smaller than the wavelength of light but larger than the wavelengths of X-rays or neutrons (~1 to 10 Å).

We shall assume, for simplicity, that we are dealing with a very dilute solution of identical particles with random positions and orientations. Scattering data, however, can also be interpreted for polydisperse solutions and in cases in which there are interparticle interactions. The intensity scattered in the forward direction by a solution of particles of concentration c_2 is given by (Eisenberg, 1981):

$$I(0) = (\text{constant})M_2 c_2 (\partial \rho_x / \partial c_2)_\mu^2 \quad (10)$$

The constant depends on the radiation used, as does $(\partial \rho_x / \partial c_2)_\mu$, which is the refractive index increment for light, the electron density increment for X-rays, and the neutron scattering density increment for neutrons.

At a scattering angle 2θ, the scattered intensity is related to the structure of the particle by the Debye formula,

$$I(Q) = (\text{constant}) \sum \Delta \rho_i \Delta \rho_j \sin Q r_{ij} / r_{ij} \quad (11)$$

where $Q = (4\pi \sin \theta / \lambda)$, $\Delta \rho_i$, and $\Delta \rho_j$ are the excess scattering density or contrast (scattering density minus the scattering density of the solvent) at positions i and j in the particle, respectively, and r_{ij} is the distance between the two positions. The summation is over all ij pairs. We see that the intensity is a function of the moduli of the interatomic vectors in the particle (and not of their orientations) and of the contrast distribution. For the same particle, the contrast distribution will, in general, be different for different types of radiation and for different solvents. This is the basis of so-called contrast variation experiments in X-ray and neutron scattering that can be used to greatly increase the amount of information derived from solution scattering. X-Ray and neutron solution scattering experiments have been reviewed by Luzzati and Tardieu (1980) and Zaccai and Jacrot (1983), respectively.

At small values of Q, the Debye formula reduces to the Guinier approximation,

$$\ln I(Q) = \ln I(0) - (1/3)R_g^2 Q^2 \quad (12)$$

where R_g is the radius of gyration of excess scattering density in the particle. In practice the forward scattering is obtained from extrapolation of the data at small Q by using the Guinier approximation.

The morphological information on the particle obtained from a solution scattering curve has been discussed carefully by Luzzati and Tardieu (1980). For a particle in which the scattering density distribution is about constant, apart from the information contained in $I(0)$ and R_g, the surface area and total volume of the particle can be calculated uniquely. Because the same value can be obtained for the right-hand side of Eq. (11) from different structures, all of these, in principle, are compatible with the data, provided they have the same values of radius of gyration, surface area, and volume.

F. Thermodynamics of Protein–Salt–Water Interactions and Structural Models

In Section IV,E we saw how to obtain experimental data that are related to the molar mass of a particle and the appropriate density increment of the solution. We shall assume that both these parameters have been measured and discuss how they can be interpreted. In certain cases, the mass density increment could be measured directly by weighing calibrated volumes of the solution at dialysis equilibrium. These, however, are difficult experiments to perform in high salt concentrations.

The thermodynamics approach to macromolecules in solution and their interactions with solvent components have been reviewed by Eisenberg (1990). Briefly, the density increment of a solution containing three components—(1) water, (2) macromolecules, and (3) salt—due to an increase in the concentration of component 2 at constant chemical potential of the other components, is given by

$$(\partial \rho / \partial c_2)_\mu = 1 + \xi_1 - \rho^\circ (\bar{v}_2 + \xi_1 \bar{v}_1) \tag{13}$$

where ρ^0 is the density of the solvent, $\bar{v}_2$ and $\bar{v}_1$ are the partial specific volumes of macromolecule and water, respectively, and ξ_1 is an interaction parameter (grams of water per gram of protein). Because of solvent interactions, the chemical potential of solvent components will change on addition of the macromolecule; if the solution were dialyzed against a large volume of solvent (of the original composition), ξ_1 would be the mass of water (per gram of macromolecule) that enters the dialysis bag in order to reestablish the initial values for the chemical potentials of water and salt in the solvent. Equation (13) can equally well be written in terms of ξ_3 and $\bar{v}_3$, corresponding values for the salt component, with ξ_1 and ξ_3 related by $\xi_1 = -\xi_3/w_3$.

Eisenberg (1981) has developed a unified approach for the interpretation of ultracentrifugation and forward scattering. In a radiation scattering experiment, the scattering density increment is written similar to Eq. (13). For example, for neutron scattering,

$$(\partial \rho_N/\partial c_2)_\mu = b_2 + b_1\xi_1 - \rho_N^\circ(\bar{v}_2 + \xi_1\bar{v}_1) \tag{14}$$

where the subscript N denotes neutron scattering length densities, and b_2 and b_1 are values of neutron scattering length per gram of macromolecule and per gram of water, respectively. There is a similar equation for X-ray scattering with electron densities in the place of neutron scattering length densities.

Consideration of a model in which 1 g of macromolecule binds B_1 grams of water and B_3 grams of salt yields $\xi_1 = B_1 - B_3/w_3$, where the solvent composition is w_3 grams of salt per gram of water. In the model example of the macromolecule with associated water and salt molecules, it can be shown that

$$(\partial \rho/\partial c_2)_\mu = 1 + B_1 + B_3 - \rho^\circ(\bar{v}_2 + \bar{v}_1 B_1 + \bar{v}_3 B_3) \tag{15}$$

$$(\partial \rho_N/\partial c_2)_\mu = b_2 + B_1 b_1 + B_3 b_3 - \rho_N^\circ(\bar{v}_2 + \bar{v}_1 B_1 + \bar{v}_3 B_3) \tag{16}$$

The complementarity between ultracentrifugation and radiation scattering experiments is obvious from Eqs. (15) and (16), provided b_2, b_1, and b_3 are different from each other, as is the case for the neutron scattering amplitudes of protein, water, and salt. This complementarity has been discussed in detail by Zaccai et al. (1986a). In particular, when the density increment versus solvent density plot is a straight line (implying that $V_T = \bar{v}_2 + \bar{v}_1 B_1 + \bar{v}_3 B_3$ could be treated as constant), an invariant particle model could be considered for these conditions (Tardieu et al., 1981), and constant values of B_1, B_3, and V_T derived from a comparison of the density and scattering density plots.

G. Halophilic Malate Dehydrogenase

Malate dehydrogenase from *H. marismortui* (*h*MDH) is the halophilic protein that has been studied most by solution structure methods. A molar mass of 87 kg/mol was determined for the native enzyme. It is stable at high concentrations of NaCl or KCl and unfolds and dissociates below 2.5 M salt. Pundak and Eisenberg (1981) first measured values for the solvent interactions of *h*MDH and found that, in contrast to nonhalophilic globular proteins in similar conditions ($B_1 \sim 0.2$–0.3 g/g, $B_3 \sim 0.01$ g/g), the halophilic protein bound extraordinary amounts of water and salt ($B_1 \sim 0.8$–1.0 g/g, $B_3 \sim 0.3$ g/g). Zaccai *et al.* (1986a, 1989) and Calmettes *et al.* (1987) have used neutron scattering and ultracentrifugation to derive B_1 and B_3 and $\bar{v}_2$ as functions of the corresponding solvent density. Some of these data are shown in Fig. 6. The data show that different solution particles are formed in the different salt solutions. The parameters calculated are given in Table III. Only in potassium phosphate buffer, pH 7, does hMDH have a solution structure similar to a nonhalophilic globular protein. In $MgCl_2$, it binds as many water molecules as in NaCl or KCl, but significantly fewer salt molecules. An interesting and important observation is that the exceptional solvent binding of hMDH is associated with its quaternary structure (Pundak *et al.*, 1981; Zaccai *et al.*, 1986b). The experiments on the dissociated, unfolded protein in 1 M NaCl are consistent with ~ 0.2 g of water and negligible salt associated with 1 g of unfolded protein.

The interpretation of X-ray and neutron scattering curves of hMDH (Reich *et al.*, 1982; Zaccai *et al.*, 1986a; Calmettes *et al.*, 1987) suggested the solution particle to be globular with a relatively large surface-to-volume ratio. The schematic structure model presented by Zaccai *et al.* (1986a) results from the scattering curves to large angles

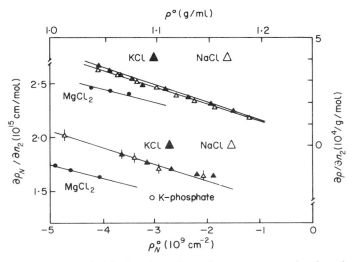

FIG. 6. Mass (top and right-hand y axis) and neutron scattering length (bottom and left-hand y axis) density increments for hMDH as functions of the respective solvent densities. From Zaccai *et al.* (1989), with permission.

TABLE III
Composition Parameters of the hMDH Solution Particles, Calculated from Neutron-Scattering and Mass Density Increments

hMDH in	KCl (1–4 M)	MgCl$_2$ (0.5–1.0 M)	Potassium phosphate (1.5 M)
Protein (M_2 g/mol)	87,000	87,000	87,000
Particle specific volume (V_{tot} cm^3/g)	1.79	1.71	
Particle volume (Å^3) ($M_2 V_{tot}/N_A$)	259,000	247,000	166,000[b]
Protein specific volume ($\bar{v}_2$ cm^3/g)	0.76 ± 0.01	0.74 ± 0.01	0.75[c]
"Bound" water N_1 (mol water/mol protein)	4080 ± 400	4000 ± 400	2000 ± 1000
"Bound" salt N_3 (mol salt/mol protein)	520 ± 60	120 ± 30	Negligible
Molality of "bound" salt ($N_3 \times 1000)/N_1 \times M_1$)	7.2	1.6	—

[a]From Zaccai *et al.* (1989), with permission.
[b]The sum of the protein and hydration volumes.
[c]Because mass density increments could not be determined for potassium phosphate solutions, a value of $\bar{v}_2 = 0.75$ was assumed in order to interpret the neutron scattering density increment.

and the radius of gyration values as a function of the contrast between the particle and the solvent. Such "contrast variation" experiments have allowed an estimate of the radius of gyration of the protein moiety of the particle and of the radius of gyration of the associated water and salt moiety. The model has a protein core with domains extending outward, on which the solvent interactions take place. These domains provide the larger surface area of hMDH when compared to MDH.

The gene coding for hMDH has recently been isolated and sequenced (Cendrin *et al.*, 1992). The gene coding for hMDH was also cloned into an *E. coli* expression vector and expressed from the strong promoter of phage T7. The resulting polypeptide appeared to fold properly and the enzyme became active when the salt concentration was increased. The molar mass of the polypeptide was calculated to be 33 kg/mol from the amino acid sequence. A similar value was obtained for the native hMDH subunit from mass spectrometry, and neutron scattering curves of the recombinant and native enzymes were indistinguishable. The molar mass of a dimer calculated from the sequence (66 kg/mol), however, is distinctly lower than the values (84–87 kg/mol) obtained from solution studies. Symmetry considerations from X-ray diffraction studies in progress on hMDH crystals are consistent with both dimer and tetramer structures for the enzyme (O. Dym and J. Sussman, private communication). L-Lactate dehydrogenase, with which hMDH shows the strongest homology, is a tetramer. The molecular mass of an hMDH tetramer based on its sequence would be 132 kg/mol, too high when compared to the solution studies value. Nevertheless, the experimental value could be accounted for by a mixture of approximately 75% by weight dimers and 25% by weight tetramers. However, such an inhomogeneity has not been observed in any of the physical studies (analytical ultracentrifugation, light, X-ray, and neutron scattering), which were undertaken over a broad concentration range and with different native (as well as recombinant) enzyme preparations. However, in equilibrium sedimentation of hMDH in CsCl density gradients (Eisenberg *et al.*, 1978), a stable additional high-molecular-weight component (of different density than the original enzyme band) was found, but this interesting observation was not followed up. These results are presently subject to further investigation, and may lead to a reevaluation of previous data concerning the molar mass and solvent interactions.

H. Model for Halophilic Protein Stabilization

A stabilization model proposed for hMDH by Zaccai *et al.* (1989) accounts for all of the observations on the stability of this protein and on its solution structure. It is based on the fact that the hMDH solution particles are different in the different solvents in which the enzyme is active, and on the reasonable assumption that the bound water and salt molecules are not associated separately with the polypeptide but as *hydrated salt ions*.

At low salt concentrations, the protein is unfolded because the hydrophobic interaction in these conditions is not sufficient to stabilize the folded structure. In molar concentrations of strongly salting-out potassium phosphate, however, the hydrophobic interaction is sufficient to stabilize the folded protein; at very high concentrations of this salt, it forms crystals (Harel *et al.*, 1988). The hydration of hMDH in potassium phosphate is similar to that of a nonhalophilic protein. Zaccai *et al.* (1989) suggested, therefore, that hMDH at low salt concentrations and in potassium phosphate could be understood in terms of the hydrophobic interaction dominating its stabilization, but with a shift toward higher concentrations of salting-out ions when compared to nonhalophilic globular proteins. The temperature dependence of stabilization in potassium phosphate (Zaccai *et al.*, 1989) is in agreement with this description, the curves with the plateaus near room temperature suggesting a dominance of the entropy term of the hydrophobic interaction. Recent experiments on the temperature dependence of stability in D_2O solutions support the validity of this description because they show the additive salting-out effects of D_2O and the phosphate ions (F. Bonnete and G. Zaccai, unpublished data).

The picture for *h*MDH in solvents with NaCl, KCl, or $MgCl_2$ is qualitatively different from that in strongly salting-out conditions (Fig. 7). The model suggests that the protein in NaCl, KCl, and $MgCl_2$ solvents is stabilized by the formation of the solution complexes given in Table III. The *h*MDH dimer in NaCl or KCl interacts with about 500 Na^+ ions and about 500 Cl^- ions hydrated by about 4000 water molecules. In $MgCl_2$, the complex contains approximately the same number of water molecules but much fewer salt ions. This is in agreement with the proposal that solvation is via hydrated ions because Mg^{2+} can coordinate more water molecules than either Na^+ or K^+ (Enderby *et al.* 1987). Halophilic proteins are particularly rich in acidic residues (Eisenberg and Wachtel, 1987), and carboxyl groups are the most solvated of the polar groups in amino acids. Even so, the number of carboxyl groups in the protein is not sufficient to account for the solvation values observed, and it must be assumed that the interaction of the protein with hydrated salt ions is cooperative and related to its tertiary or quaternary structure. This is indeed what is observed; when the protein is dissociated and unfolded, it loses its exceptional solvation properties. It is argued that despite its composition, rich in acidic groups, the protein alone would be unable to compete with essentially saturated salt solvents, so it has evolved a tertiary and quaternary structure that can compete effectively by coordinating hydrated salt *at a higher local concentration than in the solvent*. The model can account, at least qualitatively, for all the experimental observations on *h*MDH (Zaccai *et al.*, 1989). The models for *h*MDH in different solvents are drawn schematically in Fig. 7. The primary and tertiary structures of *h*MDH should be solved; these studies are in progress (O. Dym and J. Sussman, private communication; Harel *et al.*, 1988). Sufficient data are not yet available to say if the model described here is applicable to all halophilic proteins.

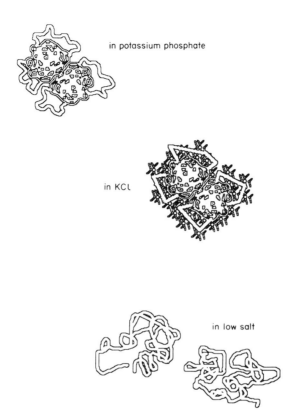

Fig. 7. Schematic representation of hMDH solution structures. The active structures have two parts: a catalytically active core, conceivably similar to that in nonhalophilic MDH, and protruding loops, required for stabilization in KCl, NaCl, and $MgCl_2$ solvents. In potassium phosphate the protein dimer is stabilized by the hydrophobicity of the core and the protruding loops are disordered. In KCl (or NaCl) the protein is stabilized by the interaction of the loops in a specific protein–water–salt hydration network. In $MgCl_2$ a similar structure exists with the same amount of water molecules coordinated by fewer salt ions. In low salt concentration, the protein is unfolded and its hydration is like that of nonhalophilic proteins. From Zaccai et al. (1989), with permission.

I. Halophilic EF-Tu and Other Halophilic Proteins under Study

The elongation factor Tu from *H. marismortui* (*h*EF-Tu), of molar mass ~43 kg/mol, has been purified (Guinet et al.,1988) and preliminary studies have shown it to have stabilization behavior and solution structures similar to hMDH (Ebel et al., 1992). The structure of its gene has been published (Baldacci et al., 1990). The derived protein sequence shows highest homology (61%) with EF-Tu from *Methanococcus viannelii*, another archaebacterium. The excess of acidic amino acids in hEF-Tu results mainly from mutations in basic and neutral polar amino acids. A crystal structure is available only for *E. coli* EF-Tu (Nyborg and LaCour, 1989). When the *E. coli* and halophilic EF-Tu sequences are aligned, the residues corresponding to the excess acidic groups—which are quite far apart in the primary structure—appear grouped together in patches on the surface of the tertiary structure, similar to the case of hDHFR (Section III,D). This is very interesting because, if the model were correct, each patch could contribute to cooperatively coordinating the hydrated ion solvation network on the protein surface. The interactions of hEF-Tu with aminoacyl-tRNA in halophilic conditions are now being studied (C. Ebel, private communication).

Halophilic proteins whose solution structures are currently under study include glyceraldehyde-3-phosphate dehydrogenase from *Haloarcula vallismortis* (Krishnan and Altekar, 1990) and a heme-binding catalase–peroxidase from *H.marismortui* (F. Cendrin, H. Jouvre, and G. Zaccai, private communication).

J. Conclusions

Macromolecular structures from halophilic bacteria are studied in order to improve our understanding of the molecular mechanisms of adaptation to high salt concentration environments, but also because in a number of cases they have advantages as models for more general biological processes. Purple membranes, which occur naturally as well-ordered two-dimensional crystals, have been of great importance for high-resolution diffraction studies of membrane proteins. The results of their study might be of even greater interest because the α-helical structure of bacteriorhodopsin is likely to be related to that of the family of receptors in higher organisms associated with G proteins (such as rhodopsin). Halophilic ribosomes are studied because they form crystals of high quality—probably because of their halophilic nature. Halophilic malate dehydrogenase might serve as a model for protein stabilization as a function of solvent environment because of the broad range of conditions in which it is active and stable. Because hMDH is active in potassium phosphate and in KCl solutions, for example, it can be assumed that its active site is identical in these two solvents. The protein, therefore, could be considered to be made up of two parts: a constant active site and a part of the structure that can adapt to different environments and is necessary for its stabilization. An interesting view is that hMDH behaves like a family of proteins from different organisms but with the same structure that may have different features related to their environments but a constant active site. An interesting difference is that, with hMDH, it is the same primary structure that yields different tertiary structures.

V. Molecular Genetic Aspects

In the earlier stages of halobacterial research very little attention was paid to the study of genome organization. This situation was completely changed when Woese and colleagues included the extremely halophilic bacteria in their pioneering publication (Fox *et al.*, 1980), including them as members of the newly defined kingdom of archaebacteria. This recognition that the extremely halophilic bacteria are archaebacteria has helped to attract many molecular geneticists to study the genome organization of these microorganisms as well as to isolate and sequence numerous genes. Review material on the molecular genetics of halobacteria can be found, therefore, both in articles dealing with halophilic microorganisms (Kushner, 1985; Pfeifer, 1988) and in review articles dealing in the molecular genetics of archaebacteria (Dennis, 1986; Brown *et al.*, 1989). In the present review special emphasis will be paid to the contribution of the molecular genetics of halobacteria to understanding the adaptation of halophilic proteins to extreme salinities.

A. *Genome Organization*

Until recently only two major groups of extremely halophilic bacteria have been recognized. The first group included the rod-shaped and pleomorphic halobacteria and the second group included the coccoid halococci. It became apparent, however, that these two groups that comprised the taxonomic family of Halobacteriaceae should be further subdivided. The main problem in the taxonomic identification of halobacteria is the lack of readily identifiable phenotypic markers. Using several new criteria, for instance, membrane lipid composition and numerical taxonomy methods (Torreblanca *et al.*, 1986), resulted in the creation of new taxonomic relationships. Currently it is accepted that the rod-shaped and pleomorphic nonalkalophilic halobacteria are divided into three genera: *Halobacterium, Haloferax,* and *Haloarcula*. In spite of the dissimilarities among the different halobacteria, expressed in the need for this division, the tendency exists to generalize observations made in members of one group and to include members of other groups. As we shall see later, in a great majority of cases these generalizations seem to hold true, indicating that in spite of the differences among groups, major characteristics are shared by all halobacteria.

The genome size of halobacteria, based on renaturation rate curves (C_0t curves), was estimated to be 4100 kilobase pairs (kbp) (Moore and McCarthy, 1969b), which is similar to the size of the eubacterial genome. A common feature of halobacterial DNA is the existence in the bacterial cells of several components having different G + C composition (Moore and McCarthy, 1969a), a characteristic that was used at one point in the taxonomic description of the halobacteria. When halobacterial DNA is fractionated by CsCl density gradient (Moore and McCarthy, 1969a) or by malachite green bisacrylamide gel (Ebert and Goebel, 1985; Pfeifer and Betlach, 1985), the DNA is separated into two fractions having different G + C compositions. The main fraction (F I) comprises 70–89% of the total and has an average G + C content of 68%. The other fraction (F II) comprises 11–30% of the DNA and has significantly lower G + C composition of 55–58%. This F II component includes mostly plasmid DNA but also some chromosomal DNA that seems to be located in defined regions of the chromosome termed "islands" (Pfeifer and Betlach, 1985). By using DNA fragments derived from F I as hybridization probes it was possible to demonstrate that F I is composed mainly of unique sequences. On the other hand, the DNA of the F II component contains many repetitive sequences and "insertion elements."

Many of the early genetic studies were done on *H. halobium* and its related strains. The popularity of these strains stemmed from the fact that several interesting spontaneous mutations could be readily detected. Among the most characterized mutations are those that affect the production of the protein part of the purple membrane, the heavily studied light-driven proton pump bacteriorhodopsin. Spontaneous mutations occur at a frequency of 10^{-4}. Analysis of these mutations showed that in almost every case a "foreign" DNA sequence was introduced into the bacterioopsin *(bop)* structural gene or into sequences surrounding it.

A number of insertion sequences (termed ISH elements) have been isolated from Bop⁻ mutants of *H. halobium* (Simsek *et al.*, 1982; Pfeifer, 1988; DasSarma, 1989). Among the most characterized are ISH1, ISH2, and ISH26. ISH1 is 1118 bp long and has a specific integration site in *bop* (Simsek *et al.*, 1982; Betlach *et al.*, 1983). ISH2, on the other hand, seems to integrate into *bop* in a nonspecific manner and is 520 bp long (DasSarma *et al.*, 1983; Pfeifer *et al.*, 1984). ISH26 (1384 bp long) occurs in the genome of *H. halobium* in at least seven copies (Ebert *et al.*, 1987). One strand of this element contains two partially overlapping "open reading frames." At the ends of the element there are 16-bp terminal inverted repeats that are almost identical to the inverted repeats of ISH2. On integration, ISH26 causes an 11-bp duplication of the target DNA. The insertion sequences ISH27 and ISH28 were also isolated from Bop$^-$ mutations but so far have been only partially characterized. A special case is ISH24, which causes the Bop$^-$ phenotype by integrating at a locus 1407 bp upstream of the *bop* gene. This mutation could be phenotypically "reverted" as a result of an additional insertion of 588 bp distal to ISH24 and the *bop* gene. Unlike the other ISH elements that have the typical structural features of eubacterial transposable elements, including terminal inverted repeats and causing duplication of the target DNA on integration, this new insertion, termed ISH25, lacks terminal inverted repeats and the integration does not cause a duplication of the target DNA (Pfeifer *et al.*, 1984).

All the ISH elements mentioned above as causing the Bop$^-$phenotype can be found in the wild-type *H. halobium* genome as sites in the chromosome or in the plasmids. The existence of one or more copies of these elements as well as the existence of additional unidentified elements are probably the reason for the very dynamic rearrangements that occur in parts of the chromosome of *H. halobium*. Sapienza and Doolittle (1982) estimated that the chromosome of *H. halobium* contains at least 50 families of repeated elements, each of which may have 2 to 20 members. These repeated sequences were shown to affect or to be affected by spontaneous genomic rearrangements (Sapienza *et al.*, 1982). Quantitative analyses showed that such rearrangements occur at frequencies greater that 4×10^3 events per family per cell generation. Under these circumstances the authors calculated that the chance of two daughter cells produced by a single division to bear identical genomes is only 80%. An insight into the molecular basis for the high degree of genetic instability of *H. halobium* was gained by studying the population dynamics of *H. halobium* plasmids (Pfeifer *et al.*, 1988). It was found that the major 150-kb pHH1 plasmid underwent a series of deletions and rearrangements that led to the fusion of noncontiguous sequences. A detailed analysis of one derivative of pHH1–pHH4 (36 kb in size) showed that when cells harboring this plasmid were plated and analyzed it was possible to detect rearrangements of pHH4 due to transpositions and deletions.

The picture evolving for the *H. halobium* genome organization is that numerous ISH elements exist in multiple copies, and although most of them are clustered in particular loci in the chromosome (the "islands") or in the plasmids, some of these elements transpose at rather high frequencies into the other regions of the chromosome and cause mutations in the sites of integration. This picture of the *H. halobium* genome might explain the fact that this species has many nutritional requirements and can therefore grow only in complex media, rendering *H. halobium* unsuitable for formal genetic analysis.

Some halobacteria are known, however, to be prototrophic, requiring for growth only simple carbon and nitrogen sources, for instance, glycerol and ammonia, respectively. Considerable progress has been made in an attempt to map the chromosome of one of these prototrophic halobacteria, *H. volcanii* (Charlebois *et al.*, 1989). In this attempt a cosmid bank of partial *Mlu*I restriction fragments was made. Overlapping cosmids were recognized by a strategy that made use of the distinctive *Mlu*I restriction fragments that contain relatively rare restriction sites. Each site recognized by these infrequently cutting restriction enzymes became a "landmark" from which different regions of the chromosome could be identified. The overlapping cosmids were ordered into "contigs"—continuous chromosomal region. The gaps between the contigs were filled by "chromosomal walking" using sequences located at the end of the contig as hybridization probes to rescreen the cosmid library. In the most recent publication of this group (Lam *et al.*, 1990), it is reported that 154 cosmids have been mapped and linked into 11 large map fragments. It is estimated that these 154 cosmids cover more than 95% of the *H. volcanii* chromosome. The availability of this linked cosmid bank enabled, as will be mentioned later, the isolation of two metabolic operons and will unquestionably facilitate the determination of the structure of the *H. volcanii* genome.

B. Genetic Tools

It is indispensable for any formal genetic analysis to be able to isolate mutants defective in genes coding for important cellular functions. In addition to the spontaneous mutations mentioned above, as caused by ISH elements, several protocols of induced mutagenesis were introduced. The most efficient method so far is the use of the alkylating agent ethyl methane sulfonate to cause point mutations

(Mevarech and Werczberger, 1985; Lam et al., 1990). Using this agent, hundreds of auxotrophic mutants of *H. volcanii* were isolated and phenotypically identified. In order to facilitate the identification of the auxotrophic mutants, an elegant method was introduced by Rosenshine (1990). The idea behind this method is to plate the mutagenized suspension of cells (after allowing proper time for phenotypic expression) on minimal plates supplemented with trace amounts of yeast extract. Colonies of auxotrophic mutants can be easily distinguished by their small size. In a different protocol of mutagenesis (Soppa and Oesterhelt, 1989), X-ray irradiation was used as the mutagenizing agent and mutants were enriched by several cycles of exposures to 5-bromo-2'-deoxyuridine followed by UV irradiation, which killed the growing cells but did not kill cells that did not incorporate the bromodeoxyuracil into their chromosome.

The availability of auxotrophic mutants enabled the identification of a natural genetic transfer system in *H. volcanii* (Mevarech and Werczberger, 1985). This genetic system has several interesting features. As in the case of "classical" bacterial conjugation, this system requires physical contact between the cells and can take place only when both the donor and the recipient cells are alice. However, unlike the classical bacterial conjugation, in the genetic transfer of *H. volcanii* it is impossible to distinguish between donor strain and a recipient strain. In this system, it seems, every cell can be a donor, a recipient, or both. This fact was demonstrated using genetic analysis (Mevarech and Werczberger, 1985) and using plasmids that served as cytoplasmic markers (Rosenshine *et al.*, 1989). Further analysis of this unusual genetic transfer system suggested that a prerequisite for the genetic transfer is the establishment of cytoplasmic continuities between the cells. These cytoplasmic continuities could be observed in electron micrographs taken of cells grown on solid surfaces (Rosenshine *et al.*, 1989).

The genetic transfer system of *H. volcanii* is so far the only one described in the halobacteria and, more generally speaking, in the archaebacteria. However, it seems that this might be largely due to the fact that suitable genetic markers are still lacking in the other archaebacteria or to the fact that not enough attempts were made to be able to identify such a system. A more general approach to the study of the genetics of halobacteria was taken by the group of W. F. Doolittle. In a series of experiments it was demonstrated that DNA can be introduced very efficiently into spheroplasts of halobacterial cells. In the first demonstration that halobacteria can accept externally introduced DNA, halobacterial phage DNA (ϕH) was introduced into spheroplasts of *H. halobium* (Cline and Doolittle, 1987). Spheroplasts were obtained simply by lowering the magnesium ion concentration in the solution used for suspending the cells. The transfection of the spheroplasts with the phage DNA was mediated by poly(ethylene glycol). This transfection method was very efficient and resulted in 5×10^6–2×10^7 transfectants per 1 μg of phage DNA. It was a little more difficult to demonstrate that this method of DNA transfection can be utilized to transform spheroplasts with plasmid DNA. The reason was that naturally occurring halobacterial plasmids do not carry selectable markers. Therefore, the first demonstrations that halobacteria can be transformed with plasmids relied on laborious screening methods (Charlebois *et al.*, 1987; Hackett and DasSarma, 1989).

A breakthrough in this area was achieved when the first useful selectable marker was discovered. It was found that the halobacteria are sensitive to the drug mevinolin, which inhibits the enzyme 3-hydroxy-3methylglutaryl-CoA reductase (HMG-CoA reductase). When *Mlu*I restriction fragments of DNA isolated from spontaneous resistant mutants of mevinolin were shotgun cloned into the *H. volcanii* endogenous plasmid pHV2, and the DNA was used to transform *H. volcanii* spheroplasts, mevinolin-resistant colonies could be obtained (Lam and Doolittle, 1989). A further trimming of the mevinolin resistance locus and the pHV2 replicon together with the introduction of the *E. coli* replicon resulted in construction of the first shuttle vector of halobacteria, pWL102 (Lam and Doolittle, 1989). Another selectable shuttle vector has been reported (Holmes and Dyall-Smith, 1990). This vector is based on the replicon of the *Haloferax* plasmid pHK2 and uses a novobiocin-resistant allele of the halobacterial gyrase gene as a selectable marker. The mevinolin and novobiocin resistance vectors have a potential to become general vectors for a great number of halobacteria, as most of the halobacteria are sensitive to these drugs. Trimethoprim, which is a competitive inhibitor of the enzyme dihydrofolate reductase, inhibits only *H. volcanii* (Rosenshine, 1990). Cells of *H. volcanii* that contain several copies of the gene coding for DHFR become resistant to trimethoprim and therefore, when the gene coding for *H. volcanii* DHFR is cloned into the multicopy plasmid pWL102, the resultant plasmid confers resistance to trimethoprim (Blecher, 1990).

An expression vector for *H. volcanii* was constructed by Nieuwlandt and Daniels (1990) by incorporating the promoter of the *H. volcanii* tRNA gene into a derivative of the *H. volcanii* shuttle vector pWL102 (Lam and Doolittle, 1989). The applicability of this vector is limited, however, to transcription of genes. Expression vectors that will enable efficient transcription and translation of protein-coding genes *in vivo* are still unavailable.

C. Isolation of Genes

Several methods have been employed for the isolation of halobacterial genes. The most straightforward method is based on the use of specific synthetic oligonucleotide probes to screen for restriction fragments or clones in plasmid or phage banks. The sequence of these oligonucleotide probes is deduced from the known protein amino acid sequence. This approach was used to purify several halobacterial genes [*bop*(Dunn *et al.*, 1981), *hop* (Hegemann *et al.*, 1987), *sop* (Blanck *et al.*, 1989), and the *sod* gene coding for *H. cutirubrum* superoxide dismutase (May and Dennis, 1989), and *H. marismortui* EF-Tu (Baldacci *et al.*, (1990)]. When the isolated *sod* gene was hybridized to a Southern blot of restriction fragments of total genomic DNA of the same organism, two different hybridizing fragments were detected. It was found that *H. cutirubrum* has another gene that is very homologous to the *sod* gene (87% DNA sequence identity and 83% protein sequence identity). This putative gene was called *slg* (superoxide-like gene). It seems that both genes are transcribed, though only *sod* is affected by the presence of paraquat (a substance that causes a high oxidation stress). Coincidentally, the *slg* gene was isolated by two groups (Salin *et al.*, 1988; Takao *et al.*, 1989) and was mistakenly identified as the homologous *sod* gene.

In the case of the genes coding for the gas vesicle protein of *H. halobium* the homologous *gvpA* gene of cyanobacteria was used as a probe (Pfeifer *et al.*, 1989); DasSarma *et al.*, 1987). This case may be exceptional because, due to the large divergence of the amino acid sequences of the halobacterial proteins, most of the eubacterial and eukaryotic genes do not hybridize to the halobacterial DNA.

Escherichia coli expression libraries were used to purify the gene coding for the surface glycoprotein of *H. halobium* (Lechner and Sumper, 1987), the genes coding for *H. halobium* flagellins (Gerl and Sumper, 1988), and the genes coding for the larger subunits of RNA polymerase of *H. halobium* (Leffers *et al.*, 1989).

The isolation of the gene coding for *H. volcanii* DHFR (Rosenshine *et al.*, 1987) deserves special mention. As mentioned above, *H. volcanii* is very sensitive to the antifolate drug trimethoprim. Spontaneous resistant colonies arise at frequencies of 10^{-10}–10^{-9}. The molecular basis for the resistance in all the resistant mutants studied so far is an amplification of the chromosomal region coding for DHFR. This gene amplification induces, in turn, the overproduction of DHFR, thus overcoming the inhibitory effect of the competitive inhibitor. The isolation of the gene coding for DHFR from the amplified DNA region then became straightforward (Zusman *et al.*, 1989).

It seems, however, that the approach introduced recently to isolate halobacterial genes by complementing mutations in *H. volcanii*, by cosmids derived from the cosmid bank, is going to become in the near future a very rich source of knowledge of the structure of halobacterial genes. This approach is based on the observation that auxotrophic mutations of *H. volcanii* can be transformed to prototrophs by DNA isolated from wild-type strains (Cline *et al.*, 1989). The efficiency of transformation was of the order of 5×10^4 transformants per 1µg DNA. This relatively high transformation efficiency (four orders of magnitudes above reversion rate) testified not only to the fact that the DNA is accepted efficiently by the halobacteria, it is also recombined efficiently into the chromosome. Taking advantage of this observation, the *hisC* gene of *H. volcanii* was isolated by complementing a His⁻ mutant of this microorganism by cosmids derived from the ordered cosmid bank (see above) (Conover and Doolittle, 1990). Moreover, it was possible to locate the complementary information within the large cosmid insert by purifying the smaller restriction fragment using agarose gel electrophoresis and transforming the mutant with these gel purified fragments. Fragments as small as 2.7 kb could efficiently transform the mutant to wild type. Using the same approach, genes coding for enzymes in tryptophan biosynthesis were purified (Lam *et al.*, 1990).

D. *Transcript Organization and Structure*

In eukaryotic cells every gene is expressed as a separate transcriptional unit. In eubacteria, on the other hand, many genes are clustered and are expressed as polycistronic mRNAs. It was therefore interesting to find out what the situation is in the halobacteria, which are members of the third phylogenetic kingdom of the archaebacteria. As more and more genes are isolated and characterized it becomes apparent that many genes in halobacteria are clustered and transcribed as polycistronic mRNAs. The genes coding for the large subunits of the RNA polymerases of *H. halobium* and *H. morrhuae* are arranged in one transcriptional unit together with two open reading frames of unknown function (Leffers *et al.*, 1989). Many of the genes coding for ribosomal proteins were also found to be clustered into polycistronic units (Shimmin *et al.*, 1989); Spiridinova *et al.*, 1989). Similar to the genes coding for the tryptophan biosynthetic pathway in eubacteria, it was shown that in *H. volcanii* these genes are also transcribed as polycistronic mRNAs, though organized differently than in *E. coli* (Lam *et al.*, 1990). Some of the genes characterized are, however, transcribed as separate transcriptional units, for instance, the gene coding for superoxide dismutase of *H. cutirubrum* (May *et al.*, 1989) and bacteriorhodopsin (*bop*) (DasSarma *et al.*, 1984).

In some of the polycistronic mRNAs the genes are very tightly linked. In the L1e–L10e–L12e ribosomal protein operon of *H. cutirubrum*, the intergenic spacers are one and five nucleotides, respectively (Shimmin *et al.*, 1989), and in the L22e–S3e–L29e ribosomal protein operon of *H. halobium*, the initiation codons of downstream genes overlap the termination codons of the previous genes (Spiridinova *et al.*, 1989). Similarly, a tight linkage between genes was observed in the polycistronic mRNA coding for the larger subunits of the RNA polymerase (Leffers *et al.*, 1989).

Archaebacterial RNA polymerases are very different from their eubacterial counterparts and more closely resemble eukaryotic enzymes both in their subunit complexity and in their amino acid sequences (for review, see Puehler *et al.*, 1989). This view is also reflected in the diversity of the DNA sequences that are used by the transcription apparatus as signals for initiation of transcription, namely, the promoters. Many attempts were made to identify a "consensus" promoter structure (Zillig *et al.*, 1988). However, as more genes are isolated and characterized, the picture becomes less coherent. Earlier identification of two upstream sequences, box A and box B, located around positions -30 and $+1$, respectively, gave way to two "elements"—DPE (distal promoter element) and PPE (proximal promoter element)—located -38 to -25 and -11 to -2, respectively (Reiter *et al.*, 1990). The DPE encompasses the box A sequence TTTA(A or T)A, but the PPE sequence seems to depend more on an (A + T)-rich sequence rather than on a specific DNA sequence.

Whereas in most of the halobacterial genes analyzed so far one can find a sequence located between -25 and -35 bp upstream from the transcription initiation site that resembles the DPE sequence, the PPE can barely be recognized. On the other hand, by comparing upstream sequences of genes that belong to the same gene family, one can find new sequence homologies that might hint of specific cis elements, for instance, when the genes coding for rRNA of *H. halobium* (Mankin *et al.*, 1984), *H. cutirubrum* (Dennis, 1985), and *H. marismortui* (Mevarech *et al.*, 1989) are compared or when the upstream sequences of the genes coding for the ribosomal proteins are compared (Shimmin et al., 1989). The evolving picture is that in the archaebacteria in general and in the halobacteria in particular the definition of the consensus promoter sequence is not as simple as in eubacteria. It is possible that different gene families use different transcription initiation factors (transacting elements) and therefore transcription initiation is promoted by different DNA sequences (cis elements). The definition of these elements must probably wait till *in vitro* transcription assays are available. *In vitro* transcription assays might help in the identification of the *cis* elements, as was the case for the 16S/23S rRNA encoding the DNA of *Sulfolobus* (Reiter et al., 1990). Also, they might help to identify protein factors (*trans*-acting elements) that are required for specific transcription initiation (Huedepohl et al., 1990).

The complex picture of the halobacterial promoters is reflected also in the lack of knowledge of the mode of regulation of halobacterial genes. Although it is clear that the expression of some of the characterized genes is regulated by physiological conditions, for instance, oxygen tension and light [bacteriorhodospin (Sumper and Herrmann, 1978)] and salt concentration [*Haloferax mediterranei* vacuoles (Englert et al., 1990)], so far nothing is known about the mode of this regulation. It is even not clear whether the expression is regulated positively or negatively. Expression of the *bop* gene was shown to be effected by two additional genes, *brp* and the *bat*, both located upstream of the *bop* gene and transcribed in the opposite orientation (Betlach et al., 1989). A complicated transcription regulation scheme was also demonstrated for the expression of the *H. halobium* phage ɸH (Gropp et al., 1989).

The length of the 5' untranslated mRNA varies greatly among the halobacterial transcripts. In many cases the transcript starts very few base pairs (2–3 bp) upstream of the first transcribed codon [for instance, *bop* (Dunn et al., 1981) and *sod* (May and Dennis, 1989)]. In other cases the 5' untranslated region is very long, for instance, in the gene coding for the cell surface glycoprotein in which the first translated codon resides 111 bp downstream from the initiation of transcription (Lechner and Sumper, 1987).

The existence of the extremely short 5' untranslated mRNA of many halobacterial mRNAs raises the question of the location of the signals for the initiation of translation (equivalent to the Shine–Dalgarno sequences). In some cases it is possible to detect a sequence complementary to the 3' end of the 16S rRNA at the 5' untranslated transcripts [as in the cases of *sop* (Blanck et al., 1989), the genes coding for the large subunits of RNA polymerase (Leffers et al., 1989), and the genes coding for ribosomal proteins (Shimmin et al., 1989; Spiridinova et al., 1989)]. In other cases these putative Shine–Dalgarno sequences could be found in the coding regions of the genes [as in the cases of *hop* (Blanck and Oesterhelt, 1987) and the genes coding for flagellins (Gerl and Sumper, 1988)]. In the *sod* gene no Shine–Dalgarno sequence is evident upstream or downstream of the initiation codon. It is therefore left to be shown, probably by analyzing the translation of transcripts using *in vitro* translation systems, how ribosomes recognize the sites for the initiation of translation.

Termination of transcription occurs usually at a stretch of T residues preceded by a GC-rich region (Dennis, 1986). Sometimes this GC-rich region is involved in a short, inverted repeat (May et al., 1989).

E. What Can We Learn about Halophilic Enzymes?

Molecular genetic tools are extremely useful in the quick analysis of protein sequences. Moreover, when predictions can be made about the functional significance of particular amino acids, these predictions can be tested experimentally by site-specific mutagenesis of these amino acids and functional analysis of the mutated proteins. Examination of the amino acid sequences of the halobacterial proteins can be made at different levels. The first level is phylogenetic. By comparing amino acid sequences of halobacterial proteins to those of homologous proteins from eubacterial and eukaryotic organisms, it is possible to draw conclusions regarding the phylogenetic position of the halobacteria. The second level is functional. A comparison of the amino acid sequences of homologous proteins from different sources can help to determine which regions of the proteins are conserved and therefore might have functional significance. The third level is to determine the specific properties of the halophilic proteins that might be significant for their adaptation to function at extreme salinities.

These three levels were demonstrated very nicely in the analysis of the primary structure of the large subunits of the RNA polymerases of *H. halobium* and *H. morrhuae* (Leffers et al., 1989). Analysis of the gene cluster coding for these subunits identified the amino acid sequences of four polypeptides: B'', B', A, and C. When these sequences were aligned with the sequences of known RNA polymerases, the B'' + B' subunits fitted the eukaryotic polymerase IIB subunit of yeast and the β subunit of the *E. coli* polymerase, where they corresponded to the N- and C-terminal parts, respectively. However, whereas the sequence alignment with the yeast polymerase IIB subunit demonstrated a high level of similarity throughout the sequence, the level of similarity to the B subunit of the *E. coli* enzyme was much lower and was confined to discrete regions. Similarly, the combined A + C subunits correspond to the N- and C-terminal regions of the A subunit of the eukaryotic RNA polymerase I, II, and III and the *E. coli* β' subunit. Again, the similarity to the eukaryotic enzymes was higher than to the eubacterial enzyme. When the sequence alignments were used to construct phylogenetic trees, it was found that the *H. halobium* sequences branch out from the tree at the same position as the sequences of the RNA polymerase of the extremely thermophilic archaebacterium *Sulfolobus acidocaldarius*. This confirmed the monophyletic nature of the archaebacterial kingdom.

Very little is known about the structural domains that are responsible for the multiple functions of RNA polymerase. Therefore, comparison of the conserved amino acid sequences of the halobacterial and the eukaryotic enzymes might hint of the possible functional importance of these regions. For instance, the sequence alignment revealed that the zinc-binding motifs suggested for the eukaryotic enzymes are also conserved in the A and B' subunits of the *H. halobium* enzyme, but only the former motif is partially conserved in the β' subunit of the *E. coli* enzyme. Also, the splits of the B and the A subunits of the eukaryotic enzymes into two parts each in the halophilic enzyme might suggest a division of these subunits into two functional domains.

The unique halophilic aspects of the halobacterial RNA polymerases are expressed in the high net negative charge of their subunits in contrast to the values found for the RNA polymerases of other organisms. The acidic amino acids tend to group throughout the primary structure, hinting at regions that are exposed to the solvent. In some halophilic proteins, clustering of some acidic residues is confined to specific loci. For instance, when the primary structures of the β subunit of the enzyme tryptophan synthase of *E. coli* and *H. volcanii* are aligned (Lam *et al.*, 1990), three insertions have to be introduced into the alignment. Half of the 16 inserted residues are acidic. In the case of the *H. halobium* ribosomal protein homologous to the *E. coli* S3, 30% of the residues are aspartic and glutamic acids. The distribution of these amino acids is not even but shows a specific clustering in the C-terminal region of the protein (in which the acidic amino acids comprise 50% of the sequence). This region seems to be "extra" when aligned to the *E. coli* S3 protein. A similar observation was made when the amino acid sequences of the halobacterial ferredoxins of *H. halobium* (Hase *et al.*, 1977) and *H. marismortui* (Hase *et al.*, 1980) were compared to ferredoxins of nonhalophilic organisms. In these cases, however, the "extra" acidic sequences reside in the N-terminal part of the molecule. Almost all the amino acid sequences of halophilic proteins determined so far confirm the old observations made by amino acid analyses, i.e., that the halophilic proteins are acidic.

Because the three-dimensional structures of the halobacterial proteins have not yet been determined, it is still not clear how the negative charges are distributed in space. Baldacci *et al.* (1990) assumed that the structure of the *H. marismortui* elongation factor EF-Tu is homologous to that of the *E. coli* EF-Tu and then superimposed the amino acid sequence of the halophilic protein on the nonhalophilic structure. As a result of this rough approximation it was possible to demonstrate that in several regions of the molecule the negative charges are clustered spatially, producing charged patches.

So far the only case in which structural information led to predictions on the relation between structure and function, which, in turn, was tested experimentally by site-directed mutagenesis, is the case of the halobacterial light-driven proton pump bacteriorhodopsin. Structural analyses of this very interesting system together with those of the other two retinal proteins, the light-driven chloride pump halorhodopsin and the sensory rhodopsin, are beyond the scope of this review (see Oesterhelt and Tittor, 1989; Blanck *et al.*, 1989; Popot *et al.*, 1989; Henderson *et al.*, 1990). The results of the very elaborate site-directed mutagenesis program of bacteriorhodopsin have been summarized in a series of publications by the group of Khorana (for a list of references, see Mogi *et al.*, 1989). In this review we would like to mention only some of the methodological obstacles described in these studies.

Because no system exists for high-level expression of cloned DNA in halobacteria, the authors tried to express the *bop* gene in *E. coli* using several of the known high-level expression vectors (Dunn *et al.*, 1987); Karnik *et al.*, 1987). It was found that the production of bacteriorhodopsin in *E. coli* from strong promoters is deleterious to the cells. Therefore, only promoters that are tightly regulated could be efficiently used. The other problem in efficient production was that though the level of transcription of the cloned gene was high, for an unknown reason the level of translation was very low. In order to improve the translation levels the nucleotide sequence in the region coding for the N-terminal part of the protein was modified to increase the A + T content. A third problem in the production was the instability of the translation products. It was estimated that the half-life of the product is 8–10 min. The addition of hydrophobic amino acids to the N-terminal region increased the stability of the bacteriorhodopsin. It was speculated that this increase in the stability is the result of incorporation of the translation products into the membrane. Once the bacteriorhodopsin was produced in *E. coli* it had to be extracted from the *E. coli* membranes and reconstituted in liposomes with retinal.

It is not clear, so far, whether problems in the expression of the bacteriorhodopsin gene in *E. coli* are specific to this gene. It is possible that these problems stemmed from the fact that this protein is a membranal protein and its production in the *E. coli* cell is specifically restricted. Recently we succeeded in expressing two halophilic genes in *E. coli* at high levels. The gene coding for hDHFR of *H. volcanii* was produced in an inactive form and could be reactivated (Blecher, 1990). The gene coding for hMDH of *H. marismortui* was, surprisingly, produced in *E. coli* in its active form (Cendrin *et al.*, 1992).

It is hoped that development of expression vectors in halobacteria, as well as development of the methodology for gene expression of halophilic proteins in *E. coli* and their subsequent reactivation, will enable the use of the site-directed mutagenesis methodology in the elucidation of structural features that are responsible for the halophilic properties of these enzymes.

ACKNOWLEDGMENTS

We would like to thank Dr. Felicitas Pfeifer for critical reading of the part of this review that deals with genetics. We would like also to acknowledge the Office of Naval Research (United States), the Centre de la Recherche Scientifique (France), the National Council for Research and Development (Israel), and the Endowment Fund for Basic Research in Life Sciences—Charles H. Revson Foundation for their support. Parts of this article were written while M. M. was on a sabbatical leave at The Max-Planck-Institute for Biochemistry in Martinsried, Germany, and H. E. was a Visiting Scientist at the National Institutes of Health in Bethesda, Maryland.

REFERENCES

Albertsson, P.-Å. (1970). *Adv.Protein Chem.* **24,** 309–341.
Andrews, J., Fierke, C. A., Birdsall, B., Ostler, G., Feeney, J., Roberts, G. C. K., and Benkovic, S. J. (1989). *Biochemistry* **28,** 5743–5750.
Altekar, W., and Rajagopalan, R. (1990). *Arch. Microbiol.* **153,** 169–174.

Baldacci, G., Guinet, F., Tillit, J., Zaccai, G., and de Recondo, A.-M. (1990). *Nucleic Acids Res.* **18,** 507–511.
Baumeister, W., Wildhaber, I., and Phipps, B. M. (1989). *Can. J. Microbiol.* **35,** 215–227.
Bayley, S. T., and Morton, R. A. (1978). *CRC Crit. Rev. Microbiol.* **6,** 151–205.
Ben-Amotz, A., and Avron, M. (1990). *Trends Biotechnol.* **8,** 121–126.
Benkovic, S. J., Fierke, C. A., and Naylor, A. M. (1988). *Science* **239,** 1105–1110.
Betlach, M., Pfeifer, F., Friedman, J., and Boyer, H. W. (1983). *Proc. Natl. Acad. Sci. U.S.A.* **80,** 1416–1420.
Betlach, M. C., Shand, R. F., and Leong, D. M. (1989). *Can. J. Microbiol.* **35,** 134–140.
Beverley, S. M., Ellenberger, T. E., and Cardingley, J. S. (1986). *Proc. Natl. Acad. Sci. U.S.A.* **83,** 2584–2588.
Blakley, R. L. (1985). In "Folates and Pterins" (R. L. Blakley and S. J. Benkovic, eds.), Vol. 8, pp. 191–253. Wiley, New York.
Blanck, A., and Oesterhelt, D. (1987). *EMBO J* **6,** 265)–273.
Blanck, A., Oesterhelt, D., Ferrando, E., Schegk, E. S., and Lottspeich, F. (1989). *EMBO J.* **8,** 3963–3971.
Blaurock, A., Stoeckenius, W., Oesterhelt, D., and Scherphof, G. (1976). *J. Cell. Biol.* **71,** 1–22.
Blecher, O. (1990). M. Sc. Thesis, Tel Aviv University, Israel.
Bolin, J. J., Filman, D. J., Matthews, D. A., Hamlin, R. C., and Kraut, J. (1982). *J. Biol. Chem.* **257,** 13650–13662.
Bonete, M. J., Camaco, M. L., and Cadenas, E. (1986). *Int. J. Biochem.* **18,** 785–789.
Brown, J. W., Daniels, C. J., and Reeve, J. W. (1989). *CRC Crit. Rev. Microbiol.* **16,** 287–338.
Bystroff, C., Oatly, S. J., and Kraut, J. (1990). *Biochemistry* **29,** 3263–3277.
Calmettes, P., Eisenberg, H., and Zaccai, G. (1987). *Biophys. Chem.* **26,** 279–290.
Casassa, E. F., and Eisenberg, H. (1964). *Adv. Protein Chem.* **19,** 287–395.
Cendrin, F., Chroboczek, J., Zaccai, G., Eisenberg, H., and Mevarech, M. (1992). In preparation.
Charlebois, R. L., Lam, W. L., Cline, S. W., and Doolittle. W. F. (1987). *Proc. Natl. Acad. Sci. U.S.A.* **84,** 8530–8534.
Charlebois, R. L., Hofman, J. D., Schalkwyk, L. C., Lam, W. L., and Doolittle, W. F. (1989). *Can. J. Microbiol.* **35,** 21–29.
Chen, J.-T., Taira, K., Tu, C.P., and Benkovic, S. J. (1987). *Biochemistry* **26,** 4093–4100.
Christian, J. H. B., and Waltho, J. A. (1962). *Biochim. Biophys. Acta* **65,** 506–508.
Cline, S. W., and Doolittle, W. F. (1987). *J. Bacteriol.* **169,** 1341–1344.
Cline, S. W., Schalkwyk, L. C., and Doolittle, W. F. (1989). *J. Bacteriol.* **171,** 4987–4991.
Conover, R. K., and Doolittle, W. F. (1990). *J. Bacteriol.* **172,** 3244–3249.
Danon, A., and Caplan, S. R. (1977). *FEBS Lett* **74,** 255–258.
Danson, M. J. (1988). *Adv. Microb. Physiol.* **29,** 165–231.
Danson, M. J. (1989). *Can. J. Microbiol.* **35,** 58–64.
Danson, M. J., McQuattie, A., and Stevenson, K. J. (1986). *Biochemistry* **25,** 3880–3884.
DasSarma, S. (1989). *Can J. Microbiol.* **35,** 65–67.
DasSarma, S., RajBhandary, U. L., and Khorana, H. G. (1983). *Proc. Natl. Acad. Sci. U.S.A.* **80,** 2201–2205.

DasSarma, S., RajBhandary, U., and Khorana, H. (1984). *Proc. Natl Acad. Sci. U.S.A.* **81** 125–129.
DasSarma, S., Damerval, T., Jones, J.G., and Tandeau de Marsac, N. (1987). *Mol. Microbiol.* **1**, 365–370.
Dennis, P. P. (1985). *J. Mol. Biol.* **186**, 457–461.
Dennis, P. P. (1986). *J. Bacteriol.* **168**, 471–478.
Dundas, I. E. D. (1970). *Eur. J. Biochem.* **16**, 393–398.
Dunn, R. J., McCoy, J. M., Simsek, M., Majumdar, A., Chang, S. H., RajBhandary, U. L., and Khorana, H. G. (1981). *Proc. Natl. Acad. Sci. U. S. A.* **78**, 6744–6748.
Dunn, R. J., Hackett, N. R., McCoy, J. M., Chao, B. H., Kimura, K., and Khorana, H. G. (1987). *J. Biol. Chem.* **262**, 9246–9254.
Ebel, C., Guinet, F., Langowski, C., Gagnon, J., and Zaccai, G. (1992). *J. Mol. Biol.* **223**, 361–371.
Ebert, K., and Goebel, W. (1985). *Mol. Gen. Genet.* **200**, 96–102.
Ebert, K., Hanke, C., Delius, H., Goebel, W., and Pfeifer, F. (1987). *Mol. Gen. Genet.* **206**, 81–87.
Edwards, C. (1990). "Microbiology of Extreme Environments." Open University Press, Milton Keynes.
Eisenberg, H. (1976), "Biological Macromolecules and Polyelectrolytes in Solution." Oxford Univ. Press (Clarendon), London and New York.
Eisenberg, H. (1981). *Q. Rev. Biophys.* **14**, 141–172.
Eisenberg, H. (1990). *Eur. J. Biochem.* **187**, 7–22.
Eisenberg, H., and Wachtel, E. (1987). *Annu. Rev. Biophys. Biophys. Chem.* **16**, 69–92.
Eisenberg, H., Haik, Y., Ifft, J. B., Leicht, W., Mevarech, M., and Pundak, S. (1978). *In* "Energetics and Structure of Halophilic Microorganisms" (S. R. Caplan and M. Ginzburg, eds.), pp. 13–32. Elsevier, Amsterdam.
Enderby, J. E., Cummings, S., Herdman, G. J., Neilson, G. W., Salmon, P. S., and Skipper, N. (1987). *J. Phys. Chem.* **91**, 5851–5858.
Englert, C., Horne, M., and Pfeifer, F. (1990). *Mol. Gen. Genet.* **222**, 225–232.
Fierke, C. A., Johnson, K. A., and Benkovic, S. J. (1987). *Biochemistry* **26**, 4085–4092.
Filman, D. J., Bolin, J. J., Matthews, D. A., and Kraut, J. (1982). *J. Biol. Chem.* **257**, 13663–13672.
Fox, G. E., Stackebrandt, E., Hespell, R. B., Gibso, J., Maniloff, J., Dyer, T. A., Wolfe, R. S., Balch, W. E., Tanner, R., Magrum, L., Zablen, L. B., Blakemore, R., Gupta, R., Bonen, L., Lewis, B. J., Stahl, D. A., Luehrsen, K. R., Chen, K. N., and Woese, C. R. (1980). *Science* **209**, 457–463.
Gerl, L., and Sumper, M. (1988). *J. Biol. Chem.* **263**, 13246–13251.
Gibbons, N. E. (1974). *In.* "Bergey's Manual of Determinative Bacteriology" (R. E. Buchanan and N. E. Gibbons, eds.) 8th ed., pp. 269–273. Williams & Wilkins, Baltimore, Maryland.
Ginzburg, M., Sachs, L., and Ginzburg, B. Z. (1970). *J. Gen. Physiol.* **55**, 187–207.
Glatter, O., and Kratky, O. (1982). "Small Angle X–ray Scattering." Academic Press, London.
Gropp, F., Palm, P., and Zillig, W. (1989). *Can. J. Microbiol.* **35**, 182–188.
Guinet, F., Frank, R., and Leberman, R. (1988). *Eur. J. Biochem.* **172**, 687–694.
Guinier, A., and Fournet, G. (1955). "Small Angle Scattering of X–rays." Wiley, New York.
Hackett, N. R., and Das Sarma, S. (1989). *Can. J. Microbiol.* **35**, 86–91.
Harel, M., Shoham, M., Frolow, F., Eisenberg, H., Mevarech, M., Yonath, A., and Sussman, J. L. (1988). *J. Mol. Biol.* **200**, 609–610.

Hartmann, R., Sickinger, H.-D., and Oesterhelt, D. (1980). *Proc. Natl. Acad. Sci. U. S. A.* **77,** 3821–3825.
Hase, T., Wakabayashi, S., Matsubara, H., Kerscher, L., Oesterhelt, D., Rao, K. K., and Hall, D. O. (1977). *FEBS Lett.* **77,** 308–310.
Hase, T., Wakabayashi, S., Matsubara, H., Mevarech, M., and Werber, M. M. (1980). *Biochim. Biophys. Acta* **623,** 139–145.
Hecht, K., and Jaenicke, R. (1989a). *Biochemistry* **28,** 4979–4985.
Hecht, K., and Jaenicke, R. (1989b). *Eur. J. Biochem.* **183,** 69–74.
Hegemann, P., Blanck, A., Vogelsang-Wenke, H., Lottspeich, F., and Oesterhelt, D. (1987). *EMBO J.* **6,** 259–264.
Henderson, R., and Unwin, N. (1975). *Nature (London)* **257,** 28–32.
Henderson, R., Baldwin, J. M., Ceska, T. A., Zemlin, F., Beckmann, E., and Downing, K. (1990). *J. Mol. Biol.* **213,** 899–929.
Hochstein, L. I. (1988). *In* "Halophilic Bacteria" (F. Rodriguez-Valera, ed.), Vol. 2, pp. 67–83. CRC Press, Boca Raton, Florida.
Hochstein, L. I., and Dalton, B. P. (1973). *Biochim. Biophys. Acta* **302,** 216–228.
Holmes, M. L., and Dyall-Smith, M. L. (1990). *J. Bacteriol.* **172,** 756–761.
Holmes, P. K., and Halvorson, H. O. (1965). *J. Bacteriol.* **90,** 312–315.
Howell, E. E., Villafranca, J. E., Warren, M. S., Oatley, S. J., and Kraut, J. (1986). *Science* **231,** 1123–1128.
Howell, E. E., Foster, P. G., and Foster L. M. (1988). *J. Bacteriol.* **170,** 3040–3045.
Hubbard, J. S., and Miller, A. B. (1969). *J. Bacteriol.* **99,** 161–168.
Huedepohl, H., Reiter, W.-D., and Zillig, W. (1990). *Proc. Natl. Acad. Sci. U.S.A.* **87,** 5851–5855.
Javor, B. J. (1984). *Appl. Environ. Microbiol.* **48,** 352–360.
Javor, B. J. (1988). *Arch. Microbiol.* **149,** 433–440.
Kamekura, M., and Kates, M. (1988). *In* "Halophilic Bacteria" (F. Rodriguez-Valera, ed.), Vol. 2, pp. 25–54. CRC Press, Boca Raton, Florida.
Karnik, S. S., Nassal, M., Doi, T., Jay, E., Sgaramella, V., and Khorana, H. G. (1987). *J. Biol. Chem.* **262,** 9255–9263.
Kerscher, L., and Oesterhelt, D. (1981a). *Eur. J. Biochem.* **116,** 587–594.
Kerscher, L., and Oesterhelt, D. (1981b). *Eur. J. Biochem.* **116,** 595–600.
Kerscher, L., and Oesterhelt, D. (1982). *Trends Biochem. Sci.* **7,** 371–374.
Kessel, M., and Klink, F. (1981). *Eur. J. Biochem.* **114,** 481–486.
Kessel, M., Wildhaber, I., Cohen, S., and Baumeister, W. (1988). *EMBO J.* **7,** 1549–1554.
Khorana, H. G. (1988). *J. Biol. Chem.* **263,** 7439–7442.
King, T. P. (1972). *Biochemistry* **11,** 367–371.
Koenig, H., and Stetter, K. O. (1986). *Syst. Appl. Microbiol.* **7,** 300–309.
Koike, M., and Koike, K. (1976). *Adv. Biophys.* **9,** 187–227.
Krishnan, G., and Altekar, W. (1990). *J. Gen. Appl. Microbiol.* **36,** 19–32.
Kushner, D. J. (1985). *In* "The Bacteria" (C. R. Woese and R. S. Wolfe, eds.), Vol. 8, pp. 171–214. Academic Press, Orlando, Florida.
Lam, W. L., and Doolittle, W. F. (1989). *Proc. Natl. Acad. Sci. U.S.A.* **86,** 5478–5482.
Lam, W. L., Cohen, A., Tsouluhas, D., and Doolittle, W. F. (1990). *Proc. Natl. Acad. Sci. U.S.A.* **87,** 6614–6618.
Langworthy, T. A., and Pond, J. L. (1986). *Syst. Appl. Microbiol.* **7,** 253–257.
Lanyi, J. K. (1974). *Bacteriol. Rev.* **38,** 272–290.
Lanyi, J. K., and Silverman, M. P. (1972). *Can. J. Microbiol.* **18,** 993–995.
Larsen, H. (1981). *In* "The Prokaryotes" (M. P. Starr, H. Stolp, H. G. Trouper, A.

Balows, and H. G. Schlegel, eds.), Vol. 1, pp. 985–994. Springer-Verlag, Berlin and New York.
Larsen, H. (1986). *FEMS Microbiol. Rev.* **39**, 3–7.
Lechner, J., and Sumper, M. (1987). *J. Biol. Chem.* **262**, 9724–9729.
Leffers, H., Gropp, F., Lottspeich, F., Zillig, W., and Garrett, R. A. (1989). *J. Mol. Biol.* **206**, 1–17.
Leicht, W. (1978). *Eur. J. Biochem.* **84**, 133–139.
Leicht, W., and Pundak, S. (1981). *Anal. Biochem.* **114**, 186–192.
Leicht, W., Werber, M. M., and Eisenberg, H. (1978). *Biochemistry* **17**, 4004–4010.
Luzzati, V., and Tardieu, A. (1980). *Annu. Rev. Biophys. Bioeng.* **9**, 1–29.
Madon, J., and Zillig, W. (1983). *Eur. J. Biochem.* **133**, 471–474.
Madon, J., Leser, U., and Zillig, W. (1983). *Eur. J. Biochem.* **135**, 279–283.
Mankin, A. S., Tetrina, N. L., Bubtsov, P. M., Baratova, L. A., and Kagramanova, V. K. (1984). *Nucleic Acids Res.* **12**, 6537–6546.
May, B. P., and Dennis, P. P. (1987). *J. Bacteriol.* **169**, 1417–1422.
May, B. P., and Dennis, P. P. (1989). *J. Biol. Chem.* **264**, 12253–12258.
May, B. P., and Tam, P., and Dennis, P. P. (1989). *Can. J. Microbiol.* **35**, 171–175.
Mayhew, S. G., and Howell, L. G. (1971). *Anal. Biochem.* **41**, 466–470.
Mescher, M. F., and Strominger, J. L., (1976). *J. Biol. Chem.* **251**, 2005–2014.
Mevarech, M., and Neumann, E. (1977). *Biochemistry* **16**, 3786–3792.
Mevarech, M., and Werczberger, R. (1985). *J. Bacteriol.* **162**, 461–462.
Mevarech, M., Leicht, W., and Werber, M. M. (1976). *Biochemistry* **15**, 2383–2387.
Mevarech, M., Eisenberg, H., and Neumann, E. (1977). *Biochemistry* **16**, 3781–3785.
Mevarech, M., Hirsch-Twizer, S., Goldmann, S., Yakobson, E., Eisenberg, H., and Dennis, P. P. (1989). *J. Bacteriol.* **171**, 3479–3485.
Minton, A. P. (1983). *Mol. Cell. Biochem.* **55**, 2095–2101.
Mogi, T., Marti, T., and Khorana, H. G. (1989). *J. Biol. Chem.* **264**, 14197–14201.
Moore, R. L., and McCarthy, B. J. (1969a). *J. Bacteriol.* **99**, 248–254.
Moore, R. L., and McCarthy, B. J. (1969b). *J. Bacteriol.* **99**, 255–262.
Mullis, K. B., and Faloona, F. A. (1987). *In* "Methods in Enzymology" (R. Wu, ed.). Vol. **155**, Part F, pp. 335–350. Academic Press, San Diego.
Newton, G. L., and Javor, B. (1985). *J. Bacteriol.* **161**, 438–441.
Nieuwlandt, D. T., and Daniels, C. J. (1990). *J. Bacteriol.* **172**, 7104–7110.
Nissenbaum, A. (1975). *Microb. Ecol.* **2**, 139–161.
Nyborg, J., and LaCour, T. (1989). *NATO ASI Ser.*, A**65**, 3–14.
Oefner, C., D'Arcy, A., and Winkler, F. K. (1988). *Eur. J. Biochem.* **174**, 377–385.
Oesterhelt, D., and Kripphal, G. (1983). *Ann. Inst. Pasteur/Microbiol.* **134B**, 137–150.
Oesterhelt, D., and Tittor, J. (1989). *Trends Biochem. Sci.* **14**, 57–61.
Oren, A. (1983). *Lymnol. Oceanogr.* **28**, 33–41.
Papadopoulos, G., Dencher, N. A., Zaccai, G., and Büldt, G. (1990). *J. Mol. Biol.* **214**, 1519.
Pfeifer, F. (1988). *In* "Halophilic Bacteria" (F. Rodriguez-Valera, ed.), Vol. 2, pp. 105–133. CRC Press, Boca Raton, Florida.
Pfeifer, F., and Betlach, M. (1985). *Mol. Gen. Genet.* **198**, 449–455.
Pfeifer, F., Friedman, J., Boyer, H. W., and Betlach, M. (1984). *Nucleic Acids Res.* **12**, 2489–2497.
Pfeifer, F., Blaseio, U., and Ghahraman, P. (1988). *J. Bacteriol.* **170**, 3718–3724.
Pfeifer, F., Blaseio, U., and Horne, M. (1989). *Can. J. Microbiol.* **35**, 96–100.
Popot, J. L., Engelman, D. M., Gurel, O., and Zaccai, G. (1989). *J. Mol. Biol.* **210**, 829–847.

Puehler, G., Leffers, H., Gropp, F., Palm, P., Klenk, H.-P., Lottspeich, F., Garrett, R. A., and Zillig, W. (1989). *Proc. Natl. Acad. Sci. U. S. A.* **86**, 4569–4573.
Pundak, S., and Eisenberg, H. (1981). *Eur. J. Biochem.* **118**, 463–470.
Pundak, S., Aloni, H., and Eisenberg, H. (1981). *Eur. J. Biochem.* **118**, 471–477.
Reich, M. H., Kam, Z., and Eisenberg, H. (1982). *Biochemistry* **21**, 5189–5195.
Reiter, W.-D., Huedepohl, U., and Zillig, W. (1990). *Proc. Natl. Acad. Sci. U.S.A.* **87**, 9509–9513.
Richey, B., Cayley, D. S., Mossing, M. C., Kolka, C., Anderson, C. F., Farrar, T. C., and Record, M. T., Jr. (1987). *J. Biol. Chem.* **262**, 7157–7164.
Rimerman, R. A., and Hatfield, G. W. (1973). *Science* **182**, 1268–1270.
Rodriguez-Valera, F., ed. (1988). "Halophilic Bacteria," Vols. 1 and 2. CRC Press, Boca Raton, Florida.
Rosenshine, I. (1990). Ph. D. Thesis, Tel Aviv University, Israel.
Rosenshine, I., Zusman, T., Werczberger, R., and Mevarech, M. (1987). *Mol. Gen. Genet.* **208**, 518–522.
Rosenshine, I., Tchelet, R., and Mevarech, M. (1989). *Science* **245**, 1387–1389.
Russel, A. J., and Fersht, A. R. (1987). *Nature (London)* **328**, 496–500.
Salin, M. L., Duke, M. V., Oesterhelt, D., and Ma. D.-P. (1988). *Gene* **70**, 153–159.
Sapienza, C., and Doolittle, W. F. (1982). *Nature (London)* **295**, 384–389.
Sapienza, C., Rose, M. R., and Doolittle, W. F. (1982). *Nature (London)* **299**, 182–185.
Schachman, H. K. (1959). "Ultracentrifugation in Biochemistry." Academic Press, New York.
Shimmin, L. C., Newton, C. H., Ramirez, C., Yee, J., Downing, W. L., Louie, A., Matheson, A. T., and Dennis, P. P. (1989). *Can. J. Micorbiol.* **35**, 164–170.
Simsek, M., DasSarma, S., RajBhandary, U. L., and Khorana, H. G. (1982). *Proc. Natl. Acad. Sci. U. S. A.* **79**, 7268–7272.
Soppa, J., and Oesterhelt, D. (1989). *J. Biol. Chem.* **264**, 13043–13048.
Spiridinova, V. A., Akhmanova, A. S., Kargamanova, V. K., Koepke, A. K. E., and Mankin, A. S. (1989). *Can. J. Microbiol.* **35**, 153–159.
Stammers, D. K., Champness, J. N., Beddell, C. R., Dann, J. G., Eliopoulos, E., Geddes, A. J., Ogg, D., and North, A. C. T. (1987). *FEBS Lett.* **218**, 178–184.
Stoeckenius, W., and Bogomolni, R. (1982). *Annu. Rev. Biochem.* **52**, 587–616.
Sumper, M. (1987). *Biochim. Biophys. Acta* **906**, 69–79.
Sumper, M., and Herrmann, G. (1978). *Eur. J. Biochem.* **89**, 229–235.
Sumper, M., Berg, E., Mengele, R., and Strobel, I. (1990). *J. Bacteriol.* **172**, 7111–7118.
Sundquist, A. R., and Fahey, R. C. (1988). *J. Bacteriol.* **170**, 3459–3467.
Sussman, J. L., Brown, D. H., and Shoham, M. (1986). *In* "Frontiers of Iron-Sulfur Protein Research" (H. Matsubara, Y. Katsube, and K. Wada, eds.), pp. 69–82. *Jpn. Sci. Soc.*, Tokyo, and Springer-Verlag, Berlin.
Takao, M., Kobayashi, T., Oikawa, A., and Yasui, A. (1989). *J. Bacteriol.* **171**, 6323–6329.
Tardieu, A., Vachette, P., Gulik, A., and Le Maire, M. (1981). *Biochemistry* **20**, 4399–4406.
Thillet, J., Adams, J. A., and Benkovic, S. J. (1990). *Biochemistry* **29**, 5195–5202.
Torreblanca, M., Rodriguez-Valera, F., Juez, G., Ventosa, A., Kamekura, M., and Kates, M. (1986). *Syst. Appl. Microbiol.* **8**, 89–99.
Villafranca, J. E., Howell, E. E., Voet, D. H., Strobel, M. S., Ogden, R. C., Abelson, J. N., and Kraut, J. (1983). *Science* **222**, 782–788.

Volz, K. W., Matthews, D. A., Alden, R. A., Freer, S. T., Hansen, C., Kaufman, B. T., and Kraut, J. (1982). *J. Biol. Chem.* **257,** 2528–2563.
von Boehlen, K., Makowski, I., Hansen, H. A. S., Bartels, H., Zayzsev-Bashan, A., Meyers, S., Paulke, C., Franceschi, F., and Yonath, A. (1991). *J. Mol. Biol.* **222,** 11–15.
von der Haar, F. (1976). *Biochem. Biophys. Res. Commun.* **70,** 1009–1013.
von Hippel, H. P., and Schleich, T. (1969). *In* "Structure and Stability of Biological Macromolecules" (S. N. Timasheff and G. D. Fasman, eds.), pp. 417–574. Dekker, New York.
Wagner, C. R., and Benkovic, S. j. (1990). *Trends Biotechnol.* **8,** 263–270.
Werber, M. M., and Mevarech, M. (1978). *Arch. Biochem. Biophys.* **187,** 447–456.
Wilkansky (Volcani), B. (1936). *Nature (London)* **138,** 467.
Woese, C. R., and Fox, G. E. (1977). *Proc. Natl. Acad. Sci. U. S. A.* **74,** 5088–5090.
Yonath, A., and Wittmann, H. G. (1988). *In* "Methods in Enzymology" (H. F. Noller, Jr., and K. Moldave, eds.), Vol. **164,** pp. 104–110. Academic Press, San Diego.
Yonath, A., and Wittmann, H. G. (1989). *Trends Biochem. Sci.* **14,** 329–335.
Yonath, A., Bennett, W., Weinstein, S., and Wittmann, H. G. (1990). *In* "The Ribosome: Structure, Function and Evolution" (W. Hill, ed.), ASM Publ., pp. 134–147. Am. Soc. Met., Metals Park, Ohio.
Zaccai, G. (1987). *J. Mol. Biol.* **194,** 569–572.
Zaccai, G., and Eisenberg, H. (1990). *Trends Biochem. Sci.* **15,** 333–337.
Zaccai, G., and Jacrot, B. (1983). *Annu. Rev. Biophys. Bioeng.* **12,** 139–157.
Zaccai, G., Wachtel, E., and Eisenberg, H. (1986a). *J. Mol. Biol.* **190,** 97–106.
Zaccai, G., Bunick, G. J., and Eisenberg, H. (1986b). *J. Mol. Biol.* **192,** 155–157.
Zaccai, G., Cendrin, F., Haik, Y., Borochov, N., and Eisenberg, H. (1989). *J. Mol. Biol.* **208,** 491–500.
Zillig, W., Palm, P., Reiter, W.-D., Gropp, F., Puehler, G., and Klenk, H.-P. (1988). *Eur. J. Biochem.* **173,** 473–482.
Zusman, T. (1990). Ph.D. Thesis, Tel Aviv University, Israel.
Zusman, T., Rosenshine, I., Boehm, G., Jaenicke, R., Leskiw, B., and Mevarech, M. (1989). *J. Biol. Chem.* **264,** 18878–18883.

STRUCTURE AND STABILITY OF BOVINE CASEIN MICELLES

By C. HOLT

Hannah Research Institute, Ayr, KA6 5HL, Scotland

I. Caseins 63
   A. Introduction 63
   B. Nomenclature of Bovine Caseins 65
   C. Casein Concentrations in Milk 71
   D. Synthesis and Secretion of Caseins 74
II. Physicochemical Properties of Caseins 85
   A. Properties of Individual Caseins 85
   B. Association Behavior of Caseins 94
   C. Interaction of Caseins with Calcium Phosphate 103
III. Structure of Native Bovine Casein Micelles 105
   A. Appearance and Substructure of Casein Micelles 105
   B. Average Size, Size Distribution, and Voluminosity of Casein Micelles 111
   C. Location of κ-Casein in Micelles 115
   D. Structure of Micelle Surfaces 119
   E. Micellar Calcium Phosphate 123
   F. Dissociation of Casein Micelles 131
IV. Stability of Casein Micelles 133
   A. Introduction 133
   B. Forces Involved in Coagulation 134
   C. Steric Stabilization of Micelles 135
   D. Action of Chymosin on Casein Micelles 137
   References 143

I. CASEINS

A. Introduction

The casein micelle is an important and characteristic macromolecular assembly of mammalian biology, occurring in all milks that have been examined in sufficient detail. Its functions, insofar as they are known, are to form a coagulum in the stomach of the nursling, allowing the slow release of nutrients down the digestive tract, and to act as a means of transporting calcium and phosphate in a readily assimilable form from mother to young. As well as providing a source of amino acids, enzymatic cleavage of casein polypeptide chains can produce various types of biologically active peptides, such as the casomorphins, though whether any are physiologically important is not known (Maubois and Léonil, 1989). Another recent speculation is that the caseins have a role in protecting the mammary gland against ectopic calcification (Holt and Sawyer, 1988a,b), a hazard that is common to all tissues in contact with supersaturated calcium solutions. In this regard, the flexibility and relative lack of secondary structure in the caseins can be seen as essential features contributing to their biological function.

Elucidation of the structure and dynamics of the casein micelle presents researchers with a formidable challenge. A typical micelle contains on the order of 10^4 polypeptide chains of four basic types, together with about 3×10^3 microgranules of an amorphous calcium phosphate. The micelles are polydisperse in size and molecular weight, variable in composition, and, within the particles, long-range order arising from regular structural features has not been discerned. Moreover, there is a dearth of structural information on the individual caseins, none of which has been crystallized.

Our understanding of casein micelle structure has been strongly influenced by the work of David F. Waugh and collaborators (Waugh, 1971), who elaborated two principal ideas. The first was that one of the caseins (κ-casein) has a size-determining role with the micelle, forming a surface coat over a core of the three other types of casein. The second idea was that casein aggregates produced by the binding of Ca^{2+} provide a model of the native casein micelle. Indeed, calcium caseinate aggregates have sometimes been called casein micelles even though they contain no calcium phosphate. The ideas of the Waugh school have been fruitful but in the light of subsequent research have required amendment. In particular, whereas the size-determining role of κ-casein is still accepted, no conclusive evidence exists that it is exclusively located in the peripheral regions of the micelle. Moreover, mechanisms for controlling the size of micelles can be envisaged that do not require the κ-casein to be exclusively on the external surface of the particles. Other work has emphasized the role of calcium phosphate rather than Ca^{2+} ions in maintaining the structural integrity of micelles.

Another controversial and evolving idea concerning casein micelle structure is the concept of the submicelle. That there is some substructure to the micelle can hardly be denied, because all of the appropriate techniques have revealed some inhomogeneities over distances of 5–20 nm. Proponents of submicellar models of casein micelle structure interpret this evidence in terms of spherical particles of casein, the submicelles, joined together, possibly, by the calcium phosphate (Schmidt, 1982; Walstra, 1990). An alternative view (depicted schematically in Fig. 13, which will be discussed later) is that the micelle is an inhomogeneous and partly mineralized gel in which there are no discrete subunits.

Historically, ideas of casein micelle structure and stability have evolved in tandem. In the earlier literature, discussions of micellar stability drew on the classical ideas of the stability of hydrophobic colloids. More recently, the "hairy" micelle model has focused attention more on the hydrophilic nature of the micelle and steric stabilization mechanisms. According to the hairy micelle model, the C-terminal macropeptides of some of the κ-casein project from the surface of the micelle to form a hydrophilic and negatively charged diffuse outer layer, which causes the micelles to repel one another on close approach. Aggregation of micelles can only occur when the hairs are removed enzymatically, e.g., by chymosin (EC 3.4.23.4) in the renneting of milk, or when the micelle structure is so disrupted that the hairy layer is destroyed, e.g., by heating or acidification, or when the dispersion medium becomes a poor solvent for the hairs, e.g., by addition of ethanol.

Though the steric stabilization mechanism of the hairy micelle model represents an advance on early theories, it takes little account of the proteinaceous nature of the colloid and the often specific and stereoregular nature of many protein–protein interactions. Almost nothing is known about the conformational state of caseins in the casein micelle or of the nature of the interaction of these phosphoproteins with the calcium phosphate. We have no depth of understanding of why the caseins do not fold into a compact conformation or the related matter of why they readily form gels. Likewise, dynamic aspects of the casein micelle, such as the rates of exchange with the serum and the flexural motions of the polypeptide chains, have not been investigated sufficiently. As a result we presently have an overwhelmingly static picture of the micelle. There is therefore a great deal of work still to be done before a satisfying molecular understanding of casein micelle structure, dynamics, and stability is reached.

B. Nomenclature of Bovine Caseins

In this review, attention is mainly given to the caseins derived from the Western breeds (e.g., Friesian, Holstein, and Ayrshire) of *Bos taurus*. Data on other species are included only when results for the cow are not available or are not as detailed or as well established.

A theme running through the long history of research on casein has been the increasing recognition of its heterogeneity. The casein of Hammarsten was first shown, unambiguously, to be a heterogeneous substance by Linderstrøm-Lang and Kodama (1929), who demonstrated clearly that fractions prepared by acid–ethanol extraction of the acid casein had different physical and chemical properties. The modern nomenclature can be traced back to the moving boundary electrophoresis of casein at alkaline pH, which showed two main peaks designated α-casein and β-casein and a lesser peak, γ-casein, in order of decreasing mobility (Mellander, 1939). The β-casein peak is now known to be a single protein and γ-casein is recognized to be largely, if not entirely, formed from three products of the proteolytic degradation of β-casein. The α-casein fraction was shown by von Hippel and Waugh (1955) and Waugh and von Hippel (1956) to comprise a fraction "sensitive" to precipitation by Ca^{2+} ($α_S$-casein) and a fraction that could not be precipitated by relatively large concentrations of $CaCl_2$. The latter fraction, though heterogeneous, is a single protein species, κ-casein, whereas $α_S$-casein has been further separated by starch and polyacrylamide gel electrophoresis into the $α_{S1}$-casein and $α_{S2}$-casein families.

Caseins were once defined as the phosphoproteins that precipitate from milk at pH 4.6 but are now defined by their amino acid sequences. In the elucidation of the primary structures of the caseins a French group has taken a prominent part (Ribadeau Dumas et al., 1975). The four polypeptide chains of bovine casein are the $α_{S1}$-, $α_{S2}$-, β-, and κ-caseins. Of these, the two $α_S$-caseins and κ-casein are each phosphorylated to different extents and κ-casein also exists in a number of glycosylated forms.

The protein nomenclature adopted in this review is partly that recommended by a committee of the American Dairy Science Association in their two most recent reports (Eigel et al., 1984; Whitney et al., 1976) and partly the nomenclature of the source literature. The relation of the new nomenclature to that in the earlier literature has been summarized, with some simplifications, in Fig. 1.

1. $α_{S1}$-Caseins

At least five genetic variants of $α_{S1}$-casein have been identified, but in Western breeds of cattle the B variant is predominant. The protein comprises a single polypeptide chain of known amino acid sequence (Fig. 2) which is found in milk as a major and a minor fraction. The major fraction, previously designated $α_{S1}$-casein, contains 8 mol P/mol ($α_{S1}$-CN-8P), whereas the minor component (previously $α_{S0}$-casein) has an additional phosphorylated serine residue at position 41

STRUCTURE AND STABILITY OF BOVINE CASEIN MICELLES

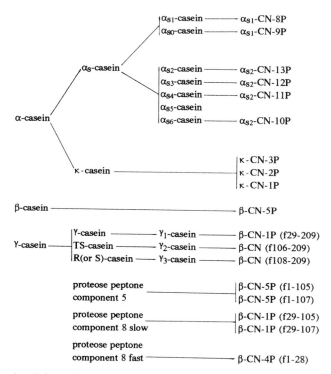

FIG. 1. Schematic summary of developments in casein nomenclature.

```
1                                                            30
R P K H P I K H Q G L P Q E V L N E N L L R F F V A P F P E
31                       •   •                               60
V F G K E K V N E L S K D I G S E S T E D Q A M E D I K Q M
61    •   •   •                 •                            90
E A E S I S S S E E I V P N S V E Q K H I Q K E D V P S E R
91                                              •           120
Y L G Y L E Q L L R L K K Y K V P Q L E I V P N S A E E R L
121                                                         150
H S M K E G I H A Q Q K E P M I G V N Q E L A Y F Y P E L F
151                                                         180
R Q F Y Q L D A Y P S G A W Y Y V P L G T Q Y T D A P S F S
181                       199
D I P N P I G S E N S E K T T M P L W
```

FIG. 2. Primary structure and sites of phosphorylation of α_{s1}-CN B-8P (Eigel *et al.*, 1984; Stewart *et al.*, 1984).

and is now designated α_{S1}-CN-9P. According to the rules of nomenclature (Eigel *et al.*, 1984), the variant is specified by a latin letter before the number of P atoms/mole: thus, the major B variant is identified as α_{S1}-CN B-8P and α_{S1}-CN B-9P, whereas the rare D variant, which contains a phosphorylated threonine residue at position 53, is α_{S1}-CN D-9P and α_{S1}-CN D-10P.

The largely hydrophobic residues 14–26 (underlined in Fig. 2) are deleted in the A variant, resulting in a marked change in its physicochemical properties compared to α_{S1}-CN B.

2. α_{S2}-Casein

Four genetic variants have been identified, with variant A the most common in Western commercial breeds. The complete amino acid sequence of α_{S2}-CN A-11P is shown in Fig. 3. Although the sites of phosphorylation are not all identified, it is thought that α_{S2}-casein exists, at least predominantly, in four phosphorylated forms containing 10–13 mol P/mol.

3. β-Casein

A total of seven genetic variants of β-casein are known, of which three (the A^1, A^2, and A^3 variants) predominate in Western breeds of cattle. The complete amino acid sequence of β-CN A^2-5P is shown in Fig. 4.

The γ-casein fractions and proteose peptone fractions 5, 8-slow, and 8-fast have amino acid sequences corresponding to parts of the sequence of β-casein (Fig. 1). Their occurrence in milk could be due to partial gene duplication but identical fractions can be produced

```
1              • • •            •                           30
K N T M E H V S S S E E S I I S Q E T Y K Q E K N M A I N P
31                                              • • •       60
S K E N L C S T F C K E V V R N A N E E E Y S I G S S S E E
      •                                                     90
S A E V A T E E V K I T V D D K H Y Q K A L N E I N Q F Y Q
91                                                         120
K F P Q Y L Q Y L Y Q G P I V L N P W D Q V K R N A V P I T
121         • •                                 •          150
P T L N R E Q L S T S E E N S K K T V D M E S T E V F T K K
151                                                        180
T K L T E E E K N R L N F L K K I S Q R Y Q K F A L P Q Y L
181                                                 207
K T V Y Q H Q K A M K P W I Q P K T K V I P Y V R Y L
```

FIG. 3. Primary structure and tentative sites of phosphorylation of bovine α_{S2}-CN A-11P (Eigel *et al.*, 1984; Stewart *et al.*, 1987).

FIG. 4. Primary structure and sites of phosphorylation of β-CN A^2-5P, showing the principal points of cleavage by plasmin (∇) (Eigel et al., 1984; Stewart et al., 1987).

by incubation of β-casein with the natural milk proteinase plasmin (EC 3.4.21.7; Eigel et al., 1984) and this appears a more reasonable explanation for their presence.

4. κ-Casein

Two genetic variants are known, A and B, with the former prevalent in most breeds of Western cattle. The complete amino acid sequence of κ-CN A-1P is shown in Fig. 5.

Chymosin, the principal proteinase of calf rennet, preferentially hydrolyzes the Phe^{105}-Met^{106} bond to produce the para-κ-casein fragment (f1–105), which is the same whatever the degree of phosphorylation or glycosylation, and the macropeptide fragment (f106–169),

FIG. 5. Primary structure and site of phosphorylation of κ-CN A-1P, and the site of cleavage by chymosin (∇). The N-terminal amino acid is pyroglutamate (Eigel et al., 1984; Thompson et al., 1985).

containing all the actual or potential sites of phosphorylation and glycosylation and all the expressed sites of genetic variation apart from the rare C variant. Normally only Ser-149 is phosphorylated and the seven or more bands of κ-casein seen in alkaline urea gel electrophoresis in the presence of 2-mercaptoethanol originate mainly from glycosylation through O-glycosidic linkages between N-acetylgalactosaminyl residues and either threonyl or seryl residues. A second site of phosphorylation (Ser-127) was identified by Vreeman et al. (1977) and there is a small degree of phosphorylation at a third, unknown, site.

Glycosylation has been observed at Thr-131, Thr-133, Thr-135 (or Thr-136), and Ser-141 (or Thr-142). The carbohydrate side chains are incompletely defined but include structures **I–III** (van Halbeek et al., 1980).

$$\text{NeuNAc}\alpha 2 \to 3\text{Gal}\beta 1 \to 3\text{GalNac}\beta 1 \to$$

(I)

$$\text{Gal}\beta 1 \to 3\text{GalNAc}\beta 1 \to$$
$$\overset{\displaystyle 6}{\underset{\displaystyle \text{NeuNAc}\alpha 2}{\uparrow}}$$

(II)

$$\text{NeuNAc}\alpha 2 \to 3\text{Gal}\beta 1 \to 3\text{GalNAc}\beta 1 \to$$
$$\overset{\displaystyle 6}{\underset{\displaystyle \text{NeuNAc}\alpha 2}{\uparrow}}$$

(III)

Because structures **I** and **III** normally predominate, it is thought that the terminal sialic acid group is added relatively rapidly. Zevaco and Ribadeau Dumas (1984) showed that there was considerable heterogeneity in side chain composition and point of substitution of the most sialylated forms of κ-casein. More abundant, complex, and variable carbohydrate side chains have been recognized in colostral κ-casein. According to Vreeman et al. (1977, 1986), the microvariants of κ-CN B include unglycosylated κ-CN-1P, κ-CN-2P, and κ-CN-3P, together with κ-CN-1P species containing from one to nine sialic acid residues/mole. Takeuchi et al. (1985a) fractionated milk from an individual Holstein cow, homozygous for κ-casein A, and found nine microvariants by chromatographic fractionation, but none was multiply phosphorylated. In general, the abundance of the fractions decreased with increasing number of side chains (zero to four) and, for a given number, the more sialylated form was present in greater amounts. Vreeman *et al.* (1986) proposed that the distribution of glycosylated fractions follows a Poisson distribution, which would be expected for a sequence of independent glycosylation events in which side chains are attached to the protein, irrespective of whether it has previously been glycosylated. In contrast, phosphorylation of Ser-127 appears to inhibit completely subsequent glycosylation.

C. Casein Concentrations in Milk

Ion-exchange chromatography in buffers containing dissociating agents has been the technique of choice for the quantification of casein composition. Davies and Law (1977a) have described a chromatographic separation of whole casein on DEAE-cellulose, which appears to give results superior to any other. A representative chromatogram is shown in Fig. 6a and illustrates the division of the casein into fractions and subfractions. Gel electrophoresis patterns (Fig. 6b), together with the effect on the patterns of chymosin treatment, allowed all fractions, apart from 2b, to be identified readily. Fraction 1a consists almost entirely of γ_2-casein, β-CN (f106–209), and fraction 1b contains largely γ_3-casein, β-CN (f108–209), with possibly some γ_1-casein, β-CN (f29–209). Fractions 2a, 3, 4a, and 4b were, respectively, the κ-, β-, α_{S2}-, and α_{S1}-caseins. The identification of the contents of fraction 2b presented some difficulty and its true nature was not established: however, it does not normally amount to much more than 5% of the total casein. Only about 80% of the total κ-casein was recovered in fraction 2a, the remainder being present in fraction 2b together with about an equal weight of some other component. As a result, Davies and Law (1977a) considered that the estimate of κ-casein obtained by summing the 2a and 2b fractions could be 1–4% larger than the true weight percent of κ-casein in whole casein. In later work κ-casein was determined independently by the gel filtration method of Yaguchi *et al.* (1968) and a comparison of the results indicated that the non-κ-casein content of fraction 2b amounted to about 2% of total casein (Davies and Law, 1980).

In Table I the average weight percent and mole percent of the four caseins and the γ-casein fragments are given for bulk milks from creameries in southwest Scotland (Davies and Law, 1980). Approximately, the mole fraction of β-casein plus γ-casein equals the mole fraction of α_{S1}-casein, and these are about four times larger than the

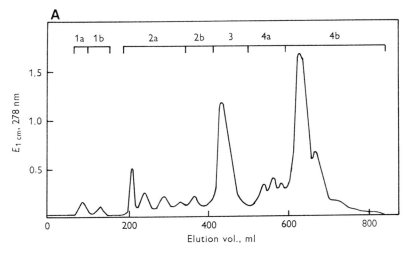

FIG. 6. (a) Ion-exchange chromatogram of whole casein according to the method of Davies and Law (1977a) and (b) alkaline PAGE of the fractions. WC, Whole casein. Reproduced by permission of Cambridge University Press from Davies and Law (1977b).

mole fraction of α_{S2}-casein. Even if about 2% is subtracted from the weight fraction of κ-casein, its mole fraction still appears to be a little larger than that of α_{S2}-casein. The figures in Table I are typical of many of the milk samples from healthy cows in midlactation and each

TABLE I
Average Casein Composition in Bulk Milk Samples[a]

Casein type	Weight %	Mole %	Casein/ skim milk (g/liter)	Casein/ skim milk (mmol/liter)	Casein P/skim milk (mmol/liter)
α_{S1}	38.1	36.4	10.25	0.434	3.52
α_{S2}	10.2	9.0	2.74	0.108	1.25
β	35.7	33.6	9.60	0.400	2.00
κ	12.8	14.8	3.45	0.176	0.19
γ	3.2	6.2	0.88	0.075	0
Total	100.0	100.0	26.92	—	6.96

[a]Data of Davies and Law (1980) and mole fractions calculated from assumed or calculated molecular weights: 23,600 (α_{S1}); 25,300 (α_{S2}); 24,000 (β); 19,500 (κ); ~11,700 (γ).

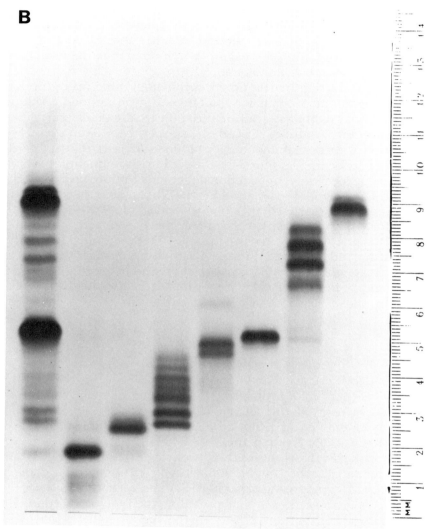

FIG. 6. (continued)

is probably accurate to within 1–2% of whole casein, as indicated by the precision of the method. The mole fractions do not appear to have a general significance, because, in other species, very different proportions are found. For example, in human milk the main casein fraction is a β-casein homologue.

The introduction of fast liquid chromatographic procedures for the separation of proteins has, so far, not given as good resolution as traditional methods of all of the individual caseins, particularly the α_S-caseins (Andrews et al., 1985; Aoki et al., 1986b). However, alkylation with cysteamine and cystamine increases the charge difference between α_{S1}- and α_{S2}-caseins and leads to quantitatively satisfactory agreement with traditional methods for most samples (Davies and Law, 1987). Davies and Law (1977b) applied their method to the variation of casein composition with stage of lactation and found that, for the greater part of lactation, the proportions of the individual caseins were effectively constant. In the beginning of lactation and at the end, however, there are significant changes, particularly in the β-, γ-, and κ-casein fractions, where the proportions of κ- and γ-caseins are negatively correlated with the proportion of β-casein.

D. Synthesis and Secretion of Caseins

1. Introduction

Although there are many gaps in our knowledge of the synthesis and secretion of milk proteins by the lactating mammary gland, the essential features at the cellular level bear a close resemblance to pathways identified in many other eukaryotes (Mercier and Gaye, 1983; Kelly, 1985; Keenan and Dylewski, 1985).

The caseins are synthesized as preproteins, with an N-terminal signal peptide, in the rough endoplasmic reticulum and are able to interact with signal recognition particles and receptors in the membrane (Walter et al., 1984), thus providing the topological conditions for the vectorial transfer of nascent polypeptide chains across the membrane. In the Golgi apparatus the signal peptides are excised, the caseins are phosphorylated, and (in the cow) κ-casein only is glycosylated as the proteins move from the proximal to the distal regions of the Golgi apparatus. Vesicles, containing casein and other components of the aqueous phase of milk, bleb off the distal face of the Golgi stack and move to the apical membrane of the cell, where exocytosis occurs and the contents of the vesicles pass into the alveolar lumen. In full lactation, only about 30 min elapse from synthesis to secretion (Foster, 1979; Boulton et al., 1984). However, the precise timings, the sequence of events, and their spatial locations are imperfectly defined. Likewise, the pH and other ion and protein concentrations prevailing during the formation of casein micelles in secretory vesicles are almost completely unknown. The concentrations of Ca^{2+} and H^+ are of particular interest, together with the stage at which the calcium phosphate is formed, because of the strong influence these factors can have on the association behavior of caseins.

2. Structure of Casein Genomes and mRNAs

The existence of allelic genes for all four bovine caseins has allowed classical Mendelian genetic studies to be made of their mode of inheritance (Grosclaude, 1979). The finding is that the four types of casein are under the control of a cluster of four autosomal genes, and from the nonrandom assortment of the alleles, a close linkage between the loci of α_{S1}-Cn (i.e., the gene coding for α_{S1}-CN) and β-Cn and between β-Cn and κ-Cn was established. Probably, there is also a close linkage of α_{S2}-Cn with the other casein genes and the hypothesis of Grosclaude (1979) is that the loci are in a sequence such as α_{S1}-Cn . . . β-Cn . . . α_{S2}-Cn . . . κ-Cn or possibly one in which the κ-Cn and α_{S2}-Cn loci are interchanged. Long-range restriction analysis of bovine casein genes has confirmed that they all lie on a single chromosome in the order α_{S1}-Cn . . . β-Cn . . . α_{S2}-Cn . . . κ-Cn. The loci lie within a sequence estimated to be 300 kb (Ferretti et al., 1990) or less than 200 kb (Threadgill and Womack, 1990) in length. In the cow and the rat, about 90 and 150 kb, respectively, of cloned sequences containing casein genes have been sequenced but no overlapping sequences have yet been identified (Bonsing and Mackinlay, 1987; Gellin et al., 1985).

Jones et al. (1985) and Yoshimura and Oka (1989) have established the complete exon/intron structure of the rat and mouse β-casein genes respectively, and, as there is evidence of conservation of exon/intron junctions in the casein gene family, this work is likely to be directly relevant to bovine β-Cn.

Jones et al. (1985) sequenced phage clones of genomic DNA containing rat β-Cn and showed it to contain nine exons with eight intervening introns. The 5' noncoding region is encoded mostly by exon I and partly by exon II and comparisons of the sequence with the 5' noncoding sequences of other casein mRNAs indicate three conserved regions that qualify as possible regulatory elements, including a putative steroid hormone-binding site. Exons VIII and IX encode the C-terminal valyl residue and the 3' noncoding region, which includes a canonical poly(A) sequence (AATAAA). Most of the hydrophobic C-terminal region of rat β-casein is encoded in a single exon (exon VII). The ability of caseins to undergo association reactions as part of the process of forming casein micelles and the importance of hydrophobic bonding in polymerization suggested to Jones *et al.* (1985) that exon VII therefore encodes a functional domain of the protein, as expected from the hypothesis of Gilbert (1978) and Blake (1978) on the significance of the exon/intron structure of eukaryotic genes. Exon II encodes the entire signal peptide sequence and the first two amino acid residues of the mature protein, and also, presumably, corresponds to a functional domain of the protein. Unlike exon VII, however, the region encoding the signal peptide was found to have the most highly conserved sequence in the rat casein mRNAs. The interpretation of Jones *et al.* (1985) of the exon structure in the remaining coded region of the gene is also given in terms of the functional domain hypothesis, but becomes more complicated. Exons IV and V as well as III and VI appear to have originated from an exon duplication event. They are quite similar in sequence, particularly at the 3' ends, both to each other and to the conserved phosphorylation sites of other casein mRNAs. This is less surprising for exons III and IV, which also encode phosphorylation sites, than for exons V and VI, which do not. Jones *et al.* (1985) postulated that the latter pair of exons originated from an earlier duplication event. Jones *et al.* (1985) distinguish between a minor phosphorylation site Ser-X-Glu and a major phosphorylation site Ser-X-Glu-Glu, where X is any other amino acid. The conserved sequences of the major phosphorylation site are split between two exons, with the Ser-Ser-Glu, the equivalent of a minor site of phosphorylation, at the 3' end of one exon and the final glutamyl residue encoded by a 5' GAA codon on a separate exon. A minor site is converted to a major site by the juxtaposition, through splicing, of glutamyl codons at the 3' and 5' termini. Thus, the presence of seryl residues close to exon/intron junctions could, according to the proposal of Craik *et al.* (1983), facilitate their phosphorylation by membrane-bound casein kinases, because they would then be preferentially located at the protein surface. Jones *et al.* (1985) identify the major phosphorylation site as a functional domain of the protein but formed from more than one exon.

Studies of the structure of the κ-casein gene (Alexander *et al.*, 1988) have supported earlier findings (Jollès *et al.*, 1978; Thompson *et al.*,

1985) that it is not closely related to the Ca^{2+}-sensitive casein gene family and that it shows some sequence similarities to the γ chain of fibrinogen, especially near exon/intron junctions in the fibrinogen gene. Alexander et al. (1988) observed that in spite of considerable variation in the coding sequences and the number and positions of the exons encoding the mature protein, there has been complete conservation of the positions, within the Ca^{2+}-sensitive casein genes, of the first two and the last two exons (Alexander et al., 1988). By these criteria there is no discernible relation between κ-casein and the other caseins. The κ-casein gene contains five exons coding, approximately, for (1) the 5' untranslated region, (2) the signal peptide, (3) the final two residues of the signal peptide together with the first nine residues of the mature protein, (4) the rest of the mature protein and a short length of 3' untranslated sequence, and (5) the remainder of the 3' untranslated sequence. The postulated functional division of κ-casein into a hydrophilic C-terminal macropeptide and a hydrophobic N-terminal para-κ-casein domain does not appear to be reflected in the structure of the modern gene.

Complete cDNA coding sequences of the four bovine caseins have been reported (Stewart et al., 1984, 1987; Nagao et al., 1984; Jimenez-Flores et al., 1987). Like the rat α-, β-, and γ-casein mRNAs, the bovine $α_{S1}$-casein mRNA has a sequence in the 5' noncoding region that has been identified as a putative steroid hormone regulatory element. Other conserved sequences are found around the major phosphorylation sites and, particularly, in the region coding for the signal peptide. Two 24-nucleotide repeats encoding a minor phosphorylation site at their 3' ends were found, which, if corresponding to exons, might indicate that the major phosphorylation site is formed by a splicing event, as in the rat β-casein mRNA. In order to optimize the alignment of the bovine $α_{S1}$-casein with the rat α-casein mRNAs, Stewart et al. (1984) proposed that a number of insertion/deletion events have occurred, one of which encloses the region of the large deletion (residues 14–26) in the bovine $α_1$-CN A variant. A second event largely involves insertions in the rat casein mRNA of a 10-times repeated 18-nucleotide block and coincides in position with the 24-nucleotide repeat in the bovine sequence. Thus, some parts of the sequences appear particularly prone to evolutionary change whereas others (e.g., the signal peptide and major phosphorylation site sequences) are very highly conserved.

When homologous $α_{S2}$-casein sequences were compared, even more insertion/deletion events were required to optimize homology (Stewart et al., 1987). These features contrast with the findings on the β-

and κ-caseins where, in spite of many amino acid substitutions, the overall architecture of the proteins appears to be conserved.

It has been recognized for many years that the casein gene family exhibits an apparently high rate of divergence, as calculated from changes in the amino acid and nucleotide sequences. The high rate of divergence in the Ca^{2+}-sensitive caseins has been taken as evidence supporting the view that these caseins generally lack secondary structural elements, which are considered unnecessary in fulfilling their role as calcium- and phosphate-transporting proteins (Stewart et al., 1984, 1987). Conservation of hydrophobic binding domains still allows scope for considerable amino acid substitutions, deletions, and insertions (Hobbs and Rosen, 1982; Stewart et al., 1984, 1987). To this may be added the consideration that the caseins are required by their biological function to form rather open, gellike structures rather than folding into a compact state, and this does not impose a severe restraint on variation in many parts of the sequence of the mature protein.

In contrast to the homologies recognized in the 5' noncoding regions of all the Ca^{2+}-sensitive casein mRNAs, the corresponding region of the κ-casein mRNA shows appreciable differences. There is, for example, no sequence corresponding to the putative steroid hormone regulatory element. Similarly, the length of the 3' noncoding region is shorter and the highly conserved signal peptide sequences of the Ca^{2+}-sensitive casein mRNAs are markedly different from that of κ-casein mRNA. These observations, taken with those on the genomic structures of the caseins, support the contention that the κ-casein gene ancestry differs from that of the other casein genes, as was originally proposed on the basis of amino acid sequence comparisons (Gaye et al., 1977; Jollès et al., 1978).

3. Secondary Processing of Nascent Casein Chains

Just as the control and coordination of casein gene transcription and mRNA processing determine the extent of translation of individual casein proteins, control and coordination are evident at every subsequent stage, including the gross sorting of proteins into secreted and nonsecreted types (Kelly, 1985) and, probably, more subtle sorting mechanisms along the secretory pathway. In many eukaryotic cells, translation and proteolytic excision of the signal peptide appear to occur concurrently. This aspect is of some importance for self-associating proteins because as long as the preprotein remains membrane bound, formation of secondary structure can occur without the complication of the association of the proteins, possibly leading to "wrong" conformational states, as has been observed by *in vitro* experiments with some oligomeric enzymes. The high proline content of the caseins suggests that they might be particularly sensitive to wrong conformational states, dependent on the sequence of folding and association steps. The formation of the correct structure of a multimeric system may require an assembly program involving some degree of cellular segregation and overall coordination of the various stages of synthesis, folding, secondary processing, and association (Jaenicke, 1982). These stages should not necessarily be regarded as either independent or successive because there is considerable scope for cooperative folding/processing/association reactions leading to the correct product. However, at present only sketchy and indirect evidence can be gleaned of the existence of such an assembly program in the secretion of milk proteins.

a. Excision of Signal Peptides. The signal peptides of caseins fall into two groups, as summarized in Fig. 7, reflecting the evolutionary development of casein genes. Thus, the Ca^{2+}-sensitive caseins all have signal peptides that are closely homologous and form one group (Fig. 7), whereas the κ-casein signal peptide is longer and, among other differences of amino acid sequence, does not contain a cysteine residue (Mercier and Gaye, 1983). The presence of cysteine is a common feature of signal peptides, suggesting that a disulfide bond might be involved in their recognition and excision. Likewise, the predominance of hydrophobic amino acid residues, and one or two basic residues adjacent to the first residue, suggest that these additional structural features are significant for the function of the peptide. Also, an interspecific comparison of the sequences of casein mRNAs in the region coding for the signal peptides revealed an unusually high degree of conservation, which suggested to Hobbs and Rosen (1982) a possible additional functional role for this region of the mRNA, but this remains a speculation.

```
(a)   M K V L I L T C L V A L A L A
            L F     F A     L     A V
            F I                       V
                                      T

(b)   M M R N F I V V M N I L A L T L P F L A A
              K S     F L   V T                 G
```

FIG. 7. Consensus sequences of the signal peptides of (a) the three Ca^{2+}-sensitive caseins and (b) the κ-caseins, taken from sequences for the cow, sheep, man, guinea pig, rat, mouse, and rabbit.

b. Phosphorylation. Phosphorylation of caseins occurs as newly synthesized proteins move along the secretory pathway and interact with at least two different membrane-bound kinases (Bingham, 1987). Mercier and Gaye (1983) have summarized the evidence that phosphorylation is a posttranslational event which occurs in the Golgi apparatus and this finding has been confirmed by recent work. Thus, the pulse-chase experiments of Boulton *et al.* (1984), in the lactating guinea pig mammary gland, support the idea of a precursor–product relationship between [^{35}S]methionine-labeled preproteins and phosphorylated forms: no phosphorylated early or intermediate forms of the caseins were identified, whereas excision of the signal peptide could be demonstrated 10–20 min before the incorporation of ^{32}P into newly synthesized protein. Secretion occurred 10–20 min later, as demonstrated by the appearance, then, of fully phosphorylated caseins in the incubation medium. Immunocytochemical studies at the light level have shown that casein kinase and galactosyltransferase have a similar subcellular distribution in lactating mammary gland cells (Moore *et al.*, 1985).

The mammary gland casein kinases are highly specialized and specific enzymes. Apart from the caseins, a number of other examples are known where protein phosphorylation occurs in an ordered sequence of steps, catalyzed by different kinases (Fiol *et al.*, 1990). Membrane-bound, serine-specific, cyclic AMP-independent kinases requiring a divalent cation for maximum activity have been found in the rat, cow, and mouse (Leiderman *et al.*, 1985). The polyamines spermine and spermidine are able to influence the selective phosphorylation of caseins at physiological concentrations. Of the two cyclic nucleotide-independent kinases isolated from lactating mouse mammary gland, the activity of only one was stimulated by polyamines, particularly at relatively high (>10 mM) concentrations of Mg^{2+} (Leiderman *et al.*, 1985). An interesting feature of mammary gland casein kinase activity is that the optimum Mg^{2+} concentration is about 10 mM, but this can be replaced by Ca^{2+} to give conditions likely to cause the substrate to aggregate extensively. Phosphorylation and micelle formation driven by a divalent cation concentration in the millimolar range could therefore be concurrent processes.

Brooks and Landt (1984) and Brooks (1987) isolated a calmodulin-dependent kinase from rat and bovine mammary acini which selectively phosphorylated dephosphorylated κ-casein rather than dephosphorylated $α_{S1}$-casein or β-casein.

In all the species that have been investigated, casein phosphorus is in the form of phosphomonoesters of seryl and, to a much lesser extent, threonyl residues. The primary recognition site is invariably of the form Ser/Thr-X-Y, where Y is either a glutamyl or occasionally an aspartyl residue and X may be any amino acid. Phosphorylation can then proceed in the direction of the N terminus, provided a suitable hydroxyamino acid residue is available in the sequence at position n and an acidic residue (Glu/SerP/Asp) is in the $n + 2$ position. Nevertheless, not all possible sites are phosphorylated completely, or even partly. Mercier (1981) has constructed a hierarchy of effectiveness of the primary phosphorylation sites:

$$\text{Ser-X-Glu} > \text{Thr-X-Glu} > \text{Ser-X-Asp} > \text{Thr-X-Asp}$$

If X is basic or particularly bulky, it may reduce the extent of phosphorylation: the best example is provided by the α_{S2}-casein C variant in which Ile replaces Thr-130 to reduce the extent of phosphorylation of Ser-129 and Ser-131. Although the triplet recognition site is necessary for phosphorylation by mammary Golgi kinases, not all sites are actually phosphorylated, so there may be a further topological requirement, such as surface location for the site or even a β-turn conformation. Some evidence for this has come from studies on the kinetics of phosphorylation of synthetic model substrates for mammary gland and other casein kinases (Meggio *et al.*, 1989).

c. Glycosylation. Only the κ-casein component of bovine casein is glycosylated, by the *O*-glycosyl rather than the more complex N-glycosylation mechanism, probably in the Golgi apparatus. Evidence that the three enzymes involved in glycosylation of κ-casein are all located in the Golgi apparatus of the lactating mammary gland was provided by Takeuchi *et al.* (1984). The activities of *N*-acetylgalactosaminyltransferase, galactosyltransferase, and sialyltransferase were all concentrated more than 10-fold, compared to cell homogenates, in a fraction enriched in membranes from the Golgi apparatus. No such concentration of activity was found in fractions enriched in Golgi vesicle membranes or other subcellular components. The relative order of phosphorylation and glycosylation of caseins in the Golgi apparatus is not known with any certainty, but there is some indirect evidence to indicate that phosphorylation precedes glycosylation. Takeuchi *et al.* (1984) found that dephosphorylation of κ-casein A 1-P gave a substrate that was glycosylated at about the same rate as the native substrate. In contrast, phosphorylation of dephosphorylated κ-casein containing three carbohydrate side chains was reduced compared to dephosphorylated, unglycosylated κ-casein. The inhibitory effect of the carbohydrate chains was still present after desialylation, indicating that it was due to the presence of a GalNAc or Gal moiety, rather than to the charged residues. Thus, phosphorylation could proceed as far as producing the major, singly phosphorylated, species of κ-casein, without affecting subsequent glycosylation, whereas, if glycosylation preceded phosphorylation, the latter process would be inhibited. It may be significant, though, that the doubly and triply phosphorylated forms of κ-casein characterized by Vreeman *et al.* (1977, 1986) were not glycosylated, indicating that phosphorylation at Ser-127 could, possibly, reduce or eliminate subsequent glycosylation.

The enzyme responsible for the initial glycosylation step, UDP-*N*-acetyl-D-galactosamine:κ-casein polypeptide *N*-acetylgalactosaminyltransferase, was purified and characterized by Takeuchi *et al.* (1985b). The enzyme was stimulated by Mn^{2+} up to a concentration of 4–10 mM and, to a lesser extent, by other divalent cations, including Ca^{2+}. Uridine nucleotides had a strong inhibitory effect, with UDP being the strongest followed by UTP and UMP, suggesting that glycosylation could be inhibited by lactose synthesis. The substrate specificities of the solubilized enzyme allowed Takeuchi *et al.* (1985b) to account for the *in vivo* pattern of glycosylation of κ-casein. Briefly, they found that the more carbohydrate chains there were on the substrate, the lower was the rate of transfer of GalNAc, due to the lower number of sites for glycosylation. The subsequent steps, to produce one or the other of the three known side chains, could then proceed, but the general reduction in the proportions of κ-casein with more carbohydrate side chains is determined by the initial step. This explanation appears to differ in principle from that of Vreeman *et al.* (1986) in which initial glycosylation events at two or more sites in a molecule are proposed to be random and independent processes, but the practical difference will be small.

d. Formation of Casein Micelles. Secretory vesicles are abundant in the apical region of the mammary secretory cell (Fig. 8) and appear to be derived from the distal face of the Golgi apparatus. The impression gained from electron micrographs is that premicellar particles, identified as casein, are present in immature, smaller vesicles and subsequently form casein micelles in mature vesicles prior to exocytosis (Keenan and Dylewski, 1985). Because preparations of Golgi membranes have lactose synthase activity (Kuhn and White, 1975; Keenan *et al.*, 1979) and can generate and maintain a large gradient in calcium ion activity across the vesicle membrane (Baumrucker and

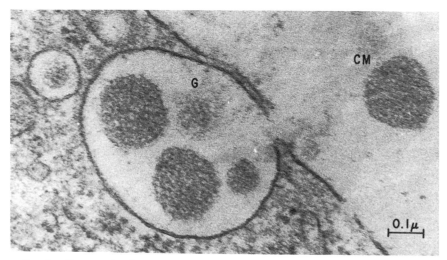

FIG. 8. Exocytosis of casein micelles (CM) from Golgi vesicles (G) in the apical region of a bovine mammary gland secretory cell. From Thompson and Farrell (1973), reproduced with permission.

Keenan, 1975; Neville et al., 1981; Watters, 1984; Virk et al., 1985), the presumption is that the formation of casein micelles is orchestrated with the transport of ions, the phosphorylation and glycosylation of the caseins, and lactose synthesis, such that the intravesicular ionic environment and casein concentration change continuously during the 20 min or so required for micelle assembly. Patton and Jensen (1975) observed, in electron micrographs, the same density of micellar particles in the alveolus as in mature vesicles, suggesting that, by this stage, the vesicular concentrations are virtually identical to those in the aqueous phase of milk.

The enzymes involved in casein phosphorylation, glycosylation, and lactose synthesis can tolerate free Ca^{2+} concentrations in the millimolar range, suggesting that the accumulation of Ca^{2+} may begin at an early stage of the secretory pathway and continue in the Golgi vesicles, by which time its presence may be demonstrated by electron microprobe analysis (Wooding and Morgan, 1978; O'Brien and Baumrucker, 1980). The mechanism of Ca^{2+} accumulation bears some resemblance to the active transport process in skeletal muscle sarcoplasmic reticulum. Golgi membrane preparations contain a Ca^{2+}-stimulated Mg-ATPase (Baumrucker and Keenan, 1975), which can sequester Ca^{2+} to intravesicular millimolar concentrations in the presence of ATP and Mg^{2+} when the extravesicular free Ca^{2+} concentration is submicromolar (West, 1981; Neville et al., 1981). The dependence of lactose synthase and Golgi kinase activity on free Ca^{2+} concentration has been used to estimate the free ion concentration within the vesicles; in rat mammary Golgi vesicles, studies of the rate of casein phosphorylation indicated the concentration to be about 0.1 mM (West and Clegg, 1981), whereas in mouse mammary Golgi vesicles a value of about 2–4 mM was indicated (Neville and Staiert, 1983). These estimates are of an order similar to values for the free Ca^{2+} concentrations in the aqueous phase of rat and cow milks, respectively (Holt and Jenness, 1984; Holt et al., 1981; Geerts et al., 1983), possibly indicating that a similar concentration is sustained throughout the assembly of the micelles.

It has been suggested that the accumulation of calcium within Golgi vesicles may be intimately linked to the mechanism of lactose synthesis in the complex, coordinated process of milk secretion (Holt, 1983). The lactose synthase complex (EC 2.4.1.22) comprises two subunits, galactosyltransferase and α-lactalbumin, and is located on the luminal surface of Golgi membranes. Galactosyltransferase has a high-affinity site for Mn^{2+} and a low-affinity divalent metal ion-binding site that is half saturated at about 1 mM free Ca^{2+}, so lactose synthesis could be initiated and modulated by the free Ca^{2+} concentration. The enzyme substrates, glucose and UDP-galactose, enter the Golgi lumen, presumably through membrane carriers, and form UDP and lactose, neither of which can permeate the Golgi membrane, but the former is hydrolyzed by a membrane-bound Ca^{2+}-dependent nucleoside diphosphatase to produce UMP and phosphate, both of which can equilibrate with the cytosol (Kuhn and White, 1975; Navaratnam et al., 1986). It appears reasonable to suppose, however, that part of the phosphate generated within the lumen by lactose synthesis is incorporated in the casein micelle as calcium phosphate (Holt, 1983). A further consequence of lactose synthesis is the dilution of Golgi vesicle contents by osmotic swelling. It has long been recognized that lactose synthesis regulates the volume of the aqueous secretion (Taylor and Husband, 1922) and Linzell and Peaker (1971) suggested that this occurs at the level of the Golgi vesicle. In effect, proteins and, specifically, caseins present in the Golgi dictyosome are diluted during transit to the alveolar lumen so that casein micelle formation occurs at some casein concentration intermediate between the high concentration present in the dictyosome and that in the aqueous phase of milk (Holt, 1983).

Uncontrolled precipitation of calcium phosphate would present a hazard to the functioning of any cell. The binding of the calcium phosphate produced in Golgi vesicles to the casein phosphoproteins offers a means of controlling the precipitation process and preventing the formation of relatively insoluble crystalline forms of calcium phosphate. In effect, limiting the growth of calcium phosphate to the size of the micellar microgranules prevents a highly cooperative first-order phase transition. Holt and Sawyer (1988a,b) suggested that this may be one of the biological functions of casein and Holt and van Kemenade (1989) have discussed the more general question of the physicochemical properties to be expected of phosphoproteins involved in the control of calcification—in particular, the advantages of a flexible solution conformation in being able to bind rapidly to calcium phosphate nuclei, forming spontaneously from a supersaturated biological fluid. This speculation may go some way to explaining why the caseins have such an open and flexible conformation.

It is argued here that the formation of casein micelles is a highly controlled process, producing a macromolecular complex with a specific structure and function. This point needs to be stressed because of a long-standing view of caseins as "random coil-type proteins" which associate to produce a largely random coil complex having only a nutritional function.

II. Physicochemical Properties of Caseins

The subject matter of this section has been treated, for the most part, in some earlier reviews (Swaisgood, 1982; Schmidt, 1982; Payens and Vreeman, 1982; Farrell and Thompson, 1988). The coverage here is highly selective, reviewing mainly the more recent findings. However, because of their relevance to our understanding of the structure and stability of native casein micelles, studies dealing with the interactions of the caseins with calcium phosphate are considered more fully.

A. Properties of Individual Caseins

1. Secondary Structures of Caseins

Caseins are usually isolated from milk by methods involving urea or some similar denaturant in high concentration. The native conformation of these proline-rich proteins may therefore not be fully recovered within a reasonable time after removal of the urea. However, Graham *et al.* (1984) were unable to demonstrate any appreciable difference between the circular dichroism spectra of β-casein isolated with and without exposure to urea, but the possibility of more subtle conformational differences remains, perhaps associated with special conformational states adopted during micelle assembly, or prior to phosphorylation, or on binding to the colloidal calcium phosphate. Certainly, phosphorylation can induce well-defined conformational transitions in globular proteins (Sprang *et al.*, 1988; Hurley *et al.*, 1989). Studies on the isolated caseins may therefore not reveal the original conformation or range of conformations that allowed at least the initial stages of aggregation to occur in the Golgi apparatus.

a. α_{S1}-Casein. Bovine α_{S1}-casein B, at near-neutral pH, is expected to carry a net negative charge of about -22, largely arising from the clusters of phosphoseryl and glutamyl residues between positions 45 and 75. Prolyl residues occur with a high frequency (17 out of 199), but of these, only two are found in the region containing most of the sites of phosphorylation. Early studies by optical rotatory dispersion (ORD), circular dichroism (CD), and infrared spectroscopy, suggested that α_{S1}-casein has a low proportion of α-helical structure in neutral solution and more recent investigations have confirmed this finding. Byler and Susi (1986) obtained deconvolved and second-derivative infrared spectra in the amide I region of a sample identified as α-casein. The strongest component corresponded to that expected to arise from unordered conformations. Likewise, the CD studies of Creamer *et al.* (1981) showed a low proportion of periodic structure in neutral buffers but more than in 4.5 M guanidinium hydrochloride or 6 M urea. A reduction in pH to 1.5 gave only a small increase in the calculated fractions of α helix and β strand, which contrasts with the behavior of the somewhat similar protein phosvitin, where a significant increase in β-strand content was observed on reducing the pH to 2 (Renugopalakrishnan *et al.*, 1985). In the spectral region between 250 and 320 nm, Creamer *et al.* (1981) found little change was produced by treating the α_{S1}-casein with carboxypeptidase to remove the C-terminal Trp-199 residue, indicating that Trp-164 and the adjacent Tyr-165 and Tyr-166 are in a relatively stable conformation. Using the structure-predicting program of Garner *et al.* (1978), Creamer *et al.* (1981) were able to match approximately the calculated fractions of α helix and β strand with values obtained from the analysis of the CD spectra. In the absence of satisfactory information from known protein structures, it was assumed that phosphoseryl residues had zero values for the directional parameters determining the propensity to form periodic or reverse-turn conformations. Nevertheless, the sequence SerP-Ile-SerP-SerP-SerP-Glu (residues 64–69) was found to lie at one end of an α-helical region (residues 50–69) in the middle, hydrophilic, part of the protein. Other evidence, however, suggests that the phosphorylated residues are in flexible regions of the protein (see Holt and Sawyer, 1988a,b). The sequence of aromatic residues identified as having a stable conformation (residues 164–166) was predicted to be in a β-strand conformation, extending from residues 164 to 168. This region of primary structure is also relatively well conserved between species (Holt and Sawyer, 1988a,b), suggesting some functional role. The peptide bonds, which are more sensitive to cleavage by chymosin, Phe^{23}-Phe^{24} and Phe^{24}-Val^{25}, lie in a region of predicted β-strand conformation (residues 22–25), as might be expected given the preference for a β-strand conformation around the susceptible bond in the substrates of aspartic proteinases (Davies, 1990).

  b. α_{S2}-Casein. Bovine α_{S2}-casein A contains between 10 and 13 phosphorylated residues in three main clusters, or phosphate centers, namely, 8–16, 56–61, and 129–131 (Fig. 3). It is the most hydrophilic of the Ca^{2+}-sensitive caseins and, with β-casein, the main substrate for the plasmin in milk. The primary structure suggests five domains, namely, an N-terminal phosphopeptide of approximately 30 residues, another acidic peptide containing the second phosphate center in the N-terminal half of the molecule, this time about 50 residues long, followed by a predominantly hydrophobic domain of low charge density and low net charge in the middle of the chain. The third phosphate center occurs in another acidic peptide (approximately residues 125–150) and the fifth domain is a C-terminal basic peptide. Sites of preferential cleavage by plasmin are found near the boundaries of each of the domains, apart from the junction between the second phosphate center and the central domain of low charge density. Only one of the 10 prolyl residues is found in the N-terminal half of the protein and none is near a phosphate center.

  Haga *et al.* (1983) have measured the ORD and CD spectra of α_{S2}-casein in Ca^{2+}-free phosphate buffer, pH 7.2, and estimated the content of α helix to be about four times as large as in α_{S1}-casein, which they attribute to the lower proline content of the more phosphorylated protein. There are marked differences in the primary and predicted secondary structures of α_{S2}-type caseins from different species. One notable constant feature, however, is a predicted α helix (residues 70–83 in the cow), duplicated at residues 152–165, which is consistently found in spite of many amino acid substitutions. These α helices could be part of α-helix–loop–α-helix motifs, centered on major sites of phosphorylation, though other evidence indicates that the solution conformation is flexible (Holt and Sawyer, 1988a,b).

c. *β-Casein*. Bovine β-casein A^2 carries a net charge of about -12 near neutral pH, largely arising from a hydrophilic N-terminal peptide of about 47 residues, which includes all of the phosphorylated seryl residues but only one of the 35 prolyl residues. The remaining prolyl and nearly all the hydrophobic residues are in the remainder of the molecule so that in terms of sequence and also in terms of physicochemical properties, β-casein resembles a surfactant with a polar head and nonpolar tail (Berry and Creamer, 1975). This division of the β-casein molecule into functional domains has its parallel in the exon/intron structure of the β-casein gene (Jones *et al.*, 1985; Yoshimura and Oka, 1989). In the largely hydrophobic tail, two remarkable lengths of uncharged residues occur, 51–90 and 149–168, containing, respectively, 30 and 25% prolyl residues.

In its hydrodynamic properties, β-casein behaves like an expanded coil of low axial ratio, rather than as a compact globular protein (Payens and Vreeman, 1982). Indeed, its radius of gyration calculated from the intrinsic viscosity, assuming a random coil structure, is only a little larger (5.1 nm) than the value found from low-angle X-ray scattering (4.6 nm) and is not appreciably changed in 6 M guanidinium hydrochloride. However, there is a considerable amount of evidence that there are, indeed, elements of regular secondary structure in the β-casein molecule.

Both ^{1}H and ^{31}P nuclear magnetic resonance spectroscopy show that there is a high degree of side chain mobility in the β-casein molecule as measured by spin relaxation times. The ratio of the intensities of the aliphatic and aromatic protons agrees fairly closely with that expected from the amino acid composition (Andrews *et al.*, 1979; Sleigh *et al.*, 1983). Nevertheless, there is a decrease in the absolute intensities in going from trifluoroacetic acid solution to 8 M urea and a further decrease in going to ^{2}H$_2$O solution, which argues that a proportion of the side chains are in fairly fixed conformational states (Andrews *et al.*, 1979). The backbone structure of β-casein has been investigated by ORD and CD spectroscopy and generally interpreted in terms of α-helical, β-strand, β-turn, and aperiodic conformations, although a polyproline II-type component has also been postulated (Garnier, 1966; Andrews *et al.*, 1979). Graham *et al.* (1984), were, however, able to simulate the CD spectrum of β-casein very precisely using the method of Provencher and Glöckner (1981) in which a linear combination of the CD spectra of 16 proteins of known X-ray structure is used, without postulating a polyproline helix. They found 7% α helix, 33% β structure, 19% β turns, and 42% aperiodic structure. Other methods of calculation, employing different reference proteins, have given different results, notably less of the extended β and more of the aperiodic structure. Under the solution conditions used by Graham *et al.* (1984), β-casein is likely to be monomeric, at least from 4 to 25°C. They noted a decrease in the magnitude of the CD band at 220 nm on raising the temperature to 60°C, which could be interpreted as an increase in β-strand content at the expense of aperiodic structure, with a constant proportion of α helix. Another interesting and somewhat surprising property of β-casein is the apparent persistence of some secondary structure in the presence of high concentrations of urea or guanidinium chloride (Creamer *et al.*, 1981; Graham *et al.*, 1984).

A variety of secondary structure prediction methods has been applied to β-casein. Regions of α helix around residues 24, 94, and 133 and β strands near residues 83, 147, and 190 are widely predicted (Creamer *et al.*, 1981; Graham *et al.*, 1984; Holt and Sawyer, 1988a,b). The predicted α helix in the N-terminal phosphopeptide region may only be stable at low pH, causing the increase in apparent helix content at pH 1.5, compared to neutrality (Creamer *et al.*, 1981).

^{31}P nuclear magnetic resonance studies have been reported on $α_{S1}$- and β-caseins, and phosphopeptides derived therefrom (Humphrey and Jolley, 1982; Sleigh *et al.*, 1983). The spectrum of β-casein at milk pH, and in the absence of paramagnetic ions, which can broaden considerably the ^{31}P resonances, contains four peaks in the intensity ratios 1:1:2:1, accounting for all five of the phosphoseryl residues. From the measured pK values of the resonances and their chemical shifts, individual assignments could be made, consistent with the idea that the ionization of each residue influences its neighboring amino acids most strongly. Thus, the middle phosphoseryl residue (SerP-18) of the cluster of three resonated at the highest field strength at milk pH and had the largest pK. Its two neighbors, each with a single adjacent phosphoseryl residue, formed the peak of double height. The low-field resonance was absent in the spectrum of a β-casein phosphopeptide, β-CN A^2 (f1–25), and was therefore due to SerP-35, leaving SerP-15 to be assigned to the remaining peak. The relaxation times (T_1 and T_2) of the ^{31}P nuclei were relatively long and unequal, consistent with rapid segmental motions that are largely independent of the rotation and translation of the molecule as a whole. In contrast to the findings of Humphrey and Jolley (1982), Sleigh *et al.* (1983) found differences of chemical shift between the ^{31}P resonance peaks of β-casein and those of its N-terminal phosphopeptide, amounting to a downfield shift of about 1 ppm in the peptide. Likewise, the resonance positions of two peaks in a phosphopeptide from α_{S1}-casein were downfield of their assigned positions in the spectrum of the whole protein. Sleigh *et al.* (1983) interpreted these observations to mean that there are specific conformations of the whole β- and α_{S1}-casein proteins which the shorter peptides cannot adopt and which alter the pK values of phosphoseryl residues by bringing hydrophobic or other amino acid side chains into their vicinity.

d. κ-Casein. The primary structure of κ-casein reveals a hydrophilic C-terminal peptide containing many hydroxyamino acid residues and a predominantly hydrophobic remainder, containing, for example, all the tyrosine residues, many of which are highly conserved. Similarly, the C-terminal peptide has a high negative charge density, particularly when glycosylated, whereas at near-neutral pH the remainder bears a small net positive charge. The chymosin-sensitive bond is at the boundary between these two domains so that cleavage brings about a profound change in the functional properties of the molecule. κ-Casein is unique among the caseins in having no clusters of phosphoseryl residues and in milk is linked by intermolecular disulfide bonds to form oligomers of various degrees of polymerization (Swaisgood, 1982).

Analysis of the CD spectrum has yielded values of 14% α helix and 31% β strand, with a possible increase in helix content observed with increase of temperature (Loucheaux-Lefebvre *et al.*, 1978). In a more recent study (Ono *et al.*, 1987), a lower fraction of α helix was calculated, but the results vary with the method of calculation. Structure prediction methods have also been applied to this protein and have given results that encourage the view that κ-casein has a number of stable conformational features. Loucheaux-Lefebvre *et al.* (1978) applied the Chou and Fasman (1974) method and predicted an α-helical content of 23%, with 31% β strand and 10% β turns. Raap *et al.* (1983) preferred the method of Lim (1974) to predict α-helix and β-strand content, because the method of Chou and Fasman, as published in 1974, was considered to overpredict these elements (Lenstra, 1977). They also tested their predictions for the structure about the chymosin-sensitive bond using the later boundary analysis method of Chou and Fasman (1978), as both an extended β structure and an α helix were considered possible conformations for this region by Loucheaux-Lefebvre *et al.* (1978). The conclusion of Raap *et al.* (1983) was that an extended β structure for residues 103–108 was slightly favored over an α helix for residues 101–108. A similar conclusion was also reached by Holt and Sawyer (1988a,b), who applied a combination of prediction methods to all the known κ-casein sequences and found that a β-turn–β-strand–β-turn motif about the chymosin-sensitive bond was commonly predicted. Visser *et al.* (1987) have proposed that the sequence His-Pro-His-Pro-His (residues 98–102), on the N-terminal side of the chymosin-sensitive bond, and Lys-111, on its C-terminal side, are involved in electrostatic binding on either side of the active-site cleft in chymosin, because synthetic substrates containing these sequences are bound more strongly and tend to be hydrolyzed more readily by chymosin. The results of other studies with synthetic substrates have been comprehensively reviewed by Visser (1981).

Raap *et al.* (1983) found that in the sequence 28–79, there were three predicted lengths of extended β structure and an α helix, which could be arranged in a βαβ configuration (Levitt and Chothia, 1976), forming a hydrophobic core, resistant to enzymatic digestion, as observed experimentally (Jollès *et al.* 1970). In the C-terminal part of κ-casein, the high charge density is likely to militate against any stable conformational states. Experimental confirmation of this view has come from the high-resolution ^{1}H NMR spectroscopic studies of Rollema *et al.* (1984) on monomeric κ-casein and micelles of κ-casein. The spectrum of the latter appeared to be a superposition of a well-resolved component on a broadened component. A good fit to the experimental spectrum could be obtained by taking linewidths of 25 Hz for the C-terminal residues 91(approx.)–169 and 300 Hz for the remainder. If the side chain mobility is accompanied by flexure of the polypeptide backbone, the predicted extended β structure around the chymosin-sensitive bond may, therefore, only be adopted on binding to the proteinase.

2. *Binding of Calcium Ions to Caseins*

Many studies have been made of the binding of Ca^{2+} to the caseins because of the undoubted importance of cation binding in the aggregation of these proteins. Binding occurs in the millimolar range of metal ion concentrations and appears to involve the phosphate moiety of phosphoseryl residues in many of the binding sites. In this regard and also in the high content of prolyl residues and degree of structural flexibility, the Ca^{2+}-sensitive caseins resemble certain other highly phosphorylated phosphoproteins found in bone, teeth, and saliva: all are thought to be involved in binding to the surfaces of calcium phosphate structures or in controlling the stability of supersaturated calcium phosphate solutions (Holt and van Kemenade, 1989).

The highly acidic tooth phosphoprotein, phosphophoryn, contains very many phosphoseryl residues in sequences that are also rich in aspartyl and glutamyl rsidues (Lee et al., 1977). Using Mn^{2+} as a paramagnetic probe, Cookson et al. (1980) demonstrated binding of the ion on phosphoseryl residues with high afifnity but with fast exchange effects, as measured by ^1H NMR spectroscopy. This led Cookson et al. (1980) to propose a "runway" for Mn^{2+}, comprising neighboring phosphorylated residues, which allows a long lifetime for the bound ion but also allows it to hop from one phosphate group to the next. The analogy with the Ca^{2+}-sensitive caseins appears particularly close, because they all contain at least one cluster of phosphoseryl and glutamyl residues along the sequence capable of acting as a Ca^{2+} runway.

Addition of $CaCl_2$ to α_{S1}- or β-casein in neutral solutions, up to a concentration of 3 mM, caused no change in the CD spectrum below 250 nm (Creamer et al., 1981). At higher Ca^{2+} concentrations, aggregation of the α_{S1}-casein caused too much light scattering to make measurements possible. Ono et al. (1976) also found no change in the CD spectrum of α_{S1}-casein on additions of $CaCl_2$ up to 5 mM and Ono et al. (1980) reported no difference in the CD spectra of κ-casein solutions at neutral pH, with and without 40 mM $CaCl_2$. There is a similar lack of effect of Ca^{2+} on the CD spectrum of phosvitin in neutral solution, but addition of Ca^{2+} to phosphophoryn in neutral solution was reported to increase the content of β-strand structure (Lee et al., 1977). The absence of any profound change in the backbone conformation of the relatively flexible caseins and the low binding affinity argue against the formation of any highly specific binding sites for Ca^{2+}, such as the EF hand conformation.

There can be little doubt that phosphoseryl and glutamyl (or aspartyl) residues are involved in the binding of Ca^{2+}, with a progressive increase in the contribution of carboxyl groups as more ions are bound (Ono et al., 1976, 1980; Sleigh et al., 1983). Binding capacities of the individual caseins increase with their content of phosphorylated residues (Dickson and Perkins, 1971; Imade et al., 1977) and, for a given casein, are increased by further phosphorylation and decreased by dephosphorylation (Dickson and Perkins, 1971; Yoshikawa et al., 1981). However, the most convincing evidence comes from infrared measurements. For example, Ono et al. (1976) showed that with α_{S1}-casein, the $-PO_3$ stretch at 980 cm^{-1} was shifted by complexation of Ca^{2+}. Infrared measurements have also indicated the involvement of carboxyl groups (Byler and Farrell, 1989).

The binding of Ca^{2+} by α_{S1}-casein also induces changes in ultraviolet absorbance (Ono et al., 1976), long-wavelength circular dichroism (Creamer et al., 1981), and fluorescence (Dalgleish, 1973) spectra, consistent with a shift of chromophores to a more hydrophobic environment, but whether these changes are brought about by a conformational change in the protein or by its association is not clear. Certainly, the C-terminal region, containing the affected chromophores, is essential for association.

Binding isotherms for α_{S1}-, β-, and κ-caseins have been obtained and the effects of pH, ionic strength, and temperature studied (Waugh et al., 1971; Imade et al., 1977; Dickson and Perkins, 1971; Dalgleish and Parker, 1980; Parker and Dalgleish, 1981). The binding of Ca^{2+} to casein can involve also the displacement of H^+ (Waugh et al., 1971; Slattery, 1975), monovalent cations such as Na^+ and K^+ (Ho and Waugh, 1965), and Mg^{2+} (C. Holt and J. Upton, unpublished results). However, a view of calcium binding as a simple reaction between the protein and the metal ion has been adopted in many studies without considering multiple equilibria. In general, the strength with which a Ca^{2+} ion is bound decreases with the number of such ions already bound, as might be expected if the binding sites can interact, as in the runway hypothesis, or are independent but of different affinities, or some combination of the two. Dalgleish and Parker (1980) considered all the binding sites to be equivalent, initially, and that there was an indefinitely large number of them. The successive association constants were then assumed to be related to each other through a linear free energy relationship such that the association constant for binding the jth ion is

$$K_j = K_0^j \, N^{j(j-1)/2} \qquad (1)$$

where K_0 is the association constant for binding the first ion and N is a substitution parameter ($0 < N < 1$) whose size determines the degree of anticooperativity of binding. Smaller values of N give greater downward curvature in the binding isotherm and hence less binding at a given free ion concentration. Values of K_0 and N required to fit

TABLE II
Parameters for Binding of Ca^{2+} to α_{S1}- and β-Caseins at pH 7

Temperature (°C)	Added NaCl (mM)	K_0 (mM^{-1})		N	
		α_{S1}-CN	β-CN	α_{S1}-CN	β-CN
4	50	1.6	1.6	0.72	0.53
10	50	1.8	—	0.53	—
14	50	1.8	—	0.73	—
20	50	2.1	2.1	0.73	0.55
30	50	2.4	2.4	0.74	0.55
40	50	2.6	2.6	0.75	0.56
20	100	1.5	1.5	0.69	0.43
20	150	1.2	1.2	0.64	0.31
20	200	1.0	1.0	0.56	0.19

the binding isotherms for α_{S1}-casein (Dalgleish and Parker, 1980) and β-casein (Parker and Dalgleish, 1981) are summarized in Table II.

Both K_0 and N were found to increase with increase of pH and temperature and decreased with increase of ionic strength. For α_{S1}- and β-casein, under comparable conditions, K_0 and the van 't Hoff enthalpy were the same, but N for β-casein was smaller, giving less binding at the same free Ca^{2+} concentration. Likewise, the average binding constant for binding the first five Ca^{2+} ions to β-casein was found to be very similar to the average binding constant for binding the first eight Ca^{2+} ions to α_{S1}-casein, under all conditions, emphasizing a similarity of the two caseins when scaled according to their content of phosphoseryl residues. Interestingly, the enthalpy of binding the first Ca^{2+} ion to α_{S1}- or β-casein is +10 kJ/mol, showing that the reaction is entropically driven.

B. Association Behavior of Caseins

In protein interactions, it is seldom possible unambiguously to separate electrostatic effects from hydrophobic bonding and the effect of changes in the hydration of hydrophilic groups, particularly those contributing to protein charge. The interface between proteins that are bonded together sometimes shows constellations of charged groups on the two surfaces that are complementary in sign, thus effecting a stereochemical constraint on the interaction. Furthermore, there may be hydrophobic regions that facilitate the expression of water molecules from between the two surfaces as they approach. This exclusion of water can strengthen the electrostatic attraction by reducing the effective dielectric constant of the intervening medium. The synergy of electrostatic and hydrophobic factors could be essential for protein aggregation because hydration forces might otherwise be completely dominant over the electrostatic attraction between hydrophilic surfaces at small separations (Israelachvili and Pashley, 1982; Rau et al., 1984; Parsegian, 1982). All of the caseins can undergo association reactions, either with themselves or with other caseins, to give a wide variety of multimers. The relatively open conformations of the caseins and the high voluminosities of the aggregates suggest that the protein–protein interface is, possibly, more likely to be defined by isolated elements of the secondary structure than is the case for globular proteins, where tertiary structure can create the surfaces of interaction.

1. Association in the Absence of Calcium

a. α_{S1}-Casein. The association behavior of α_{S1}-casein has been reviewed by Schmidt (1982), thus only the briefest outline will be given here. It associates in a series of consecutive steps rather than in the monomer–micelle manner of β- and κ-casein, in a way that depends strongly on pH and ionic strength, but temperature has almost no effect.

Thurn et al. (1987b), on the basis of static and dynamic light scattering measurements, concluded that at pH 6.7 and 35°C, α_{S1}-casein associates to form very long, wormlike chains, having a relative molecular mass per unit length of 2239 nm^{-1}. A freely jointed chain of the same radius of gyration and total number of subunits would have a segment length of about 130 nm, indicating considerable rigidity. At higher protein concentrations, there was evidence that the chains could associate laterally to create even larger aggregates.

b. α_{S2}-Casein. Snoeren et al. (1980) found by light scattering and viscosity measurements that α_{S2}-casein can undergo indefinite self-association to form spherical particles. The intermolecular association was described by an isodesmic model in which there is a constant change in free energy for the addition of a monomer to any i-mer, though the association depended strongly on ionic strength, passing through a maximum at about 0.2 M, in buffers of pH 6.7. The calculated hydrodynamic volume was found to decrease below an ionic strength of 0.15, contrary to what might be expected from increased electrostatic repulsion. Snoeren *et al.* (1980) thought that this might be explained by the unusual charge distribution along the amino acid sequence, with a net negative charge near the N terminus and a net positive charge near the C terminus. Increased intramolecular electrostatic attraction, for example, between lysyl and phosphoseryl side chains, could then allow a tighter packing of monomers at low ionic strength.

c. β-Casein. In an excellent review of the association properties of β- and κ-caseins, Payens and Vreeman (1982) argue that these caseins behave like simple surfactants in producing spherical, soaplike micelles above a critical micelle concentration. Various findings support a monomer $\rightleftharpoons$ *n*-mer equilibrium model, with *n* fairly large, such as the good separation between monomer and micelle peaks seen in the analytical ultracentrifuge, the agreement of the calorimetric and van 't Hoff enthalpies of association, and the existence of a sharp critical micelle concentration. The association is entropically driven and can be reversed by increasing pressure. Though the individual association steps are cooperative, a limit to growth must exist, which Kegeles (1979) has proposed to be a steric limit in the packing of head groups in a shell. This allowed Tai and Kegeles (1984) to calculate the distribution of intermediates between monomer and the largest possible spherical micelle, in good agreement with the bimodal distribution of sizes found by electron microscopy (Buchheim and Schmidt, 1979). The polymerization reaction is very sensitive to temperature (Takase *et al.*, 1980; Payens and Vreeman, 1982), ionic strength, and protein concentration, indicating a delicate balance of forces. At pH 7, ionic strength 0.2, and 21°C, the average degree of polymerization is about 23 and the average polymer radius about 17 nm.

From a combination of small-angle neutron scattering and static and dynamic light scattering measurements, Thurn *et al.* (1987a) concluded that the β-casein micelle has a starlike structure, somewhat like a spherical soap micelle, but with stiffer arms, particularly in the middle of the micelle. The average degree of polymerization and hydrodynamic radius increased with temperature (10–39.7°C), but at 15°C and pH 6.7, in 0.2 *M* phosphate buffer, the values were, respectively, 38 and 15.4 nm. Low-angle X-ray scattering measurements as a function of temperature and protein concentration led Kajiwara *et al.* (1988) to question the assumption of spherical symmetry. In all cases oblate spheroid models provided better fits to the angular variation in scattered X-ray intensity than spheres, rods, or random coils. The particles became more nearly spherical, mainly by growth of the shortest (c) axis, as the temperature was increased from 5 to 25°C. The calculated axial ratios also decreased with increase of protein concentration, particularly at 5°C. Because the hydrodynamic radii were invariably greater than those calculated from the X-ray measurements, Kajiwara *et al.* (1988) postulated that there was a surface layer of low mass density that was nevertheless able to restrict solvent drainage, as in the postulated hairy layer of casein micelles and micelles of κ-casein.

d. κ-Casein. Soaplike micelles of reduced (SH) κ-casein form above a concentration of about 5 g/liter in neutral solutions of ionic strength $0.1–0.5\ M$. Unlike the β-casein association, that of κ-casein is relatively insensitive to temperature, giving a constant average degree of polymerization of about 30 and a polymer diameter of 23 nm. Moreover, the ability to form micelles is not affected by reduction and alkylation of the intermolecular S–S bonds, through, for the most reproducible results it appears necessary to treat the protein with 6 M guanidinium chloride and dithiothreitol, prior to dialysis against buffer, to allow slow renaturation and micelle formation (Vreeman *et al.*, 1981). To explain the high specific volume of κ-casein micelles and the constancy of the degree of polymerization, Vreeman *et al.* (1981) and Payens and Vreeman (1982) proposed an ingenious model that draws on the ideas of quasi-equivalence for proteins in virus coats (Caspar and Klug, 1962). There are 60 equivalent positions on the surface of an icosahedron, so 30 subunits having pseudo-twofold symmetry could also generate such a structure. The large specific volume of the reduced κ-casein micelle can be explained by either a shell of protein with a water-filled core or by supposing that the C-terminal macropeptide of the protein extends from the surface as a random coil, trapping solvent in a "hairy layer." The ^1H NMR measurements of Rollema *et al.* (1984) are also consistent with a hairy κ-casein micelle. Note, though, that a quite different model has been proposed for κ-casein micelles formed in a simulated milk ultrafiltrate (Thurn *et al.*, 1987a).

2. Association in the Presence of Calcium Ions

a. α_{S1}-Casein. Horne, Parker, and co-workers, have described kinetic and equilibrium aspects of the reaction between α_{S1}-casein and Ca^{2+} ions. Two distinct approaches were used, one having its origin in the statistical theory of polymer condensation and the other deriving from the classical theory for the coagulation of hydrophobic colloids; both approaches have led to the proposition that net protein charge is of overriding importance in determining the nature and extent of the coagulation reactions of α_{S1}-casein.

When solutions of α_{S1}-casein and $CaCl_2$, suitably buffered, are rapidly mixed, there is at first little change in molecular weight (the lag phase) and then, after a critical time, a rapid linear increase in average molecular weight occurs. Likewise, under given solution conditions of pH, temperature, ionic strength, etc., the $CaCl_2$ concentration in a solution of α_{S1}-casein can be increased up to a critical level with little change in the proportion of the casein remaining in the supernatant after a defined centrifugation procedure (the so-called soluble fraction). Above the critical concentration, however, the soluble fraction decreases rapidly to low levels (Fig. 9). At higher salt concentrations, the proteins can be salted-in or salted-out and Farrell et al. (1988) have related solubility to ion binding properties.

Parker and Dalgleish (1977a,b) considered α_{S1}-casein to have f equally reactive and independent sites (functionalities) per monomer, enabling the casein to aggregate and form a coagulum of potentially infinite molecular weight, according to the condition

$$(f - 1)g = 1 \qquad (2)$$

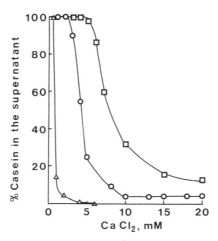

FIG. 9. Sensitivity to precipitation by Ca^{2+} at about neutral pH of α_{S2}-casein (△) at 25°C (Aoki et al., 1985), and α_{S1}-casein (□) and β-casein (○) at 35°C (Yoshikawa et al., 1981). Redrawn from the original figures and reproduced with permission.

where g is the fraction of reacted functionalities. For example, with $f = 3$, an infinite molecular weight occurs when 50% of the functionalities have reacted and when 12.5% of the monomers remain free. Thus a true gelation time can be defined unambiguously in terms of the fraction of unreacted monomers. Dalgleish and Parker (1980) showed that the critical calcium concentration for precipitation of α_{S1}-casein corresponded to a fairly constant number of bound Ca^{2+} under most of the conditions of temperature (4–40°C), pH (6.0–7.5), and concentration of added NaCl (0–200 mM) examined. The fraction of soluble casein, however, was found to depend on the total protein concentration, which led Dalgleish and Parker (1979) to propose that though casein functionality depends on the number of bound Ca^{2+} ions, there exists an equilibrium between reacted and unreacted functionalities. Under their conditions, the calculated number average functionality was always close to 2, which means that aggregating particles have a constant reactivity, giving the same linear growth in molecular weight with time as the classical Smoluchowskian mechanism, or, indeed, the simplest growth kinetics of linear condensation polymers. The appeal of the polyfunctional condensation model lies in its elegance and simplicity, yet the difficulty is in ascribing a physical significance to the derived values for the casein functionalities: the importance of net protein charge was revealed because protein functionality could be related to the number of bound Ca^{2+}, but the link between functionality and protein structure remains obscure. Subsequently, Horne and Dalgleish (1980) and Dalgleish *et al.* (1981) demonstrated that protein structure need not be considered explicitly in describing the coagulation process. It was shown that the change in turbidity with time, after mixing a solution of α_{S1}-casein with a solution containing a variable concentration of $CaCl_2$, was independent of the concentration of $CaCl_2$, provided a reduced time scale was employed. The reduced time scale was the time after mixing divided by the critical time for coagulation, CT. The CT was obtained by extrapolation of the turbidity versus time plot, in the linear growth phase, back to the point where it crosses the time axis. This successful rescaling of the turbidity–time plots demonstrates the importance of coagulation time as a characteristic of the whole coagulation mechanism. The second important finding of Horne and Dalgleish (1980) was that, at any given casein concentration, $\log(CT)$ was a linear function of Z^2, where Z is the calculated net charge on the casein at pH 7.0 (Fig. 10). Assuming no displacement of bound ions such as H^+ and Na^+ occurs when a total of n Ca^{2+} ions are bound, Z is given by $Z = -22 + 2n$. Moreover, when

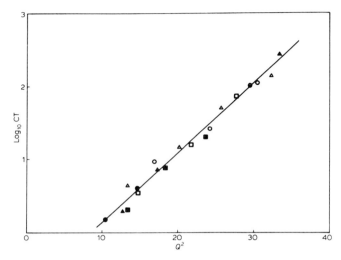

FIG. 10. Logarithm of the coagulation time as a function of the net protein charge (Q in the figure, Z in the text) for fluorescamine-treated caseins at different Ca^{2+} concentrations. From Horne (1983), reproduced by permission of the publishers, Butterworth-Heineman Ltd.

the charge on the α_{S1}-casein was modified by reaction of lysyl side chains with dansyl chloride or fluorescamine, $\log(CT)$ versus Z^2 plots were again linear and could be superimposed on the data obtained with the native protein (Horne, 1983). Because the coagulation time is inversely proportional to the rate constant for aggregation, application of the Arrhenius equation showed that the activation energy for coagulation was directly proportional to Z^2. Horne and Dalgleish (1980) elaborated on the significance of this by considering that the free energy of interaction between protein surfaces could be calculated as the sum of double-layer repulsion and dispersion force attraction—effectively treating the casein as a hydrophobic colloid. The double-layer repulsion is, for small surface potentials, proportional to Z^2, whereas the dispersion force is independent of Z. The only property of the protein considered explicitly in this approach is, therefore, the net protein charge. Solvent effects, usually described in terms of hydrophobic bonding and protein hydration, will certainly be present because the protein surface influences the free energy density of the solvent over distances of several nanometers (Parsegian, 1982; Israelachvili and Pashley, 1982; Rau *et al.*, 1984). They are, presumably, in this instance, either constant or so closely correlated with net protein charge that their influence cannot be distinguished separately. Horne and Moir (1984) investigated the coagulation of α_{S1}-casein chemically modified by iodination of tyrosyl residues. However, they found that the turbidity versus time curves for different degrees of substitution could not be superimposed successfully by using a reduced time axis, as was the case with results for the native protein and the fluorescamine and dansyl derivatives. In the Smoluchowskian phase, the rate of aggregation, calculated after rescaling, increased with the degree of iodination, suggesting a stronger net attraction between the derivatized protein particles.

Other evidence implicating solvation forces comes from studies of the A variant of α_{S1}-casein, which is generally more soluble than the B variant in solutions containing $CaCl_2$ (Thompson *et al.*, 1969; Farrell *et al.*, 1988), presumably because of the deletion, in this variant, of the mainly hydrophobic amino acid residues, 14–26. Likewise, the results of a study by Kaminogawa *et al.* (1983) emphasize the importance of the hydrophobic C-terminal region because removal of the final 23 residues generated a product that could not be precipitated by Ca^{2+} even up to a concentration of 30 mM. One of the prime candidates for hydrophobic bonding in the C-terminal region is the tyrosine-rich sequence Ala^{162}-Trp-Tyr-Tyr-Val-Pro-Leu^{168} with a predicted propensity to form a β strand. Iodination could presumably alter the interactions of this sequence, causing the nonelectrostatic effects seen by Horne and Moir (1984). On the other hand, the enzymatic treatment of Kaminogawa *et al.* (1983) reduced the propensity of α_{S1}-casein to aggregate without removing this sequence, so it is somewhat puzzling as to why the ability to precipitate was lost by the removal of only the final 23 residues.

In α_{S1}-casein, it is arguable that because the hydrophobic interaction surfaces are well separated from Ca^{2+} ion-binding sites, the electrostatic and hydrophobic free energies of association can be treated as separate and additive, leading to the Z^2 dependence of the rate of aggregation under many circumstances. Likewise, the nearly bifunctional nature of the aggregation reaction is consistent with the formation of linear polymers, as observed in the absence of Ca^{2+} (Thurn *et al.*, 1987b), and may involve the apposition of hydrophobic surfaces formed from the N- and C-terminal peptides.

b. α_{S2}-Casein. Comparatively little work has been reported on the association behavior of α_{S2}-casein, but it is clear that precipitation by Ca^{2+} at pH 7.0 occurs at lower levels of Ca^{2+} than for the other caseins, in spite of it being the most hydrophilic casein, as shown in

Fig. 9. For example, under comparable conditions, Ca^{2+} concentrations of 2 and 8 mM are required to precipitate 80% of the protein from solutions of α_{S2}- and α_{S1}-casein, respectively.

   c. β-Casein. In the Ca^{2+}-induced aggregation of β-casein (Parker and Dalgleish, 1981), a constant protein charge model was shown to be inadequate and the effects of solvation forces are more easily seen. For example, with β-casein, the number of bound Ca^{2+} ions at the critical concentration of $CaCl_2$ required for precipitation increased from 4.62 to 6.75 mol/mol casein as the temperature was reduced from 40 to 20°C, and 4°C no precipitation was seen. The aggregation of this protein induced by Ca^{2+} ions therefore shows a marked dependence on temperature, similar to the self-association behavior in the absence of Ca^{2+}. Unlike α_{S1}-casein, the electrostatic and solvation forces leading to association cannot be treated separately.

   d. κ-Casein. Though κ-casein is described as a Ca^{2+}-insensitive casein, some recent work has reported a notable different type of κ-casein aggregate in a simulated milk serum compared to those proposed to be formed in Ca^{2+}-free buffers (Thurn *et al.*, 1987a). The aggregates formed are four to five times larger in molecular weight at comparable protein concentrations and have the structure depicted schematically in Fig. 11. The angular dependence of neutron scattering was intermediate between that expected of a homogeneous sphere and that of a statistically branched polymer. In the structural model proposed by Thurn *et al.*(1987a), substructures are linked together in a network by interactions between the outer regions of

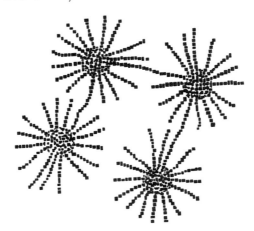

FIG. 11. Schematic diagram of the structure of aggregates of κ-casein formed in a simulated milk serum of pH 6.7. Redrawn from Thurn *et al.* (1987a).

some of the polypeptide chains. the substructures depicted in Fig.11 superficially resemble the hairy κ-casein micelles of Vreeman, Rollema, and co-workers (Vreeman *et al.*, 1981; Payens and Vreeman, 1982; Rollema *et al.*, 1984), formed from reduced (SH) κ-casein, but Thurn *et al.*, (1987a) calculate that they contain only about half the number of protein molecules. Thurn *et al.* (1987a) do not appear to have pretreated their κ-casein solutions with guanadinium hydrochloride and dithiothreitol; such pretreatments might influence the size and structure of the κ-casein aggregates apart from any possible influence of divalent cations in the simulated milk serum.

C. *Interaction of Caseins with Calcium Phosphate*

According to Bloomfield and Mead (1974), "The ultimate goal of all workers on casein is to reconstitute micelles with native properties from the separated constituents of skim-milk." This assertion reflects the large number of studies in the literature on the precipitation and association properties of the caseins.There are, however, legitimate scientific goals in this kind of work other than the creation of artificial casein micelles, such as the elucidation of the mechanisms by which phosphoproteins profoundly influence the nucleation and growth of calcium phosphate phases.

It is usually assumed that the interaction of caseins with calcium phosphate is mainly mediated through the phosphorylated residues. The experimental evidence for this is largely indirect but it would be surprising if it were proved not to be correct. Thus, the phosphate centers of caseins are resistant to attack and remain bound to the colloidal calcium phosphate when casein micelles are digested enzymatically (Holt *et al.*, 1986). The effect of different caseins on the lag time and rate of precipitation of calcium phosphates is linearly related to the content of phosphorylated residues (Van Kemenade, 1988), and there are profound effects of chemical phosphorylation and dephosphorylation of the caseins on the size and composition of complexes formed with calcium phosphate (Schmidt and Poll, 1989). Enzymatic phosphorylation of β-casein increased its binding to hydroxyapatite, particularly when the site of phosphorylation was close to the phosphate center (Sleigh *et al.*, 1979), indicating a possible synergy of action of phosphorylated residues close together in the sequence. Indeed, this finding may help to explain why the phosphorylated residues are predominantly clustered in the primary structure. Other types of residue may be involved in the interaction but no convincing evidence for the involvement of basic residues has yet been presented, though this remains a possibility meriting further investigation.

The patterns of association of the caseins induced by precipitation of calcium phosphate are quite different from those seen with Ca^{2+} alone. At a fixed total calcium concentration, the addition of increasing amounts of phosphate will, at first, hardly reduce the free Ca^{2+} concentration at about neutral pH because HPO_4^{2-} ions bind only weakly to Ca^{2+}. Thus, if the initial free Ca^{2+} concentration is high enough to induce extensive aggregation of casein, the degree of association may hardly be altered by small additions of phosphate. As the concentration of phosphate is increased, precipitation of calcium phosphate occurs and, at a fixed total concentration of calcium, the free Ca^{2+} ion concentration decreases: the formation of a casein–calcium phosphate complex can occur in which the casein acts as a protective colloid, limiting the size of the calcium phosphate particles. There can then exist calcium caseinate particles in equilibrium or quasi-equilibrium with the casein–calcium phosphate complexes, the precise balance being sensitive to solution conditions. On further increasing the phosphate concentration, two possibilities arise, each of which could destabilize the system: first, precipitation of more calcium phosphate could exhaust the supply of stabilizing casein, leading to the precipitation of a more-or-less crystalline calcium phosphate, or second, the free Ca^{2+} could fall to such a level that the caseins dissociate from the calcium phosphate. The latter possibility must be considered because the binding of casein to the calcium phosphate in the casein micelle has been shown to depend on solution free Ca^{2+} ion concentration (Holt et al., 1986). These possibilities are in accord with the observation by Horne (1982) of an intermediate range of phosphate concentrations within which stable colloidal particles of α_{S1}-casein are formed in the presence of more than 4mM $CaCl_2$.

The most thorough study of the formation of artificial casein micelles is that of Schmidt and co-workers (1977; 1979; Schmidt and Koops, 1977; Schmidt and Both, 1982; Schmidt and Poll, 1989), who not only studied the properties of the casein aggregates but also attempted to relate them to the solution conditions under which they were formed. In the precipitation of calcium phosphate from solution, the means by which solutions are mixed together is of crucial importance: Schmidt et al. (1977) described a method in which four solutions were pumped simultaneously into a reaction vessel while keeping the pH constant. As a result of careful, slow mixing, the reproducibility of the size distributions of particles, measured by electron microscopy on freeze-fractured and freeze-etched specimens, was very good. In the first series of experiments, the objective was to produce milk like concentrations of the most important ions while varying the proportions of the individual caseins. In terms of internal structure, the artificial micelles were indistinguishable by the freeze-etching method from the natural ones, though it is also true that this typical micelle substructure can be created by Ca^{2+} alone (Schmidt et al., 1974). In terms of size, the effect of increasing the proportion of κ-casein was to narrow the size distribution and reduce the average diameter. The effects cannot be attributed to changes in the free Ca^{2+} concentration because the concentrations of calcium, magnesium, and phosphate in the supernatant formed by high-speed centrifugation of the artificial micelle solution remained virtually constant. As the proportion of κ-casein increased, the amount of casein pelleted by centrifugation tended to decrease whereas the phosphate content of the sedimented micelles tended to increase. The size-regulating role of κ-casein suggested by these experiments received further support in experiments wherein κ-casein was labeled with small gold particles and was found by electron microscopy to be preferentially located on the surface of the artificial micelles (Schmidt and Both, 1982). In contrast, labeled α_{S1}- and β-caseins were found to be uniformly distributed throughout the micelles. However, an effect on the inorganic constituents in regulating the size of the artificial micelles was demonstrated in a second series of experiments which utilized whole casein (Schmidt et al., 1979). When the amount of calcium added to the mixing vessel was increased, the concentration in the centrifuged supernatant remained fairly constant but the amount of calcium phosphate formed tended to increase, as indicated by falling calcium and phosphate ion concentrations in the supernatant. There was a clearly established increase in the average diameter of the artificial micelles with the amount of colloidal calcium phosphate formed.

In their stability to dialysis, ethanol addition, heating, and applied pressure, the artificial micelles generally were more similar to native micelles than "micelles" made with Ca^{2+} alone (Schmidt and Koops, 1977; Schmidt et al., 1979).

III. Structure of Native Bovine Casein Micelles

But facts are chiels that winna ding, An' downa be disputed.
Robert Burns, A Dream

A. *Appearance and Substructure of Casein Micelles*

Although casein micelles are more or less spherical particles, their appearance in electron micrographs is not smooth but has been likened to the surface of a raspberry (Schmidt, 1980). This appearance was recognized in early electron microscopic work involving quite drastic changes in the environment of micelles (e.g., Rose and Colvin, 1966; Carrol *et al.*, 1968) and by the freeze-etching method (Schmidt, 1982; Kalab *et al.*, 1982). In freeze-fractured and freeze-etched specimens, sections also have an uneven appearance on a scale of 8–20 nm (Figs. 12a and 12b), and this substructure has been ascribed to submicelles (Buchheim and Welsch, 1973; Knoop *et al.*, 1973; Farrell, 1973; Creamer and Berry, 1975; Schmidt and Buchheim, 1976; Schmidt, 1982).

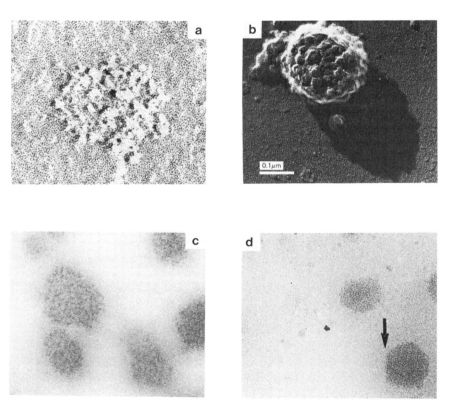

FIG. 12. Electron micrographs of bovine casein micelles obtained by different techniques. (a) Freeze fractured and etched (L. K. Creamer and D. M. Hall, unpublished). (b) Fixed and rotary shadowed (Kalab *et al.*, 1982). (c) Unstained, embedded thin section [reproduced from Knoop *et al.* (1979), by permission of the publishers, Cambridge University Press]. (d). Unstained, unfixed, hydrated sample (van Bruggen *et al.*, 1986).

A substructure is also seen in unstained thin sections of casein micelles (Fig. 12c), with regions of greater electron density dispersed evenly in a less dense matrix (Knoop et al., 1979). When artificial micelles were prepared by reaction of casein with Ca^{2+}, no substructure was seen by this method even though similar artificial casein micelles prepared by Schmidt et al. (1974) showed the same apparent substructure as native micelles by the freeze-fracture and freeze-etching method. When artificial micelles were prepared containing calcium phosphate, the darker regions were again seen. From the published electron micrographs (Fig. 12c) it is difficult to gauge the size or structure of the darker regions with any certainty, but appreciable variations in electron density occur on a scale of a few nanometers. Knoop et al. (1979) attributed the darker regions to the colloidal calcium phosphate. McGann et al. (1983a) estimated the diameter of the colloidal calcium phosphate microgranules to be about 2.5 nm.

In cryoelectron microscopy, the micelles can be examined in the frozen hydrated state without staining. Van Bruggen et al. (1986) in a preliminary report, showed micrographs of particles with a type of substructure similar to that seen by other methods (Fig. 12d).

When casein micelles are dissociated, spherical particles are observed with a size similar to the scale of the substructure. Moreover, the number of spherical particles formed by dissociation appears to correspond roughly to the number of substructural elements in the micelle. In electron micrographs of mammary gland secretory cells, some of the Golgi vesicles contain particles of a size similar to that of the particles formed by dissociation of micelles, whereas others contain larger particles. Buchheim and Welsch (1973) proposed that the smaller particles are not small micelles but subunits that are to be assembled into full-sized micelles. The envisaged sequence of assembly is as follows:

Casein monomers or small polymers → caseinate subunits + calcium phosphate → casein micelles

This sequence of events has its parallel in the model of casein micelle structure proposed by Schmidt (1982) and Walstra (1990) in which small calcium phosphate particles link together discrete spherical protein subunits.

As association polymers, the caseins can form aggregates in solution with a wide range of sizes, so it is difficult to identify any particular size as the units of bioassembly or dissociation of casein micelles.

Notwithstanding this, claims have been made to have isolated the putative submicelles, on the basis that the particles have a size similar to the scale of the micellar substructure. Not surprisingly, the reported molecular weights differ between methods of preparation and depend on concentration (Schmidt, 1982). Their existence as discrete structures within the micelle is more problematic. In Fig. 13, substructure is depicted in a protein gel without requiring the existence of submicelles.

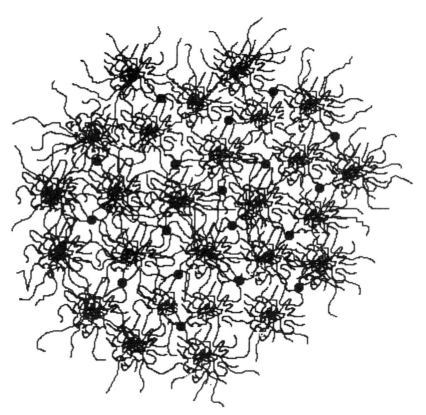

FIG. 13. Diagrammatic representation of a casein micelle showing a more or less spherical, highly hydrated, and open particle. Embedded in the protein matrix are small microgranules of calcium phosphate (●); the outermost region comprises a diffuse layer confering steric stability on the particle. The illustration is designed to give the impression of a substructure in a mineralized, cross-linked, protein gel arising from short-range but not long-range order.

Confirmation of the existence of micellar substructure has come from the application of an independent method—small-angle neutron scattering (Stothart and Cebula, 1982; Stothart, 1989). A plot of log I (I is the intensity of coherent neutron scattering) against Q ($= 4\pi \sin \varphi/\lambda$, where $\lambda = 1.2$ nm is the wavelength of the neutrons and 2φ is the total angle of scatter) is shown in Fig. 14. The inflection at $Q = 0.35$ nm^{-1} is thought to arise from the internal interference of scattered neutrons as a result of some short-range order. Stothart and Cebula (1982) were more specific about the cause of the inflection, calculating that a cluster of 12 uniform spheres, of radius 8.4 nm, in hexagonal close packing would give rise to the observed inflection. Larger clusters would give rise to a definite peak, which ultimately becomes the (110) reflection of the lattice. Stothart and Cebula (1982) also examined the particles formed when micelles were dissociated by prolonged dialysis against 70 mM NaCl, pH 6.7. Such

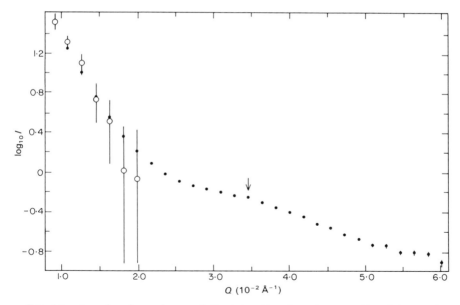

FIG. 14. Angular dependence of the intensity of neutrons scattered by casein micelles suspended in simulated milk ultrafiltrate made with two different H$_2$O/^{2}H$_2$O mixtures. The arrow marks the point of inflection ascribed to the intersubunit interference. From Stothart and Cebula (1982), reproduced with permission.

particles have been described as submicelles on the basis of their size and appearance in the electron microscope (Schmidt and Buchheim, 1976), but they will have lost most, if not all, of the calcium phosphate and possibly also some of the more soluble caseins. From a Guinier plot, radii of gyration of 6.3 nm in H_2O, 6.5 nm in ^2H_2O, and an estimated molecular weight of 3×10^5 were obtained. Assuming uniform spheres, the radius of the proposed subunit is then 8.4 nm and the specific volume is between 4 and 5.5 ml/g (Stothart and Cebula, 1982). The high voluminosity is comparable to that of whole micelles, as determined by light scattering or viscometry, whereas the micelle would be expected to have an even higher voluminosity because of the space between the spheres. Alternatively, the high voluminosity could be explained if the putative subunits have a shelllike or porous structure, having lost mass, in the form of calcium phosphate and the more soluble caseins, from the core of the particle during the dialysis procedure. Another observation that is not easy to explain on the basis of uniform spheres is that the radius of gyration was found to decrease with the reciprocal of the contrast. Most proteins show the opposite behavior because the hydrophilic residues, with higher scattering length densities, are concentrated on the outside.

Stothart (1989) reported a more complete study on whole micelles, using contrast variation with $^2H_2O/H_2O$ mixtures. He found that the inflection at $Q = 0.35$ nm^{-1} was still present in a mixture of 74% ^2H_2O, close to the calculated neutron scattering match points for two basic, anhydrous, crystalline calcium phosphates, used as models for the colloidal calcium phosphate. Calculations for a more realistic acidic, hydrated, amorphous calcium phosphate were not given, but it is reasonable to conclude that the inflection does not arise solely from the scattering by the calcium phosphate. However, at 41% ^2H_2O, close to the average match point for casein, the inflection was surprisingly more marked, amounting to a definite peak at the same value of Q. The spatial geometry of the mineral microgranules could possibly be correlated with that of the protein subunits, but with longer range order so that a more definite peak is obtained. However, Stothart (1989) noted that there are differences in scattering length density between the individual caseins, and that spatial fluctuations in scattering length density could still dominate the scattering intensity and hence be responsible for the peak. Further interpretation of the results obtained by Stothart and Cebula (1982) and Stothart (1989) might be aided by an understanding of the scattering behavior of subunit-sized particles that have retained their calcium phosphate.

B. Average Size, Size Distribution, and Voluminosity of Casein Micelles

1. Micelle Size

Investigations of the average size and size distribution of micelles have used various electron microscopic, light scattering, and column chromatographic methods (Holt, 1985a; Creamer, 1984; Ekstrand and Larsson-Raznikiewicz, 1984). Casein particles occur in milk in a broad range of sizes, the smallest of which may not be true micelles, as judged by the criterion that they should also contain some calcium phosphate. The smallest particles are not easily counted by electron microscopic methods and so are sometimes not included in histograms of the size distribution. The error is probably not serious for weight or higher averages of size, because the smallest particles amount to only a small weight fraction, even though they may be the largest number fraction. Moreover, the size distribution has a long tail, which presents a problem in counting and sizing enough particles to form a statistical number of the largest particles. Notwithstanding these difficulties, there has been substantial agreement between the results of electron microscopy and other techniques in determining the shape of the size distribution and the average sizes computed therefrom. Fig. 15, shows a calculated volume fraction histogram obtained by

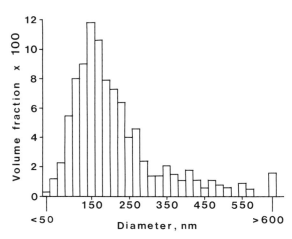

FIG. 15. Volume fraction distribution calculated, assuming homogeneous spheres of constant voluminosity, from electron micrographs of casein micelles from three cows (Holt *et al.*, 1978a, and unpublished work by the same authors).

averaging results from three samples from individual cows using freeze-fractured and freeze-etched and negatively stained thin section methods. The maximum volume fraction was at a diameter of 150 nm but particles of over 600 nm were also seen, demonstrating the broad distribution of sizes present. It is usually assumed that there is a continuous distribution of particle sizes and, in electron microscopy, such an assumption is used to correct for the distortion and smoothing caused by section thickness or etching depth on the apparent size distribution. Vreeman *et al.* (1981) challenged this orthodoxy when they observed that micelles from individual cows sedimented as a number of discrete peaks in the analytical ultracentrifuge, with sedimentation coefficients between 50 S and 500 S, and a large, rapidly broadening peak, containing the bulk of the material, sedimenting at a faster rate. The explanation offered by Vreeman *et al.* (1981) was that micelle size is governed by the requirement for quasi-equivalent symmetry for the κ-casein molecules of the coat. Only spherical shells containing $30Pi^2$ are allowed, where the class $P = 1, 3, 7, \ldots$ and i is an integer. Thus, for $P = 1$, the allowed numbers of κ-casein molecules on the surface of the micelle are 30, 120, 270, etc., and for $P > 1$, the possibilities are 90, 210, 360, etc. Vreeman *et al.* (1981) considered that only the smaller allowed sizes would be resolved by sedimentation velocity, giving the appearance of a continuous distribution in the larger micelle size classes but discrete smaller sizes, as observed experimentally. The results of Vreeman *et al.* (1981) have not been confirmed by electron microscopic or column fractionation methods of determining the size distribution, but because sedimentation velocity is much the least perturbative technique, the deficiency may well be in the other two techniques. In electron microscopy, apart from problems in sample preparation, the algorithms used to correct for the effects of sectioning on the apparent size distribution have assumed a continuous size distribution, so discrete size fractions may be smoothed out. In the column fractionation of casein micelles, insufficient resolution, micelle dissociation, and rearrangement could all prevent the recognition of discrete size fractions.

2. *Micelle Voluminosity*

The weight average molecular weight of casein micelles is generally within the range $2 \times 10^8 - 2 \times 10^9$ (Morr *et al.*, 1973; Dewan *et al.*, 1974; Holt, 1975b) and, when combined with the appropriate average radius (Holt, 1975a), yields specific hydrodynamic volumes (voluminosities) typically in range 3–7 ml/g, substantially in agreement with values obtained by viscometry (Sood *et al.*, 1976; Dewan *et al.*,

1974). However, casein pellet hydration measurements indicate lower values of 1.6–2.7 ml/g, depending somewhat on centrifugation time and speed (Thompson et al., 1969; Richardson et al., 1974). Water may be squeezed out of the pellet by ultracentrifugal compression of the micelles (Bloomfield, 1976) but, on the other hand, dissociation of micelles at the high hydrostatic pressure sometimes used (Schmidt and Buchheim, 1976; Richardson et al., 1974) could lead to an increase in apparent voluminosity, if not fully reversible. All these voluminosities are considerably higher than typical values for protein solvation and indicate that water is trapped in the micelle, moving with it as a kinetic unit due to the frictional drag of a rather open network of polypeptide chains. In this respect, the voluminosity of the micelle is to be expected from the hydrodynamic properties of the constituent monomeric proteins and indicates that, in associating to form micelles, the proteins do not fold into compact globular conformation. Walstra (1979) considered that a hairy layer on the outside of micelles would considerably increase the specific hydrodynamic volume and gave a calculation for a layer of 10nm thickness. The hydrodynamic thickness of the hairy layer has since been estimated to be about 5 nm (Walstra et al., 1981), which will reduce the specific volume of the core particle of a typical micelle by about 0.6 ml/g from that of the whole micelle. The conclusion remains, therefore, that the casein micelle is a particle with a very open structure in which the protein occupies only about a quarter of the total volume. Such a particle, depicted schematically in Fig. 13, might be expected to exhibit some properties typical of protein gels, such as compression on the application of pressure (Bloomfield, 1976), contraction in poor solvents (Horne, 1986), and swelling behavior when the solvent quality is improved or calcium is removed (Vreeman et al., 1989).

3. Variations in Micelle Properties with Size

Dewan et al., (1974) fractionated casein micelles by a rate–zone ultracentrifugation method (Morr et al., 1971) and measured the sedimentation and diffusion coefficients of the micelle fractions after dialysis against a simulated milk ultrafiltrate, to remove the sucrose used in the density gradient. The relation between the sedimentation coefficient and molecular weight, $s_{20,w} = 1.79 \times 10^{-3} M^{0.665}$, indicates that the fractionated micelles are uniform spheres of a constant, large, voluminosity. This finding provides the justification for the assumption that micelle volume is proportional to molecular weight, which is required in the calculation of the micelle size distribution and average molecular weight, from light scattering measurements on unfractionated micelles. In contrast, however, Sood et al. (1976) found a systematic difference in the voluminosity of micelles with ease of sedimentation: the largest micelles had lower voluminosities than micelles that were more difficult to sediment. Moreover, the weighted average voluminosity of the fractions was virtually the same as the micelles in the starting milk, indicating that ultracentrifugal compression and dissociation of casein from the micelles did not occur, or were fully reversible. Sood et al. (1976, 1979) followed Creamer and Yamashita (1976) in attributing the change in voluminosity with size to the systematic increase in the calcium and phosphate content of native micelles with increasing size (Ford et al., 1955; McGann and Pyne, 1960; McGann et al., 1979; Holt and Muir, 1978; Saito, 1973). Certainly, when micelles are allowed to dissociate through the progressive removal of micellar calcium and phosphate, there is an initial decrease in molecular weight without very much of a change in hydrodynamic radius (Lin et al., 1972). Sood et al. (1979), however, considered that the micelle might swell or shrink with change in its calcium content. Evidence that micelle swelling can occur on reducing the pH of milk to 5.5, where virtually all the colloidal calcium is solubilized, was obtained by Vreeman et al. (1989) using a combination of light scattering and electron microscopic methods.

It has been established in many investigations that the proportion of κ-casein increases as the average size of the micelles decreases (Sullivan et al., 1959; Yoshikawa et al., 1982; Heth and Swaisgood, 1982; Davies and Law, 1983; Donnelly et al., 1984; Dalgleish et al., 1989). The relationship between micellar size and the proportions of the other caseins is less clearly defined. For example, an increasing proportion of β-casein with increase in micelle size was found by Rose (1968), Saito (1973), Davies and Law (1983), and Dalgleish et al. (1989) by differential centrifugation, but the opposite was found by Morr et al. (1971) by rate–zone ultracentrifugation in a sucrose density gradient and by McGann et al. (1979) using chromatography on controlled pore glass at 20°C. The temperatures at which the column chromatography or centrifugal separations are carried out appear to be an important factor because Donnelly et al. (1984) were able to achieve results in substantially better agreement with the findings of Davies and Law (1983) by working at 30°C rather than 20°C, to limit the temperature-dependent dissociation of β-casein from the micelles during chromatography. Davies and Law (1983) found α_{S1}-casein to be present in about the same proportion in each of their micelle fractions, whereas the proportion of α_{S2}-casein increased sightly with ease of sedimentation. Donnelly et al. (1984) reported that both of the α_s-caseins were present in greater proportions in the larger micelles.

C. Location of κ-Casein in Micelles

The high voluminosity and the appearance of micelles in electron micrographs suggest that, rather than being simple uniform spheres, they are topologically more complex (Figs. 12 and 13). An important point to be considered in relation to such particles is the meaning to be given to measured or calculated surface areas, which increase with the resolution of the measurement and can include internal as well as external surface. Donnelly *et al.* (1984) have defined micelle surfaces by drawing a smooth closed loop around the boundary of the images of micelles in electron micrographs. Effectively, an external surface is measured at a resolution below the dimension of the putative submicelles. On the other hand, the surface area may be defined as the area that can make contact with a water molecule and include internal pores and cavities as well as external surface. Clearly, the latter could be orders of magnitude larger than the external surface area defined by Donnelly *et al.* (1984).

Much of the current debate on the location of κ-casein is concerned with whether this component is exclusively or preferentially located on or near the external surface and the relevance of this to its size-determining role. In the coat–core model (Waugh, 1971), κ-casein was placed on the surface as a way of explaining the increased proportion of κ-casein in smaller micelles and to account for the chymosin-induced aggregation of micelles. This concept has constrained the interpretation of experiments demonstrating the inverse relation between external surface-to-volume ratio and κ-casein content (Donnelly *et al.*, 1984; Dalgleish *et al.*, 1989). Other work has demonstrated that alternative models of casein micelle structure can explain the observations without requiring the κ-casein to be exclusively on the external surface. For example, in the well-known model of Slattery and Evard (1973), subunits of variable casein composition associate to form micelles. A geometric argument is used (Slattery, 1977) to allow the subunits to associate to an extent depending on the fraction of their surface area that is not occupied by κ-casein. A proportion of the κ-casein is on the internal surface of the cluster and this proportion varies inversely with the size of the cluster. When the external surface of a cluster of subunits is sufficiently rich in κ-casein molecules, further growth is terminated. In the final, stable particle, the composition of the external surface achieves a constant value, irrespective of the size of the micelle. Thus, in the model of Slattery and Evard (1973), micelle size is expected to correlate negatively with the κ-casein content, especially as the fraction in the external surface becomes small compared to the total. Essentially the same argument applies if micelle growth is regarded as a branching process in which the κ-casein acts to terminate growth of the branches (Horne et al., 1989). Depending on the average functionality, a proportion of the termini (and hence of the κ-casein) will not lie in the external surface of a cross-linked gel. In contrast, Dalgleish et al. (1989) ascribe the change in κ-casein content with size exclusively to the change in external surface-to-volume ratio and reach the conclusion that all the κ-casein is on the external surface.

That some of the κ-casein is on the external surface of micelles is indicated by the experiments of Yoshikawa et al. (1978), in which rapid coagulation of micelles was induced by the addition of wheat germ lectin. One argument that has been advanced in favor of the view that κ-casein is exclusively on the external surface is derived from the observation that a high proportion of the κ-casein in renneted milk is cleaved by the time clotting occurs (Green et al., 1978; Dalgleish, 1979). However, the chymosin molecules may be able to enter the relatively open structure of the micelles and hydrolyze the internal κ-casein with equal facility. Bringe and Kinsella (1986) observed that the proportion of κ-casein hydrolyzed by chymosin at the clotting time decreased from 71% when the micelles were suspended in 3mM CaCl$_2$ to 26% in 60mM CaCl$_2$. It appears, therefore, that the hydrolysis of a substantial fraction of the κ-casein is not required for clotting to occur and hence that the κ-casein is not necessarily predominantly located on the external surface.

In principle, the proportion of κ-casein on the external surface can be studied by using cross-linked or immobilized enzymes, antibodies, lectins, or other specific functionalities. This approach is not without its problems, however. The rate of exchange of κ-casein between the micelle and milk serum has not been established but may have a relaxation time of several hours, similar to that for β-casein (Creamer et al., 1977). If this rate is of the order of the time of the experiment, or faster, the proportion apparently in the surface will be overestimated. Likewise, solubilization of micellar κ-casein, e.g., by dissociation during the experiment, would lead to a similar error.

Heth and Swaisgood (1982) examined the preferential binding of caseins to thioester-derivatized glass beads. The micelles were first separated into micellar and serum casein fractions by chromatography at room temperature using a simulated milk ultrafiltrate as eluent to reduce dissociation during fractionation. Micelle fractions were then allowed to bind to the glass beads over a period of 12–14 hr and noncovalently linked casein was removed by treatment with a 4 M urea buffer. The covalently linked casein was then analyzed by quantitative ion-exchange chromatography. All of the major caseins (α_S-, β-, and κ-caseins) were found to be covently linked to the glass beads, but the proportion of κ-casein was much higher, β-casein was lower, and α_S-casein was much lower than in the whole casein. Control experiments indicated that the individual caseins were bound to the derivatized glass with similar affinities, so the conclusion drawn was that κ-casein is preferentially located on the external surface of the micelles but some β- and α_S-caseins also are present in the periphery. This conclusion cannot be accepted without reservation, however. In simulated milk ultrafiltrate, some dissociation of caseins from the micelle can still occur (Donnelly *et al.*, 1984): in the large particles, even as little as 1–2% dissociation could double the amount of casein apparently in the external surface, as judged by the method of Heth and Swaisgood (1982).

The micellar distribution of glyco-κ-casein was studied by Mehaia and Cheryan (1983a) using immobilized neuraminidase, incubated with pelleted casein micelles, suspended in simulated milk ultrafiltrate. The activity of the immobilized enzyme was low, so incubations were for 48 hr whereas the soluble enzyme could release 60% of the sialic acid residues in only 2 hr. After 48 hr the neuraminidase released 88% of the sialic acid groups and after making a correction for the dissociation of κ-casein from the micelles, it was estimated that about 65% of the glyco-κ-casein was on the external surface of the micelles. As acknowledged by the authors, however, proteolysis of the caseins could have occurred during these long incubation times, leading to an overestimation of the surface fraction through the solubilization of glycopeptides. A further point is that the conclusion of Mehaia and Cheryan (1983a) is not necessarily valid if there is exchange of κ-casein between the micelle and serum on a time scale of the order of the incubation time of 2 days. Similar considerations apply to the work of Chaplin and Green (1981) using pepsin and dextran–pepsin conjugates. The problem of the long incubation times with the immobilized enzymes was avoided by Mehaia and Cheryan (1983b), who used immobilized chymosin and pepsin to determine the distribution of κ-casein. About 90% of the glyco-κ-casein hydrolyzed by soluble chymosin could also be cleaved by the immobilized enzyme preparation in about the same time. No soluble enzyme was detected and only a small correction was considered necessary to allow for micelle dissociation. That the reactivity of the immobilized chymosin with micellar casein should be almost the same as with the soluble enzyme is remarkable, in view of the reduced reactivities reported in many other investigations (Beeby, 1979; Dalgleish, 1979; Chaplin and Green, 1981, 1982). If the residence time of κ-casein molecules in the micelle is long compared to the incubation time, the experiment of Mehaia and Cheryan (1983b) would support the view that a large proportion of the κ-casein lies on the external surface of the micelle.

Many attempts have been made to locate κ-casein in the casein micelle by electron microscopic methods, but the chief problems have been a lack of sensitivity in some of the methods used and doubts about the reliability of the techniques of sample preparation in not altering micellar structure.

Parry and Carroll (1969) attempted to react ferritin-labeled rabbit anti-κ-casein with glutaraldehyde-fixed micelles and found no κ-casein on the surface of the micelles, but Carroll and Farrell (1983) considered this might have been due to overfixation and reinvestigated the problem using underfixation to preserve reactive sites on the casein, and a ferritin-labeled double-antibody method for greater sensitivity. In labeled thin sections, Carroll and Farrell (1983) found the micelles to be of two types: uniformly stained particles with, on average, a smaller size, and about an equal number of somewhat larger particles with a preferential distribution of stain in the peripheral region. It can be envisaged that a section passing not through the center of a micelle but through the periphery only would give the appearance of a smaller and more uniformly stained particle than would a section exposing more of the interior. Carroll and Farrell (1983), however, considered that the smaller micelles were nevertheless more uniformly stained than the larger ones. Horisberger and Vonlanthen (1980) described a method for locating κ-casein using lectin-labeled gold particles and found a uniform staining of glutaraldehyde-fixed sections. This work was criticized by Carroll and Farrell (1983) on the grounds of poor sensitivity and detection of only the glycosylated fraction of κ-casein. Horisberger and Vauthey (1984) reinvestigated the question using rabbit anti-κ-casein and the protein A–gold method with a low-temperature embedding procedure. They confirmed the earlier findings of uniform marking of equatorial sections (Horisberger and Vonlanthen, 1980) and further noted that the density of marking tended to decrease with increasing size of the particles, in accordance with the lower weight fraction of κ-casein in larger micelles. Horisberger and Vauthey (1984) posited that the underfixation used by Carroll and Farrell (1983) allowed some reorganization of micelle structure to occur prior to staining.

On balance, electron microscopy appears to show a distribution of κ-casein throughout the micelle and, less certainly, a preferential location toward the periphery. No support is given to those experiments with immobilized reagents that appear to show that κ-casein is located overwhelmingly on the external surface of micelles.

D. Structure of Micelle Surfaces

In retrospect, it can be seen that the first convincing evidence for a diffuse surface layer came from the experiments of Scott-Blair and Oosthuizen (1961), who observed on adding rennet to milk an initial decrease in viscosity before the rapid increase associated with the aggregation of casein micelles. A decrease in hydrodynamic radius of about 5 nm, without an appreciable decrease in mass (Fig. 16), was also seen by Walstra et al. (1981) in the initial phase of rennet action and this has been confirmed in a number of other studies using quasi-elastic light scattering methods (Horne, 1984a; Holt and Dalgleish 1986; Griffin, 1987). the observations of Scott-Blair and Oosthuizen (1961) were also essentially confirmed by Griffin et al. (1989) in a viscometric study on concentrated skimmed milk.

Addition of ethanol to milk causes the protein to coagulate and the concentration required to produce visible clots within a few minutes depends on the pH and other ion concentrations (Davies and White, 1958; Horne and Parker, 1981a,b). However, Horne (1984b) has argued that the fundamental reason for coagulation is the collapse of the surface layer of the micelles, induced by the nonsolvent. Unfractionated micelles, diluted into a buffer containing $5mM$ $CaCl_2$ and up to 20% ethanol, showed a progressive and rapidly reversible reduction in hydrodynamic radius of up to 8 nm. Because ethanol can induce conformational changes in casein (Griffin et al., 1986), the contraction could be uniform throughout the micelle. Indeed, the amount of contraction was found to be an increasing function of micelle size, varying from 4 to 14 nm in different size fractions (Horne, 1986), whereas the change in hydrodynamic radius on renneting is only about 5 nm. On the other hand, Horne (1984b) found that only a small increase in turbidity accompanied the contraction (Fig. 17), which he suggested was inconsistent with uniform shrinkage because

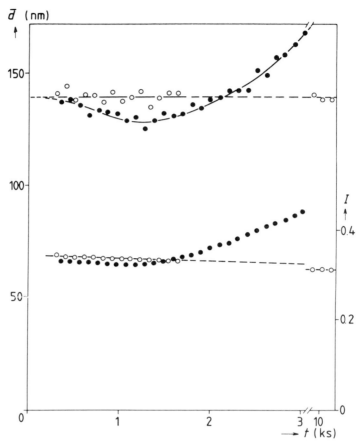

FIG. 16. Apparent average hydrodynamic diameter (upper curves) and total scattering intensity in arbitrary units (lower curves) as a function of time following the addition of chymosin (●) or controls (○). From Walstra et al. (1981), reproduced with permission.

the change in the particle scattering factor would produce a larger increase of light scattering. Additionally, at the point of maximum contraction in 16% ethanol, there was no further decrease in hydrodynamic radius after renneting, indicating that a substantial part of the overall decrease in hydrodynamic radius is due to the collapse of the hairy layer (Horne, 1984b). Environmental factors influence the extent of contraction and the critical concentration of ethanol required for coagulation (Horne and Davidson, 1986). For example,

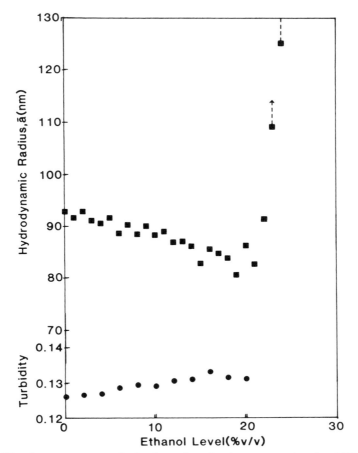

FIG. 17. Apparent average hydrodynamic radius (upper curve) and turbidity (lower curve) of casein micelles as a function of ethanol concentration. From Horne (1984b), reproduced with permission.

the overall decrease in hydrodynamic radius is reduced as the calcium concentration in the ethanol/water mixture increases, in line with the idea that the calcium ions are able to cross-link the caseins to make a more rigid gel of the protein in the micelle. Moreover, the micelles coagulate at a lower ethanol concentration, the higher the calcium level, indicating that net charge has a significant effect on the strength of the steric stabilization force.

Holt and Dalgleish (1986) elaborated an explicit model of the structure of the hairy layer based on measurements of the change in hydrodynamic radius and electrophoretic mobility of micelles on renneting. By combining equivalent theories for the hydrodynamics of particles with a partly draining outer layer, Holt and Dalgleish (1986) were able to explain the curvilinear relationship between the electrophoretic mobility and the fraction of κ-casein chains cleaved by the chymosin, a result difficult to account for if the particles have a smooth surface. They were also able to explain quantitatively the dependence of the electrophoretic mobility on free calcium ion activity and the decrease in mobility on treatment of the micelles with neuraminidase, using a model in which the macropeptide portion of only 10% of the total κ-casein in the micelles forms a hairy layer of thickness 12 nm. A decrease in hydrodynamic radius of only 5 nm, on cleavage of the hairs by chymosin, arises because of partial solvent drainage through the hairy layer. It was concluded that if all the κ-casein lies on the external surface of micelles, then the great majority of the molecules must be folded or oriented in such a way that they do not contribute to the density of hairs in the hairy layer. On the other hand, the density of hairs on the surface is about the same as would be calculated for a uniform distribution of κ-casein molecules in a typical micelle.

Other evidence of the nature of the hairs has come from ^{1}H NMR spectroscopy of κ-casein micelles and native casein micelles in ^{2}H$_2$O buffers (Griffin and Roberts, 1985; Rollema et al., 1988). The high-resolution spectra of the micelles appear to comprise a component of relatively narrow linewidth, superimposed on an envelope of overlapping, very broad bands, consistent with the idea that a proportion of the side chains are mobile and the rest are largely responsible for the unresolved features of the spectrum. In the case of the κ-casein micelles (Rollema et al., 1988), the narrow component could be unequivocally assigned to the C-terminal portion of the molecule, from about residue 90 onward. For the bovine casein micelles, the result was less clear, but the same peptide appears to be the major contributor to the narrow component, although only about half the κ-casein molecules are involved (Rollema et al., 1988). On treatment of the micelles with chymosin (Griffin and Roberts, 1985), the broad component became broader and the narrow component became narrower, without change of relative intensity or chemical shift, consistent with the hypothesis that (some of) the macropeptide side chains are mobile. Likewise, increasing concentrations of ethanol produced a progressive broadening of the narrow resonances until, at 18%

ethanol, i.e., at about the critical ethanol concentration for precipitation, they had virtually disappeared.

Thus, the ^{1}H NMR spectra support the contention that the hairy layer of micelles is formed predominantly from the C-terminal peptide of a proportion (possibly as much as half) of the κ-casein molecules. However, the mobile fraction will include contributions from any caseins that have dissociated from the micelles as a result of suspension in the ^{2}H$_2$O buffers and any mobile side chains inside the micelle, as well as the mobile external surface fraction.

E. Micellar Calcium Phosphate

Electron microscopy (Knoop et al., 1973, 1979) indicates that the inorganic ions are distributed throughout the organic matrix as ion clusters or microgranules of a few nanometers in size (Fig. 12). These ion clusters are thought to interact with the caseins through the phosphate moiety of the phosphoseryl residues to produce a structure that is unique to milk, or at least to milk and some other calcium phosphate–phosphoprotein complexes. It is convenient to distinguish between micellar calcium phosphate and colloidal calcium phosphate, the former referring to the structure in the native micelle and the latter applying to only the small ions in the structure and not including the interacting polypeptide chains. This distinction is drawn partly for historical reasons and partly because there is no clear phase boundary delimiting the structure. As a small system (Hill, 1968), the interfacial free energy cannot be neglected in attempting a thermodynamic description of micellar calcium phosphate but, equally importantly, the milk–salt system is not in a thermodynamically stable state. Thus, there are serious difficulties to be overcome in providing a structural or thermodynamic description of micellar calcium phosphate. Moreover, even its chemical composition is not easily defined, because the colloidal calcium in milk is sometimes regarded as the sum of calcium caseinate and calcium phosphate fractions (Schmidt et al., 1977; Schmidt and Knoops, 1977; Holt, 1982). Experimentally, it is very difficult to distinguish between these two types of colloidal calcium, and hence to calculate the composition of the calcium phosphate alone.

1. Chemical Characterization of Micellar Calcium Phosphate

Two approaches have been taken in attempts to establish the chemical composition of the colloidal calcium phosphate. In one, salt concentrations in milk and the milk serum are determined and the colloidal concentrations are obtained by difference, after making appropriate corrections for the excluded volume of solutes and the Donnan effect (Holt, 1985b). The other method is more direct and involves the measurement of the composition of a micellar calcium phosphate preparation. This is defined operationally as the material remaining after proteolytic digestion of casein micelles. During digestion, the material is dialyzed against a large volume of milk in order to remove the products of digestion and to prevent the calcium phosphate from dissolving (Rose and Colvin, 1966; Holt et al., 1986). A disadvantage of the first method is that colloidal ions are not necessarily all part of the colloidal calcium phosphate, whereas in the second method some change in composition could result from digestion and dialysis. When results are compared (Table III) it is seen that the micellar calcium phosphate preparation contains only some of the calcium and ester phosphate (mainly if not entirely phosphoserine) in the whole micelles: the differences in magnesium and citrate could be due to experimental error and sample differences. About 25% of the casein phosphate and 20% of the colloidal calcium in micelles appear, from this comparison, to have been lost in the digestion and dialysis procedures.

No satisfactory method has been devised to distinguish between the so-called caseinate calcium and calcium phosphate–calcium frac-

TABLE III
Salt and Ester P Composition of Casein Micelles and Micellar Calcium Phosphate

Component	Composition (mmol/g casein)	
	Casein micelles[a]	Micellar calcium phosphate[b]
Ca	787	621
Mg	69	39
P_i	369	369
Ester P	271	197
Citrate	21	44
Ca/P_i	2.13	1.63
$Ca/(P_i + \text{ester P})$	1.23	1.10

[a]Bulk milk values of White and Davies (1958) with corrected citrate values (Holt, 1985b).

[b]Recalculated from Holt et al. (1986) and scaled to make phosphate (P_i) the same in each column.

tions in casein micelles. For example, several methods have been used in which the calcium phosphate is first removed from the micelles and the resulting casein solution dialyzed against milk (Pyne and McGann, 1960; Holt, 1982; Chaplin, 1984). Caseinate calcium is then estimated from the amount of calcium bound to the casein. This approach cannot be correct because it is now realized that additional binding sites for calcium ions are generated by the removal of the colloidal calcium phosphate. Likewise, calcium caseinate cannot be calculated from the calcium binding isotherms of the individual caseins because the colloidal calcium phosphate occupies the principal binding sites and because magnesium ions in milk exert a competitive binding effect.

An attempt to circumvent these difficulties was made by Holt *et al.* (1986) by measuring the calcium binding capacity of whole casein micelles. Casein micelles were dialyzed against various buffers that were saturated with respect to the colloidal calcium phosphate but which had free calcium ion concentrations in the range 0.4–5.9 mM. The colloidal calcium content varied with the free calcium ion concentration of the dialyzed buffer in the manner expected of a binding isotherm. Colloidal phosphate was virtually constant, indicating that the colloidal calcium content of micelles, extrapolated to zero free calcium ion concentration, should give the calcium associated only with the colloidal calcium phosphate. As a means of determining the composition of the colloidal calcium phosphate, the approach of Holt *et al.* (1986) can be criticized because calcium ions could bind to both the casein and to the surface of the calcium phosphate. Because of the very high surface-to-volume ratio of the colloidal calcium phosphate, surface binding could appreciably change its apparent chemical composition.

When the rate of exchange of calcium between the micelle and milk serum is examined (Yamauchi *et al.*, 1969; Yamauchi and Yoneda, 1977; Pierre *et al.*, 1983; Wahlgren *et al.* 1990), a continuous distribution of rates appears to be present, so any division into fast- and slow-exchanging fractions that might be identified with the conceptual division into caseinate calcium and calcium phosphate calcium does not seem possible.

An independent means of establishing the chemical nature of the colloidal calcium phosphate is provided by the form and magnitude of the solubility product that governs the equilibrium between the ions in the milk serum and the solid phase. A thermodynamic model of the equilibria can be constructed as follows. First, ions in solution are supposed to be in equilibrium or quasi-equilibrium with a calcium phosphate ion cluster (i.e., the nuclei from which a precipitating phase forms),

$$Ca^{2+} + xHPO_4^{2-} + [2(1-x)/3]PO_4^{3-} \rightleftharpoons Ca(HPO_4)_x(PO_4)_{2(1-x)/3}$$
$$nCa(HPO_4)_x(PO_4)_{2(1-x)/3} \rightleftharpoons [Ca(HPO_4)_x(PO_4)_{2(1-x)/3}]_n$$

and, in a subsequent reaction, the colloidal calcium phosphate cluster binds to calcium caseinate, limiting further growth:

$$[Ca(HPO_4)_x(PO_4)_{2(1-x)/3}]_n + Ca_y\text{casein} \rightleftharpoons [Ca_{(1+y/n)}(HPO_4)_x(PO_4)_{2(1-x)/3}]_n \text{ casein}$$

Here, the minor magnesium and citrate constituents of the micelle have been neglected. The equilibrium between the ion cluster and the milk serum can then be written in the form of a solubility product, involving the activities of the individual ions comprising the cluster:

$$K_s = \{Ca^{2+}\}\{HPO_4^{2-}\}^x\{PO_4^{3-}\}^{2(1-x)/3} \qquad (3)$$

From an analysis of the ion equilibria in a large number of milk ultrafiltrate samples, covering a range of pH and other ion concentrations, Holt (1982) found an invariant ion activity product for $x = 0.7$ (i.e., nearly 80% of the phosphate ions are protonated). Milk sera were found to be supersaturated with respect to the crystalline dicalcium phosphate salts and to all the more basic calcium phosphates. Similarly, Chaplin (1984) adjusted milk pH in the range 4.0–8.0 and from a chemical analysis of the resulting equilibrated milk sera found an invariant ion activity product with the form expected of a dicalcium phosphate. Apparently, milk is in a metastable state in that nucleation and growth of thermodynamically more stable calcium phosphate phases could occur, though possibly not at a fast enough rate to be observed easily. Holt (1982) attempted to reconcile the acid calcium phosphate formula, implied by the solubility product measurements, with the apparently basic formula, implied by the high Ca/P_i value of micelles, by supposing that the solubility equilibrium applied to a surface phase on the colloidal calcium phosphate. Such an assumption now appears unnecessary because, if the casein phosphate groups are allowed to participate in the structure, electrically neutral ion clusters can be formed with a substantial proportion of esterfied or portonated phosphate groups and low $Ca/(P_i + \text{casein P})$ ratio. The acid salt formula is, however, in apparent conflict with a calculation by Pyne and Ryan (1932), based on an oxalate titration method, that only a small fraction (on average about 12%) of the phosphate groups in colloidal calcium phosphate are protonated.

Their calculation takes the premise that the casein proteins do not change their titration behavior when bound to the colloidal calcium phosphate, an assumption that appears unreasonable if the phosphoseryl residues are involved in the interaction. An alternative calculation which allows the casein phosphoseryl groups to titrate when the calcium phosphate is removed predicts, from the same titration data, that over half the orthophosphate groups in the micelles are protonated (Holt et al., 1989a).

Though there is not an exact agreement of methods used to determine the chemical nature of the colloidal calcium phosphate, the various approaches are consistent in indicating a largely acidic salt, provided the phosphate moiety of the phosphoseryl residues can be included in the structure. Whereas it was thought, at one time, that an amorphous calcium phosphate was necessarily basic in character, more recent work has shown that acidic amorphous calcium phosphates can be prepared (Holt et al., 1988, 1989b). Indeed, the C_s symmetry of the HPO_4^{2-} ion and the phosphate moiety of the phosphoseryl residue might facilitate their interaction in the micelle.

Micellar calcium phosphate preparations invariably contain phosphate esters and nitrogen, indicating that some casein peptides are bound to the inorganic material. The amino acid composition of the micellar calcium phosphate, shown in Table IV, is markedly different from that of whole casein. Over 70% of the amino acids found were derived from glutamic acid + glutamine and serine + phosphoserine, with only small amounts of hydrophobic and basic amino acids. Table IV also shows the calculated amino acid composition of a mixture of peptides derived from the regions of whole casein rich in phosphoseryl residues (residues 46–51 and 61–70 of α_{s1}-casein, 11–21 of β-casein, and 5–12, 49–61, and 126–133 of α_{s2}-casein). The calculated composition is very similar to that found experimentally, indicating that these acid peptides are, indeed, linked to the colloidal calcium phosphate and can be regarded as part of the structure of the micellar calcium phosphate. In contrast, it has been suggested (ter Horst, 1963; Visser et al., 1979) that micellar calcium phosphate interacts with casein through lysyl residues. Such interactions, if present in the native micelle, do not appear to survive the enzmatic digestion and dialysis procedures involved in the preparation of the micellar calcium phosphate, as indicated by the low content of lysine given in Table IV.

The amino acid and ionic composition of micellar calcium phosphate can be used to estimate the degree of protonation of the phosphate groups. By summing the positive and negative charges, a charge

TABLE IV
Experimental and Calculated Amino Acid Composition of Micellar Calcium Phosphate[a]

Component	Calcium phosphate sample (mole %)	Phosphate centers (mole %)
Ser + SerP[b]	37.3	36.5
Glu + Gln	35.0	38.0
Ile	7.6	6.6
Val	4.3	3.6
Leu	4.1	3.6
Ala	3.3	2.9
Asp + Asn	2.7	2.9
Gly	2.1	0.7
Thr	1.2	3.6
Lys	1.0	Absent
His	0.8	0.7
Arg	0.6	Absent
Tyr	Trace	Absent
Phe	Trace	Absent
Pro	Trace	Absent
Met	Trace	Absent
Trp	Not determined	Absent

[a]From Holt *et al.* (1986); reproduced with permission.
[b]Corrected assuming 10% loss during acid digestion.

balance is obtained with the phosphate esters all present as RPO_4^{2-}, and 80% of the inorganic phosphate ions present as HPO_4^{2-}, whereas an excess negative charge is calculated, assuming only PO_4^{3-} ions (Holt *et al.*, 1986, 1989a), confirming the other findings that a high proportion of the phosphate groups are protonated.

The diameter of micellar calcium phosphate microgranules has been estimated by electron microscopy to be about 2.5 nm (McGann *et al.*, 1983a). Such a microgranule occupies a volume in which about 66 calcium and 66 phosphate ions together with 132 water molecules could be accommodated at the same density as they would have in the lattice of dicalcium phosphate dihydrate, $CaHPO_4 \cdot 2H_2O$, the mineral brushite (Holt and Hukins, 1991). Brushite is used for this calculation as the closest model compound, in the sense that it is a hydrated acidic salt, to micellar calcium phosphate. Because about one third of the phosphate groups in the granule are derived from casein, typically about four to six polypeptide chains are required to be bound through their phosphate centers to the surface of the inorganic phase. Thus, the micogranules can be the agents responsible for crosslinking caseins in the micelle and, because the α_s-caseins contain more than one phosphate center, the microgranules can, in turn, be crosslinked by the casein (Fig. 13). Interaction of the caseins with the colloidal calcium phosphate is therefore responsible for maintaining the integrity of the casein micelle under conditions of serum pH and free Ca^{2+} ion concentration that could not, otherwise, bring about such a high degree of association (Holt et al., 1986).

2. *Physical Characterization of Micellar Calcium Phosphate*

All physicochemical measurements are in agreement in showing that there is no long-range crystalline order in micellar calcium phosphate. No sharp Bragg lines characteristic of crystalline order were found by X-ray powder diffraction analysis on freeze-dried casein micelles (Knoop et al., 1979). The amorphous nature was confirmed by Lyster et al. (1984) using high-resolution transmission electron microscopy and selected area electron diffraction methods, which could possibly detect small regions of crystalline order in an otherwise amorphous material. However, no electron diffraction spots were found. Of 21 patterns recorded with micellar calcium phosphate, 15 were wholly amorphous and the rest were essentially diffuse with some spots, sometimes showing streaking. High-resolution transmission electron microscopy on thin micelles and thin edges of the micellar calcium phosphate gave only nonperiodic images. Lyster et al. (1984) concluded that there was no evidence of order in either of the calcium phosphate preparations over distances greater than about 1.5 nm.

McGann et al. (1983b) computed the reduced radial distribution function of a micellar calcium phosphate preparation from the angular dependence of X-ray scattering. This is essentially a plot of electron density versus distance, with the oscillations determined by the atomic separations. The oscillations were strongly damped, indicating that only short-range order was present. Some differences in the reduced radial distribution curves were found between samples of amorphous calcium phosphates, which either contained or were free of magnesium. McGann et al. (1983b) concluded that micellar calcium phosphate was most like a magnesium-containing amorphous calcium phosphate isolated from mitochondria, having a relatively high Ca/P_i ratio. McGann et al. (1983b) also reported that the infrared spectrum of the micellar calcium phosphate showed no splitting of the O–P–O bending mode at 500–600 cm^{-1}, indicating an amorphous sample.

The short-range structure in amorphous materials can often be defined by studying the absorption of X-rays near an absorption edge of one or more of the atoms in the sample. The information gained takes the form of a radial average of the environment about the absorbing atoms, known as a shell structure. Holt et al. (1982) measured the spectrum of a freeze-dried micellar calcium phosphate preparation near the K absorption edge of calcium. When compared qualitatively to the spectra of crystalline samples of hydroxyapatite, β-tricalcium phosphate, brushite, monetite, and calcium tetrahydrogen diorthophosphate, and a noncrystalline calcium phosphate, the micellar calcium phosphate was judged to resemble that of brushite more closely than the other model compounds. The absence of higher frequency modulations in the spectrum of the micellar calcium phosphate confirmed, however, that only short-range order was present. In subsequent work with better model compounds, improved instrumentation, and a more complete theory, a much deeper understanding of the relation between the shell structures and the spectra of

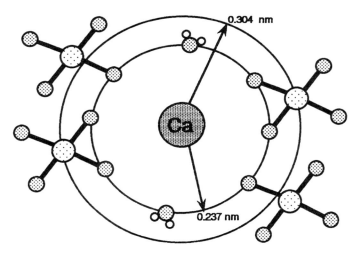

FIG. 18. Average environment of calcium ions in amorphous calcium phosphates as determined by X-ray absorption spectroscopy near the K absorption edge of calcium. The diagram shows two bidentate and two monodentate phosphate tetrahedra and two water molecules contributing to the eight oxygens of the first coordination sphere of the calcium ions. A third shell, possibly comprising the two phosphorus atoms of the monodentate phosphate ions, can be seen in some of the preparations. The positions of the protons are not established (Holt et al., 1988, 1989b; Holt and Hukins, 1991).

calcium phosphates was obtained. The effects of multiple scattering were shown to be important in influencing the spectra of hydroxyapatite, brushite, and monetite (Harries et al., 1986, 1987), and probably other crystalline and amorphous calcium phosphates. Likewise, the noncrystalline calcium phosphate used in the original work was shown to have had sufficient apatite character to render it distinguishable from more truly amorphous calcium phosphates. The spectrum of micellar calcium phosphate is nearly identical to spectra of other amorphous calcium phosphates having a wide range of chemical compositions (Holt et al., 1988, 1989b).

The approach adopted for the interpretation of the spectra of poorly crystalline calcium phosphates was to take the shell model of the crystalline phase having the greatest chemical similarity and progressively simplify and refine the model while maintaining a good fit to the observed spectrum. For the amorphous calcium phosphates, however, it was found that virtually identical shell models resulted from simplification and refinement of either the hydroxyapatite or brushite shell models to give the structure depicted in Fig. 18.

F. Dissociation of Casein Micelles

A limited degree of dissociation of micelles can be induced by cooling, whereupon β-casein in particular is solubilized (Davies and Law, 1983), or by heat treatments, which, in the first instance, appear to render the κ-casein more soluble (Aoki et al., 1975; Singh and Fox, 1985), especially after the addition of mercaptoethanol (Aoki and Kako, 1984). A more extensive dissociation can be induced by the addition of urea (McGann and Fox, 1974; Aoki et al., 1986a) or a calcium chelating agent (Lin et al., 1972; Holt et al., 1986; Griffin et al., 1988; Ono and Obata, 1989), or by acidification at low temperature to a pH of about 5 (McGann and Pyne, 1960; Dalgleish and Law, 1988; Vreeman et al., 1989). The effect of pressure in causing the dissociation of polymers of β-casein (Payens and Heremans, 1969) suggests that similar effects might be observed in the native casein micelle, but the subject has not received the attention it deserves (Ohmia et al., 1989).

The experiments with urea, in particular, demonstrate most clearly that casein micelles can be broken down into much smaller particles without rupturing the casein–calcium phosphate linkages. Hydrophobic and hydrogen bonding are therefore important in maintaining the integrity of micelles. Nevertheless, the importance of the colloidal calcium phosphate is clear, as was demonstrated by McGann and Pyne (1960). They showed that the size of the casein particles in milk was much reduced when milk was acidified to dissolve the calcium phosphate, and the serum salt composition was returned to normal by dialysis of the depleted milk against large volumes of untreated milk. Holt et al. (1986) attempted to distinguish between dissociation brought about by a reduction of free Ca^{2+} ion concentration and dissociation resulting from the solubilization of the colloidal calcium phosphate. When the colloidal calcium phosphate was slowly reduced by dialysis of the milk at 20°C against a phosphate-free buffer of pH 6.7, containing 3 mM CaCl$_2$, a progressive but selective loss of caseins occurred, such that κ- and β-caseins dissociated to the greatest extent, followed by α_{s1}-casein, which in turn dissociated more readily than α_{s2}-casein. Holt et al. (1986) interpreted the pattern of dissociation to reflect the strength of interaction of the caseins with the calcium phosphate, which is correlated with the number of phosphorylated residues in the individual caseins. However, the similarity in the dissociation behavior of κ- and β-caseins argues that some other factor, such as hydrophobic bonding, is important.

When micelles were dialyzed against buffers that were saturated with respect to the colloidal calcium phosphate but with free Ca^{2+} concentrations in the range 0.4–5.9 mM, micelle dissociation could still occur, even without dissolving the colloidal calcium phosphate, if the serum free Ca^{2+} concentration was below about 2 mM (Holt et al., 1986). The strength of binding of the individual caseins was found to be directly related to their phosphoserine content and was in the order $\alpha_{s2}>\alpha_{s1}>\beta>\kappa$. The experiments show, therefore, that a minimum free Ca^{2+} ion concentration in the serum is necessary before the phosphorylated residues of casein are able to form a stable linkage to the mineral. Moreover, this free Ca^{2+} ion concentration is of the same order as that normally found in milk, suggesting that the binding is not irreversible but is better described in terms of an equilibrium of caseins between the serum and the micelle.

Griffin et al. (1988) reported that when the colloidal calcium phosphate was depleted, by addition of a EDTA solution to a micellar dispersion, there was essentially no selective dissociation of the individual caseins. This difference from the results of Holt et al. (1986) could reflect a difference of methodology. The method of Griffin et al. (1988) could bring about an almost complete and therefore nonselective disintegration of some micelles in the immediate vicinity of the added EDTA while leaving others virtually intact. In the dialysis method of Holt et al. (1986), the free Ca^{2+} concentrations is never depressed and hence micelles dissociate only because of the solubilazation of the colloidal calcium phosphate.

Aoki *et al.* (1986a) were able to separate a higher molecular weight casein–calcium phosphate complex from lower molecular weight caseins by gel filtration of skim milk containing 6 M urea. All of the κ-casein was in the low-molecular-weight fraction and nearly all of the α_{s2}-casein was in the complexes, whereas the other two caseins were divided between the two fractions. Aoki *et al.* (1986a, 1988) considered that the high-molecular-weight fraction represented all the casein that is cross-linked by calcium phosphate in the micelle—about 67.5% of the total casein. However, urea appears to weaken the interaction of caseins with hydroxyapatite (Sleigh *et al.*, 1979) and it might also perturb the ion equilibria in the aqueous phase, which would affect the proportions of the caseins appearing in the two fractions. Nevertheless, it is clear that a high proportion of the Ca^{2+}-sensitive caseins are linked to the calcium phosphate in the casein micelle and that the strength of binding is closely related to their content of phosphorylated residues.

As milk pH is progressively reduced, the colloidal calcium phosphate dissolves, allowing some dissociation of caseins to occur, but the dissociation is not as marked as would be expected from the experiments at milk pH, because of the rise in free Ca^{2+} and H^+ concentrations, both of which tend to bind to the caseins, thereby reducing the charge and favoring association. At room temperature, dissociation is at a maximum at about pH 5.4, but over the range 4–30°C, the position of the maximum increases by about 0.17 pH units per 10°C rise in temperature (Dalgleish and Law, 1988). Of the individual caseins that dissociate on acidification, β-casein predominates but there is no evidence of dissociation in constant proportions, as might be expected of subunits of invariant composition (Dalgleish and Law, 1988). Indeed, the experiments on casein micelle dissociation, in general, have served neither to confirm nor refute concepts of micelle structure and substructure.

IV. Stability of Casein Micelles

A. Introduction

Casein micelles are remarkably stable structures. Milk may be boiled, sometimes for several hours, without coagulating the micelles. Also, the addition of $CaCl_2$ to milk does not precipitate the micelles up to concentrations greatly in excess of that required to precipitate purified whole casein. On the other hand, micelles rapidly flocculate after treatment with chymosin, at or above room temperature, and casein precipitates rapidly when milk is acidified to about pH 4.6 (so-called isoelectric precipitation). Clearly, casein has an enhanced stability through its organization into micelles.

It is at least superficially attractive to apply the theories of colloid science to describe these properties, and a number of attempts have been reported in which the classical theory of Deryaguin, Landau, Verwey, and Overbeek (DLVO theory) has been used (Payens, 1966; Green and Crutchfield, 1971; Kirchmeier, 1973). In the DLVO theory, the balance of the attractive dispersion force with a repulsion arising from the overlap of electrical double layers is used to explain many aspects of the stability of lyophobic colloids. However, application of this theory to protein solutions has proven inadequate. For example, the dispersion force is computed from the bulk dielectric properties of the disperse and continuous media and the double-layer repulsion is usually described in terms of a smeared out charge density in the surface and point charges in the double layer. These approximations are most nearly correct while the interacting particles remain far apart, but the association reactions of protein molecules often proceed in a highly specific manner, involving the close apposition of particular parts of their surfaces and, possibly, involving some change of shape or internal structure. Additionally, hydrophobic bonding or the binding of a counterion at a specific site can be of crucial importance, so the averaged properties of the protein molecules are hardly relevant. Nevertheless, the application of the theories of colloid science to casein micelles has been fruitful, particularly in introducing the idea of steric stabilization.

B. Forces Involved in Coagulation

The dispersion energy of interaction between colloidal particles increases with the degree of difference between the bulk dielectric properties of the particles and the surrounding medium, summed over all frequencies. Because casein micelles contain a high volume fraction of water (Walstra, 1979), the dispersion force is likely to be negligibly small compared to other forces acting on the particles. The repulsive potential arising from the overlap of thin diffuse double layers can be estimated from the electrophoretic mobility to be orders of magnitude larger than the dispersion force of attraction at a similar distance. Even if it is halved by the action of chymosin, the double-layer repulsion should still prevail, demonstrating the inability of the DLVO theory to explain rennet coagulation. The hydration force can become dominant over electrical repulsion at separations less than about 3 nm (Rau et al., 1984). Here, the hydration force is considered to be due to the work of removing water from the vicinity of a surface. It could arise from the spatially varying perturbation of water near a polarizing surface, since such a mechanism gives rise to the observed exponential decrease in repulsion with distance of separation (McIntosh and Simon, 1986). Israelachvili and Pashley (1982) also observed an exponentially decaying solvation force, but between hydrophobic surfaces. The force was attractive and is conventionally described as a hydrophobic interaction. In the range of 1–10 nm separation, the exponential force had a decay length of about 1 nm and in this range was about an order of magnitude stronger than the dispersion force. That the solvation force of attraction is dominant over the dipersion force is attested to by the observation that fully renneted micelles can be quite stable in cooled milk and aggregate on warming. This marked dependence of aggregation on temperature is reminiscent of that seen in β-casein micellization and can be ascribed to hydrophobic bonding, whereas no strong dependence on temperature is expected for the dispersion force.

C. Steric Stabilization of Micelles

The term steric stabilization has been used by colloid scientists to describe how a lyophilic substance, located on the surface of a lyophobic colloid, can prevent aggregation of the dispersion. The phenomenology of steric stabilization has been recognized and put to use over many millenia; one notable example is the use, by the ancient Egyptians, of casein as a steric stabilizer of carbon (lamp black) in the production of inks for writing on papyrus. Only in the last 50 years or so has a scientific understanding of steric stabilization mechanisms emerged.

When two sterically stabilized particles approach one another, entropic and enthalpic interactions between the lyophilic molecules on the surface lead to a repulsion that is sufficiently strong to overcome the dispersion force of attraction. The positive free energy change can be particularly large when the lyophilic substance is a polymer adsorbed on the surface such that loops and tails project some way into the continuous medium. Stabilization results from a negative entropy change, due to a restriction in the configurational freedom of the polymer (volume restriction) and a positive free energy change from the mixing of the polymer chains, though there may be other contributions, e.g., from polymer desorption or bridging. When the polymer is a polyelectrolyte, a further degree of stabilization results from the charges on the loops and tails. No rigorous statistical treatment of all three factors in steric stabilization by polyelectrolytes has been reported. Although the steric mechanism provides stability at large distances of separation of the core particles, it is essentially short-range in character and hence crucially dependent on the magnitude of the solvation forces around the polymer segments. In a good solvent, steric stabilization can be very strong and the repulsive force can then be treated as effectively infinite. In a Θ solvent, the free energy of mixing of the polymer chains is zero. Precipitation of sterically stabilized colloids is, therefore, often associated with solvent conditions that approximate those of a Θ solvent for the stabilizing polymer.

The evidence that a hairy layer exists on the outside of micelles, formed from the macropeptide portion of κ-casein, was reviewed in Section III,D. Prior to the accumulation of most of the experimental evidence, it was argued on theoretical grounds that the clotting of milk by rennet is due to the removal of sterically stabilizing κ-casein hairs by the action of chymosin (Holt, 1975a; Walstra, 1979). However, the evolution of the idea of a sterically stabilized casein micelle can be traced further back through the work of Hill and Wake (1969), who recognized the importance of the amphiphilic nature of κ-casein to its stabilizing action, to the seminal work of Linderstrøm-Lang and Kodama (1929) in identifying a casein fraction (Schutzcolloid) capable of protecting whole casein against precipitation.

Heating milk to elevated temperatures causes a progressive loss of κ-casein from the micelles (Aoki *et al.*, 1975; Singh and Fox, 1985) and, eventually, aggregation of the proteins. Though there are many other factors influencing the heat stability of milk, the loss of the protective colloid may turn out to be of crucial importance in determining the length of time for which milk can be heated. For example, when micelles are cross-linked by aldehydes, a considerable increase in heat stability results; it has been suggested that this is due to retention of the κ-casein in the micelle (Holt *et al.*, 1978b; Kudo, 1980; Singh and Fox, 1985). Precipitation of casein micelles by addition of ethanol to milk has been attributed to the collapse of the hairy layer by the nonsolvent (Horne, 1984b, 1986) and the so-called isoelectric precipitation at pH 4.6 can be ascribed to the disruption of micellar structure consequent on the solubilization of the colloidal calcium phosphate. Likewise, the steric stabilization mechanism has provided the basis of all recent treatments of the theory of rennet clotting of milk.

D. Action of Chymosin on Casein Micelles

As part of the traditional process of making cheese, calf rennet is added to milk to cause the clotting of casein micelles. The primary phase of the reaction in cow's milk is a specific cleavage of the Phe^{105}-Met^{106} bond of κ-casein by the chymosin in the rennet, after which the micelles are able to form a gel occupying virtually the same volume as the original milk. Indeed, there is a striking parallel between the clotting of blood and milk, in that aggregation follows a limited and specific proteolytic reaction, which is no accident of history as the similarities between κ-casein and γ-fibrinogen at the gene, RNA, and protein levels demonstrate (Jollès *et al.*, 1978; Thompson *et al.*, 1985; Alexander *et al.*, 1988).

The primary (enzymatic) phase of renneting overlaps somewhat with the secondary phase of aggregation. The gel subsequently undergoes syneresis to produce curds and whey while a slow but more general proteolysis of the caseins begins, which eventually contributes substantially to the distinctive flavor and texture of cheese. The enzymatic coagulation of milk and formation of the curd has been reviewed by Dalgleish (1987). Here, attention will be confined to parts of the subject that most clearly relate to the structure and stability of bovine casein micelles.

When rennet is added to milk during cheesemaking, there is a period (the lag time) during which there is little apparent change in fluidity and then a rapid increase in viscosity occurs at the clotting time. Holter (1932) proposed an empirical equation to describe the relation between the clotting time, τ_c, and the amount of rennet added, which also applies to clotting by chymosin alone:

$$\tau_c = A + B/[E] \tag{4}$$

where A and B are constants and $[E]$ is the chymosin concentration. The goal of many of the theoretical treatments of renneting has been to derive, as elegantly as possible, an expression for the clotting time and to show the physicochemical basis for the Holter equation.

Models of Renneting Reactions

Payens (1976, 1977) realized that in order to derive an expression for the clotting time of milk, a kinetic description of the enzyme reaction must be combined with the kinetics of aggregation of destabilized colloidal particles. In early versions of his theory, the enzymatic reaction was described by Michaelis–Menten kinetics with the assumption that the production of fully renneted micelles was at the constant rate V_{max}. This restriction was removed in the more complete theory of Payens and Wiersma (1980). In this theory, casein micelles are represented as monomers of molecular weight M_0 which form labile (fully renneted) micelles of molecular weight M_1 by loss of all their macropeptides. The rate of production of labile monomers is described by the integrated form of the Michaelis–Menten equation. The mechanism of chymosin action is envisaged as adsorption of a molecule of the enzyme to a micelle, complete hydrolysis of all the κ-casein on that micelle, followed by desorption of the enzyme. For the aggregation phase, the rate of reaction of fully renneted micelles was described by the Smoluchowski equation (von Smoluchowski, 1917) in which the net production of i-mers is the difference between the production of i-mers by combination of smaller j-mers with $(i-j)$-mers and the loss of i-mers by reaction of i-mers with all j-mers:

$$\frac{d[P_i]}{dt} = -[P_i] \sum_{j=1}^{\infty} k_{ij} [P_j] + 0.5 \sum_{j=1}^{i-1} k_{j(i-j)} [P_j] [P_{i-j}] \quad (5)$$

The k_{ij} are the elements of the reaction kernel, given in Smoluchowski's original theory for coalescing spherical particles by the diffusional collision rate as a constant, k_s, independent of i and j:

$$k_{ij} = k_s = 8kT/3\eta \quad (6)$$

For the weight-average molecular weight, M_w, as a function of time, Payens and Wiersma (1980) then derived the result

$$M_w/M_0 = 1 - M_0 (1 - f) P(t)\{2f - (1 - f) k_s [\int_0^t P(t)^2 \, dt/P(t)]/C_0\} \quad (7)$$

where $P(t)$ is the total number of reactive monomers ($= \Sigma_j [P_j]$) at time t, C_0 is the original substrate concentration, and f is the fraction of monomer mass released into solution by the enzyme action. Because it is assumed that the production of reactive monomers parallels the hydrolysis of κ-casein, $P(t)$ is obtained as the solution to the integrated Michaelis–Menten equation, thereby bringing the effects of the enzymatic reaction explicitly into the expression for the aggregation process. Dalgleish (1980a) proposed a modification of the scheme of Payens and Wiersma (1980) by supposing that micelles became fully reactive when a critical fraction of their κ-casein was hydrolyzed; under his conditions of measurement, this was only a little less, at 97%, than the total cleavage demanded by Payens and Wiersma (1980). If the rate constant for aggregation is effectively zero below the critical fractional degree of hydrolysis of κ-casein, then the time for clotting is of the order of the sum of the time required to give critical proteolysis, plus the time for this amount of material to aggregate to a supposed, critical, extent. Dalgleish (1980b) was then able to recover the empirical relationship of Holter (1932) for the clotting time of milk [Eq. (4)]. Similar reasoning has been employed elsewhere in deriving expressions for the clotting time (Payens *et al.*, 1977; Darling and van Hooydonk, 1981) but they are all unsatisfactory in that they rely on somewhat arbitrary definitions of critical extents of reaction, which must relate, in some way, to the formation of a gel without explicitly considering the structure of the aggregating micelle clusters or their mechanism of growth.

The theory of Payens and Wiersma (1980) can, therefore, simulate some of the behavior observed in the clotting of micelles, namely, a lag time during which the molecular weight of the micelles decreases slightly (Payens, 1977), followed by a rapid increase of molecular weight at the sol–gel transition. Moreover, at low substrate concentrations, τ_c is shown to be dominated by the rate of aggregation of the destabilized micelles, whereas at high substrate concentrations (e.g., in milk), the overall rate is effectively determined by the rate of proteolysis. Nevertheless, certain features of this model can be considered unsatisfactory. First, the enzyme kinetics in milk appear to be first order rather than of the Michaelis–Menten type; no particular problems are encountered in incorporating such a change in the model (e.g., Hyslop, 1989). Second, and possibly more serious, is the approach of using a step function (Payens and Wiersma, 1980; Dalgleish, 1980a) for the reaction kernel and making the temporal course of the aggregation dependent on the production of reactive particles. It can be argued that the correct approach is to devise a model for the influence of proteolysis on the reaction kernel, which allows the rate constants to increase smoothly with increasing average degree of proteolysis. However, there are other implications.

A constant reaction kernel is to be expected in the absence of enzyme reaction if the aggregation rate is determined by the Brownian motion of spherical particles which coalesce to form larger spheres. To a first approximation, the increased collisional cross-section is then compensated for by the decrease in diffusion rate (von Smoluchowski,

1917). Micelles, however, form loose clusters on aggregation, with a geometry which is related, in a complex way to their diffusion rate collision cross-section. Moreover, it might be envisaged that dissociation is more significant from loose structures than from the coalesced spheres of the original Smoluchowski mechanism and hence such reactions should be incorporated in the overall scheme. Finally, the proposed theories take little account of micelle structure, in that the effect of the broad distribution of sizes is ignored and the ideas of steric stabilization do not enter explicitly into the schemes.

The theory of particle aggregation has been developed along three different but complementary lines, namely, (1) generalizations of the Smoluchowski equation in which the reaction kernel is given a more complex form, (2) developments in the Flory–Stockmayer (Flory, 1953) statistical theory of polymer condensation, and, to an increasing extent in recent years, (3) computer simulations of the formation of aggregates constructed on a lattice. A constant kernel in Eq. (5) is justified only for the initial rate of coagulation of a dilute colloid; in general, a more complex form is required but, because the arguments $[P_j]$ appear as products, there are then few analytical solutions to Eq. (5). The three most important analytical solutions are (a) $K_{ij} = A$, (b) $k_{ij} = B(i + j)$, and (c) $k_{ij} = Cij$, where A, B, and C are constants. Case (a), the Smoluchowski result, leads to the same kinetics as the Flory–Stockmayer theory for reaction of bifunctional units because the growing polymer always has two reactive ends, independent of size. Case (c) is equivalent to a statistical theory of polymer growth in which the polymers have an infinite number of functional groups, i.e., polymers are reactive over their entire surface. In cases (a) and (b) the molecular weight increases monotonically to infinity with increasing time whereas case (c) shows a sol–gel transition at an intermediate, critical, degree of conversion. Because the surface functional groups of a growing particle may be more reactive than those in the interior, due to steric constraints, a kernel of the form $Ci^p j^q$ is suggested where p and q represent the effective fraction of reactive surface ($p + q < 2$). Computer simulations have shown that a sol–gel transition is expected if $p + q > 1$ (van Dongen and Ernst, 1985). The constant kernel model does not predict a clotting time.

Evidence of the fraction of free monomer micelles at the clotting time would help to determine the appropriate form of the reaction kernel and whether growth is limited by diffusion or by the reaction itself. The growth of polymers from polyfunctional monomers, the formation of diffusion-limiting aggregates, and many other natural phenomena can all be scale invariant fractals with a similar fractal dimension (Mandelbrot, 1982). Witten and Sander (1981) made a seminal study of the growth of fractal structures by computer simulation on a lattice and found the fractal dimension $D = 2.5$ rather than the 3.0 for Smoluchowski's coalescing spheres. Many subsequent studies have elaborated the Witten–Sander model to show the limits of the Smoluchowski equation and how it may be rescaled and generalized. For example, if clusters are formed by cluster–cluster aggregation, rather than by growth from monomers, a lower fractal dimension is to be expected. If the product of a Witten–Sander growth is allowed to rearrange at constant size (a so-called equilibrium model), the initial fractal dimension is increased (Botet and Jullien, 1985). Other simulations have explored the effects on growth of diffusion-limited aggregation at the surface of a cluster together with percolation of monomers through the cluster to internal points of attachment (Termonia and Meakin, 1985).

The rate of aggregation of fully renneted micelles is very sensitive to temperature. At room temperature it is appreciably less than the diffusional collision rate, which led Payens (1977) to consider the possibility that only a fraction of the surface is reactive (so-called hot spots). The idea of hot spots is consistent with the low fractal dimension of micelle clusters formed during renneting and leads to only a proportion of all encounters between fully renneted micelles being successful. In effect, a statistical prefactor is included in the reaction kernel to reduce the diffusion rate to a level comparable with experiment. However, Payens developed the idea of hot spots only within his theory of the aggregation of fully renneted micelles.

Darling and van Hooydonk (1981) also considered how to reduce the diffusional collision rate to obtain slow coagulation and used the classical approach of Fuchs (Reerink and Overbeek, 1954), whereby an activation energy is computed from the pair interaction free energy of the aggregating particles. The reaction kernel is given by Eq. (6) divided by the stability ratio W,

$$W = \exp(U_{max}/kT) \qquad (8)$$

where U_{max} is the maximum in the pair interaction curve. Although the stability ratio was originally applied to slow coagulation by hydrophobic sols within the context of DLVO theory, Darling and van Hooydonk (1981) considered that it could be used for a sterically stabilized hydrophilic colloid and that U_{max} would be a linear, decreasing function of the number of intact κ-casein molecules in the micelle. By this means, a temporal dependence to the reaction kernel is introduced. Darling and van Hooydonk (1981) derived an expression for a so-called aggregation time of milk by combining the linearized Michaelis–Menten equation with the bimolecular rate equation for the dimerization of casein micelles, assuming that there is a critical number concentration of monomers at the time at which clotting occurs. Very similar reaction schemes have been put forward by van Hooydonk and Walstra (1987) and Hyslop (1989) for first-order enzymes kinetics but the problems remain that no explicit account is taken of the fractal nature of the aggregating clusters of casein micelles. It is also not obvious why the interaction energy should be a linear function of κ-casein hydrolysis. Both the free energy of mixing of polymer chains and double-layer repulsion should depend primarily on the square of the density of segments in the hairy layer. The experimental evidence to support the linear function is that of van Hooydonk and Walstra (1987), who infer from viscometry that $\log(k_s)$ is proportional to α but only over a very limited range of α values (78–86%).

Dalgleish and Holt (1988) found that a statistical prefactor was required even for the dimerization reaction of partly rennetted casein micelles. They calculated the fraction of successful collisions from a geometric model of micelles stabilized by a hairy layer, using the idea (Holt, 1975a, 1985a) that the steric stabilizing force is effectively infinite, except when two bare patches make contact. Bare patches are assumed to form randomly on the surface of micelles as a result of chymosin action. The theory predicts that $\log(k_b)$ is a linear function of $\log(\alpha)$, where k_b is the rate constant for dimerization, as was observed experimentally over a wide range of α values. It proved possible to account for the dependence of the bimolecular collision rate constant on the average degree of proteolysis, using values for the hairy layer thickness and segment density obtained in an earlier study of the electrophoretic mobility of micelles (Holt and Dalgleish, 1986). The theory also predicts a dependence of aggregation rate on micelle size, with smaller particles reacting more quickly, and with a lower average functionality than larger particles (Holt, 1985a). Extension of the theory to describing the full course of the clotting process (Dalgleish, 1988), however, ignores the dependence of the reaction kernel on the geometry and hydrodynamics of the clusters and so suffers from the same disadvantages as all the other models of the clotting time. Horne (1987, 1989) and Horne and Davidson (1990) have made some progress in describing the aggregation of micelles using fractal concepts and experimental methodologies at milk concentrations, which offer some hope of a more complete and self-consistent treatment, not only of the renneting reaction, but of micelle aggregation phenomena in general.

REFERENCES

Alexander, J., Stewart, A. F., Mackinlay, A. G., Kapelinskaya, T. V., Tkach, M., and Gorodetsky, S. I. (1988). *Biochem. J.* **178**, 395–401.
Andrews, A. L., Atkinson, D., Evans, M. T. A., Finer, E. G., Green, J. P., Phillips, M. C., and Robertson, R. N. (1979). *Biopolymers* **18**, 1105–1121.
Andrews, A. T., Taylor, M. D., and Owen, A. J. (1985). *J. Chromatogr.* **348**, 177–185.
Aoki, T., and Kako, Y. (1984). *J. Dairy Res.* **51**, 439–445.
Aoki, T., Suzuki, H., and Imamura, T. (1975). *Milchwissenschaft* **30**, 30–35.
Aoki, T., Toyooka, K., and Kako, Y. (1985). *J. Dairy Sci.* **68**, 1624–1629.
Aoki, T., Kako, Y., and Imamura, T. (1986a). *J. Dairy Res.* **53**, 53–59.
Aoki, T., Yamada, N., Kako, Y., and Kuwata, T. (1986b). *Jpn. J. Zootech. Sci.* **57**, 624–627.
Aoki, T., Yamada, N., Kako, Y., and Imamura, T. (1988). *J. Dairy Res.* **55**, 189–195.
Baumrucker, C. W., and Keenan, T. W. (1975). *Exp. Cell Res.* **90**, 253–260.
Beeby, R. (1979). *N. Z. J. Dairy Sci. Technol.* **14**, 1–11.
Berry, G. P., and Creamer, L. K. (1975). *Biochemistry* **14**, 3542–3545.
Bingham, E. W. (1987). *J. Dairy Sci.* **70**, 2233–2240.
Blake, C. C. F. (1978). *Nature (London)* **273**, 267.
Bloomfield, V. A. (1976). *Biopolymers* **15**, 1243–1249.
Bloomfield, V. A., and Mead, R. J., Jr. (1974). *J. Dairy Sci.* **58**, 592–601.
Bonsing, J., and Mackinlay, A. G. (1987). *J. Dairy Res.* **54**, 447–461.
Botet, R., and Jullien, R. (1985). *Phys. Rev. Lett.* **55**, 1943–1946.
Boulton, A. P., Pascall, J. C., and Craig, R. K. (1984). *Biochem. J.* **222**, 501–510.
Bringe, N. A., and Kinsella, J. E. (1986). *J. Dairy Res.* **53**, 371–379.
Brooks, C. L. (1987). *J. Dairy Sci.* **70**, 2226–2232.
Brooks, C. L., and Landt, M. (1984). *Biochem. J.* **224**, 195–200.
Buchheim, W., and Schmidt, D. G. (1979). *J. Dairy Res.* **46**, 277–280.
Buchheim, W., and Welsch, U. (1973). *Neth. Milk Dairy J.* **27**, 163–180.
Byler, D. M., and Farrell, H. M., Jr. (1989). *J. Dairy Sci.* **72**, 1719–1723.
Byler, D. M., and Susi, H. (1986). *Biopolymers* **25**, 469–487.
Carroll, R. J., and Farrell, H. M., Jr. (1983). *J. Dairy Sci.* **66**, 679–686.
Carroll, R. J., Thompson, M. P., and Nutting, G. C. (1968). *J. Dairy Sci.* **51**, 1903–1908.
Caspar, D. L. D., and Klug, A. (1962). *Cold Spring Harbor Symp. Quant. Biol.* **27**, 1–24.
Chaplin, B., and Green, M. L. (1981). *Neth. Milk Dairy J.* **35**, 377–380.
Chaplin, B., and Green, M. L. (1982). *J. Dairy Res.* **49**, 631–643.
Chaplin, L. C. (1984). *J. Dairy Res.* **51**, 251–257.
Chou, P. Y., and Fasman, G. D. (1974). *Biochemistry* **13**, 222–242.
Chou, P. Y., and Fasman, G. D. (1978). *Adv. Enzymol.* **47**, 45–148.
Cookson, D. J., Levine, B. A., Williams, R. J. P., Jontell, M., Linde, A., and de Bernard, B. (1980). *Eur. J. Biochem.* **110**, 273–278.
Craik, C. S., Rutter, W. J., and Fletterick, R. (1983). *Science* **220**, 1125–1129.
Creamer, L. K. (1984). *J. Chromatogr.* **291**, 460–463.

Creamer, L. K., and Berry, G. P. (1975). *J. Dairy Res.* **42,** 169–183.
Creamer, L. K., and Yamashita, S. (1976). *N. Z. J. Dairy Sci. Technol.* **11,** 257–262.
Creamer, L. K., Berry, G. P., and Mills, O. E. (1977). *N.Z. J. Dairy Sci. Technol.* **12,** 58–66.
Creamer, L. K., Richardson, T., and Parry, D. A. D. (1981). *Arch Biochem. Biophys.* **211,** 689–696.
Dalgleish, D. G. (1973). *Eur. J. Biochem.* **40,** 375–380.
Dalgleish, D. G. (1979). *J. Dairy Res.* **46,** 653–661.
Dalgleish, D. G. (1980a). *Biophys. Chem.* **11,** 147–155.
Dalgleish, D. G. (1980b). *J. Dairy Res.* **47,** 231–235.
Dalgleish, D. G. (1987). *In* "Cheese: Chemistry, Physics and Microbiology" (P.F. Fox, ed.), Vol. 1, pp. 63–96. Elsevier Appl. Sci., London.
Dalgleish, D. G. (1988). *J. Dairy Res.* **55,** 521–528.
Dalgleish, D. G., and Holt, C. (1988). *J. Colloid Interface Sci.* **123,** 80–84.
Dalgleish, D. G., and Law, A. J. R. (1988). *J. Dairy Res.* **55,** 529–538.
Dalgleish, D. G., and Parker, T. G. (1979). *J. Dairy Res.* **46,** 259–263.
Dalgleish, D. G., and Parker, T. G. (1980). *J. Dairy Res.* **47,** 113–122.
Darling, D. F., and van Hooydonk, A. C. M. (1981). *J. Dairy Res.* **48,** 189–200.
Dalgleish, D. G., Paterson, E., and Horne, D. S. (1981). *Biophys. Chem.* **13,** 307–314.
Dalgleish, D. G., Horne, D. S., and Law, A. J. R. (1989). *Biochim. Biophys. Acta* **991,** 383–387.
Davies, D. R. (1990). *Annu. Rev. Biophys. Chem.* **19,** 189–215.
Davies, D. T., and Law, A. J. R. (1977a). *J. Dairy Res.* **44,** 213–221.
Davies, D. T., and Law, A. J. R. (1977b). *J. Dairy Res.* **44,** 447–454.
Davies, D. T., and Law, A. J. R. (1980). *J. Dairy Res.* **47,** 83–90.
Davies, D. T., and Law, A. J. R. (1983). *J. Dairy Res.* **50,** 67–75.
Davies, D. T., and Law, A. J. R. (1987). *J. Dairy Res.* **54,** 369–376.
Davies, D. T., and White, J. C. D. (1958). *J. Dairy Res.* **25,** 256–266.
Dewan, R. K., Chudgar, A., Mead, R., Bloomfield, V. A., and Morr, C. V. (1974). *Biochim. Biophys. Acta* **342,** 313–321.
Dickson, I. R., and Perkins, D. J. (1971). *Biochem. J.* **124,** 235–240.
Donnelly, W. J., McNeill, G. P., Buchheim, W., and McGann, T. C. A. (1984). *Biochim. Biophys. Acta* **789,** 136–143.
Eigel, W. N., Butler, J. E., Ernsrøm, C. A., Farrell, H. M., Jr., Harwalkar, V. R., Jenness, R., and Whitney, R. McL. (1984). *J. Dairy Sci.* **67,** 1599–1631.
Ekstrand, B., and Larsson-Raznikiewicz, M. (1984). *Milchwissenschaft* **39,** 591–593.
Farrell, H. M., Jr. (1973). *J. Dairy Sci.* **56,** 1195–1206.
Farrell, H. M., Jr., and Thompson, M. P. (1988). *In* "Calcium Binding Proteins" (M.P. Thompson, ed.), Vol. 2, pp. 31–51. CRC Press, Boca Raton, Florida.
Farrell, H. M., Jr., Kumosinski, T. F., Pulaski, P., and Thompson, M. P. (1988). *Arch Biochem. Biophys.* **265,** 146–158.
Ferretti, L., Leone, P., and Sgaramella, V. (1990). *Nucleic Acids Res.* **18,** 6829–6833.
Fiol, C. J., Wang, A., Roeske, R. W., and Roach, P. J. (1990). *J. Biol. Chem.* **265,** 6061–6065.
Flory, P. J. (1953). "Principles of Polymer Chemistry" Cornell Univ. Press, Ithaca, New York.
Ford, T. F., Ramsdell, G. A., and Landsman, S. G. (1955). *J. Dairy Sci.* **38,** 843–857.
Foster, R. C. (1979). *Am. J. Physiol.* **236,** C286–C292.
Garnier, J. (1966). *J. Mol. Biol.* **19,** 586–590.
Garnier, J. Osguthorpe, D. J., and Robson, B. (1978). *J. Mol. Biol.* **120,** 97–120.

Gaye, P. Gautron, J.-P., Mercier, J.-C., and Hazé, G. (1977). *Biochem. Biophys. Res. Commun.* **79,** 903–911.
Geerts, J. P., Bekhof, J. J., and Scherjon, J. W. (1983). *Neth. Milk Dairy J.* **37,** 197–211.
Gellin, J., Echard, G., Yerle, M., Dalens, M., Chevalet, C., and Gillois, M. (1985). *Cytogenet. Cell Genet.* **39,** 220–223.
Gilbert, W. (1978). *Nature (London)* **271,** 501.
Graham, E. R. B., Malcolm, G. N., and McKenzie, H. A. (1984). *Int. J. Biol. Macromol.* **6,** 155–161.
Green, M. L., and Crutchfield, G. (1971). *J. Dairy Res.* **38,** 151–164.
Green, M. L., Hobbs, D. G., Morant, S. V., and Hill, V. A. (1978). *J. Dairy Res.* **45,** 413–422.
Griffin, M. C. A. (1987). *J. Colloid Interface Sci.* **115,** 499–506.
Griffin, M. C. A., and Roberts, G. C. K. (1985). *Biochem. J.* **228,** 273–276.
Griffin, M. C. A., Price, J. C., and Martin, S. R. (1986). *Int. J. Biol. Macromol.* **8,** 367–371.
Griffin, M. C. A., Lyster, R. L. J., and Price, J. C. (1988). *Eur. J. Biochem.* **174,** 339–343.
Griffin, M. C. A., Price, J. C., and Griffin, W. G. (1989). *J. Colloid Interface Sci.* **128,** 223–229.
Grosclaude, F. (1979). *Proc. Int. Conf. Anim. Blood Groups, Biochem. Polymorphisms, 16th Leningrad, 1978* vol. 1, pp. 54–92.
Haga, M., Yamauchi, K., and Aoyagi, S. (1983). *Agric. Biol. Chem.* **47,** 1467–1471.
Harries, J. A., Hukins, D. W. L., Holt, C., and Hasnain, S. S. (1986). *J. Phys. C.* **19,** 6859–6872.
Harries, J. E., Irlam, J. C., Holt, C., Hasnain, S. S., and Hukins, D. W. L. (1987). *Mater. Res. Bull.* **22,** 1151–1157.
Heth, A. A., and Swaisgood, H. E. (1982). *J. Dairy Sci.* **65,** 2047–2054.
Hill, R. J., and Wake, R. G. (1969). *Nature (London)* **221,** 635–639.
Hill, T. L. (1968). "Thermodynamics for Chemists and Biologists" Addison-Wesley, Reading, Massachusetts.
Ho, C., and Waugh, D. F. (1965). *J. Am. Chem. Soc.* **87,** 110–116.
Hobbs, A. A., and Rosen, J. M. (1982). *Nucleic Acids Res.* **10,** 8079–8098.
Holt, C. (1975a). *Proc. Int. Conf. Colloid Surf. Sci., 1975* Vol. 1, pp. 641–644.
Holt, C. (1975b). *Biochim. Biophys. Acta* **400,** 293–301.
Holt, C. (1982). *J. Dairy Res.* **49,** 29–38.
Holt, C. (1983). *J. Theor. Biol.* **101,** 247–261.
Holt, C. (1985a). *Food Microstruct.* **4,** 1–10.
Holt, C. (1985b). *In* "Developments in Dairy Chemistry" (P.F. Fox, ed.), Vol. 3, pp. 143–181. Elsevier Appl. Sci., Barking, UK.
Holt, C., and Dalgleish, D. G. (1986). *J. Colloid Interface Sci.* **114,** 513–524.
Holt, C., and Hukins, D. W. L. (1991). *Int. Dairy J.* **1,** 151–165.
Holt, C., and Jenness, R. (1984). *Comp. Biochem. Physiol.* **77A,** 275–282.
Holt, C., and Muir, D. D. (178). *J. Dairy Res.* **45,** 347–354.
Holt, C., and Sawyer, L. (1988a). *Protein Eng.* **2,** 251–259.
Holt, C., and Sawyer, L. (1988b). *Protein Eng.* **3,** 273.
Holt, C., and van Kemenade, M. J. J. M. (1989). *In* "Calcified Tissues" (D. W. L. Hukins, ed.), pp. 175–214. Macmillan, Basingstoke.
Holt, C., Kimber, A. M., Brooker, B. E., and Prentice, J. H. (1978a). *J. Colloid Interface Sci.* **65,** 555–565.

Holt, C., Muir, D. D., and Sweetsur, A. W. M. (1978b). *J. Dairy Res.* **45**, 47–52.
Holt, C., Dalgleish, D. G., and Jenness, R. (1981). *Anal. Biochem.* **113**, 154–163.
Holt, C., Hasnain, S. S., and Hukins, D. W. L. (1982). *Biochim. Biophys. Acta* **719**, 299–303.
Holt, C., Davies, D. T., and Law, A. J. R. (1986). *J. Dairy Res.* **53**, 557–572.
Holt, C., van Kemenade, M. J. J. M., Harries, J. E., Nelson, L. S., Jr., Bailey, R. T., Hukins, D. W. L., Hasnain, S. S., and de Bruyn, P. L. (1988). *J. Cryst. Growth* **92**, 239–252.
Holt, C., van Kemenade, M. J. J. M., Nelson, L. S., Jr., Sawyer, L., Harries, J. E., Bailey, R. T., and Hukins, D. W. L. (1989a). *J. Dairy Res.* **56**, 411–416.
Holt, C., van Kemenade, M. J. J. M., Nelson, L. S., Jr., Harries, J. E., Bailey, R. T., Hukins, D. W. L., Hasnain, S. S., and de Bruyn, P. L. (1989b). *Mater. Res. Bull.* **23**, 55–62.
Holter, H. (1932). *Biochem. Z.* **255**, 160–188.
Horisberger, M., and Vauthey, M. (1984). *Histochemistry* **80**, 9–12.
Horisberger, M., and Vonlanthen, M. (1980). *J. Dairy Res.* **47**, 185–191.
Horne, D. S. (1982). *J. Dairy Res.* **49**, 107–118.
Horne, D. S. (1983). *Int. J. Biol. Macromol.* **5**, 296–300.
Horne, D. S. (1984a). *J. Colloid Interface Sci.* **98**, 537–548.
Horne, D. S. (1984b). *Biopolymers* **23**, 989–993.
Horne, D. S. (1986). *J. Colloid Interface Sci.* **111**, 250–260.
Horne, D. S. (1987). *Faraday Discuss. Chem. Soc.* **83**, 259–270.
Horne, D. S. (1989). *J. Phys. D.* **22**, 1257–1265.
Horne, D. S., and Dalgleish, D. G. (1980). *Int. J. Biol. Macromol.* **2**, 154–160.
Horne, D. S., and Davidson, C. M. (1986). *Colloid Polym. Sci.* **264**, 727–734.
Horne, D. S., and Davidson, C. M. (1990). *Milchwissenschaft* **45**, 712–715.
Horne, D. S., and Moir, P. D. (1984). *Int. J. Biol. Macromol.* **6**, 316–320.
Horne, D. S., and Parker, T. G. (1981a). *J. Dairy Res.* **48**, 273–284.
Horne, D. S., and Parker, T. G. (1981b). *Int. J. Biol. Macromol.* **3**, 399–402.
Horne, D. S., Parker, T. G., and Dalgleish, D. G. (1989). *In* "Food Colloids" (R. D. Bee, P. Richmond, and J. Mingins, eds.), No. 75, pp. 400–406. Spec. Publ. R. Soc. Chem., Cambridge.
Humphrey, R. S., and Jolley, K. W. (1982). *Biochim. Biophys. Acta* **708**, 294–299.
Hurley, J. H., Thorsness, P. E., Ramalingham, V., Helmers, N. H., Koshland, D. E., Jr., and Stroud, R. M. (1989). *Proc. Natl. Acad. Sci.* **86**, 8635–8639.
Hyslop, D. B. (1989). *Neth. Milk Dairy J.* **43**, 163–170.
Imade, T., Sato, Y., and Noguchi, H. (1977). *Agric. Biol. Chem.* **41**, 2131–2137.
Israelachvili, J., and Pashley, R. (1982). *Nature (London)* **300**, 341–342.
Jaenicke, R. (1982). *Biophys. Struct. Mech.* **8**, 231–256.
Jimenez-Flores, R., Kang, Y. C., and Richardson, T. (1987). *Biochem. Biophys. Res. Commun.* **142**, 617–621.
Jollès, J., Allais, C., and Jollès, P. (1970). *Helv. Chim. Acta* **53**, 1918–1926.
Jollès, P., Loucheaux-Lefebvre, M. H., and Henschen, A. (1978). *J. Mol. Evol.* **11**, 271–277.
Jones, W. K., Yu-Lee, L.-Y., Clift, S. M., Brown, T. L., and Rosen, J. M. (1985). *J. Biol. Chem.* **260**, 7042–7050.
Kajiwara, K., Niki, R., Urakawa, H., Hiragi, Y., Donkai, N., and Nagura, M. (1988). *Biochim. Biophys. Acta* **955**, 128–134.
Kalab, M., Phipps-Todd, B. E., and Allan-Wojtas, P. (1982). *Milchwissenschaft* **37**, 513–518.

Kaminogawa, S., Shimizu, M., and Yamauchi, K. (1983). *Proc. World Conf. Anim. Prod. 5th,* Vol. 2, pp. 681–682.
Keenan, T. W., and Dylewski, D. P. (1985). *J. Dairy Sci.* **68,** 1025–1040.
Keenan, T. W., Sasaki, M., Eigel, W. N., Morrié, D. J., Franke, W. W., Zulak, I. M., and Bushway, A. A. (1979). *Exp. Cell Res.* **124,** 47–61.
Kegeles, G. (1979). *J. Phys. Chem.* **83,** 1728–1732.
Kelly, R. B. (1985). *Science* **230,** 25–32.
Kirchmeier, O. (1973). *Neth. Milk Dairy J.* **27,** 191–198.
Knoop, A.-M., Knoop, E., and Wiechen, A. (1973). *Neth. Milk Dairy J.* **27,** 121–127.
Knoop, A.-M., Knoop, E., and Wiechen, A. (1979). *J. Dairy Res.* **46,** 347–350.
Kudo, S. (1980). *N. Z. J. Dairy Sci. Technol.* **15,** 255–263.
Kuhn, N. J., and White, A. (1975). *Biochem. J.* **148,** 77–84.
Lee, S. L., Veis, A., and Glonek, T. (1977). *Biochemistry* **16,** 2971–2979.
Leiderman, L. J., Criss, W. E., and Oka, T. (1985). *Biochim. Biophys. Acta* **844,** 95–104.
Lenstra, J. A. (1977). *Biochim. Biophys. Acta* **491,** 333–338.
Levitt, M., and Chothia, C. (1976). *Nature (London)* **261,** 552–558.
Lim, V. I. (1974). *J. Mol. Biol.* **88,** 873–894.
Lin, S. H. C., Leong, S. L., Dewan, R. K., Bloomfield, V. A., and Morr, C. V. (1972). *Biochemistry* **11,** 1818–1821.
Linderstrøm-Lang, K., and Kodama, S. (1929). *C. R. Trav. Lab. Carlsberg, Ser. Chim.* **17,** 1–116.
Linzell, J. L., and Peaker, M. (1971). *J. Physiol. London* **216,** 683–700.
Loucheaux-Lefebvre, M. H., Aubert, J. P., and Jollès, P. (1978). *Biophys. J.* **23,** 323–336.
Lyster, R. L. J., Mann, S., Parker, S. B., and Williams, R. J. P. (1984). *Biochim. Biophys. Acta* **801,** 315–317.
Mandelbrot, B. B. (1982). "Fractal Geometry of Nature." Freeman, New York.
Maubois, J. L., and Léonil, J. (1989). *Lait* **69,** 245–269.
McGann, T. C. A., and Fox, P. F. (1974). *J. Dairy Res.* **41,** 45–53.
McGann, T. C. A., and Pyne, G. T. (1960). *J. Dairy Res.* **27,** 403–417.
McGann, T. C. A., Kearney, R. D., and Donnelly, W. J. (1979). *J. Dairy Res.* **46,** 307–311.
McGann, T. C. A., Buchheim, W., Kearney, R. D., and Richardson, T. (1983a). *Biochim. Biophys. Acta* **760,** 415–420.
McGann, T. C. A., Kearney, R. D., Buchheim, W., Posner, A. S., Betts, F., and Blumenthal, N. C. (1983b). *Calcif. Tissue Int.* **35,** 821–823.
McIntosh, T. J., and Simon, S. A. (1986). *Biochemistry* **25,** 4058–4066.
Meggio, F., Perich, J. W., Meyer, H. E., Hoffmann-Porsorske, E., Lennon, D. P. W., Johns, R. B., and Pinna, L. A. (1989). *Eur. J. Biochem.* **186,** 459–464.
Mehaia, M. A., and Cheryan, M. (1983a). *J. Dairy Sci.* **66,** 390–395.
Mehaia, M. A., and Cheryan, M. (1983b). *J. Dairy Sci.* **66,** 2474–2481.
Mellander, O. (1939). *Biochem. Z.* **300,** 240–245.
Mercier, J.-C. (1981). *Biochimie* **63,** 1–17.
Mercier, J.-C., and Gaye, P. (1983). *In* "Biochemistry of Lactation" (T. B. Mepham, ed.), pp. 177–227. Elsevier, Amsterdam.
Moore, A., Boulton, A. P., Heid, H. W., Jarasch, E.-D., and Craig, R. K. (1985). *Eur. J. Biochem.* **152,** 729–737.
Morr, C. V., Lin, S. H. C., and Josephson, R. V. (1971). *J. Dairy Sci.* **54,** 994–1000.

Morr, C. V., Lin, S. H. C., Dewan, R. K., and Bloomfield, V. A. (1973). *J. Dairy Sci.* **56,** 415–418.
Nagao, M., Maki, R., Sasaki, R., and Chiba, H. (1984). *Agric. Biol. Chem.* **48,** 1663–1667.
Navaratnam, N., Virk, S. S., Ward, S., and Kuhn, N. J. (1986). *Biochem. J.* **239,** 423–433.
Neville, M. C., and Staiert, P. A. (1983). *J. Cell Biol.* **97,** 442a.
Neville, M. C., Selker, F., Semple, K., and Watters, C. D. (1981). *J. Membr. Biol.* **61,** 97–105.
O'Brien, T. L., and Baumrucker, C. R. (1980). *Trans. Am. Microsc. Soc.* **99,** 403–415.
Ohmia, K., Kajino, T., Shimizu, S., and Gekko, K. (1989). *J. Dairy Res.* **56,** 435–442.
Ono, T., and Obata, T. (1989). *J. Dairy Res.* **56,** 453–461.
Ono, T., Kaminogawa, S., Odagiri, S., and Yamauchi, K. (1976). *Agric. Biol. Chem.* **40,** 1717–1723.
Ono, T., Yahagi, M., and Odagiri, S. (1980). *Agric. Biol. Chem.* **44,** 1499–1503.
Ono, T., Yada, R., Yutani, K., and Nakai, S. (1987). *Biochim. Biophys. Acta* **911,** 318–325.
Parker, T. G., and Dalgleish, D. G. (1977a). *J. Dairy Res.* **44,** 79–84.
Parker, T. G., and Dalgleish, D. G. (1977b). *Biopolymers* **16,** 2533–2547.
Parker, T. G., and Dalgleish, D. G. (1981). *J. Dairy Res.* **48,** 71–76.
Parry, R. M., Jr., and Carroll, R. J. (1969). *Biochim. Biophys. Acta* **194,** 138–150.
Parsegian, V. A. (1982). *Adv. Colloid Interface Sci.* **16,** 49–56.
Patton, S., and Jensen, R. G. (1975). *Prog. Chem. Fats Other Lipids* **14,** 163–276.
Payens, T. A. J. (1966). *J. Dairy Sci.* **49,** 1317–1324.
Payens, T. A. J. (1976). *Neth. Milk Dairy J.* **30,** 55–59.
Payens, T. A. J. (1977). *Biophys. Chem.* **6,** 263–270.
Payens, T. A. J., and Heremans, K. (1969). *Biopolymers* **8,** 335–345.
Payens, T. A. J., and Vreeman, H. J. (1982). *In* "Solution Behaviour of Surfactants" (K. L. Mittal and E. J. Fendler, eds.), pp. 543–571. Plenum, New York.
Payens, T. A. J., and Wiersma, A. K. (1980). *Biophys. Chem.* **11,** 137–146.
Payens, T. A. J., Wiersma, A. K., and Brinkhuis, J. (1977). *Biophys. Chem.* **6,** 253–261.
Pierre, A., Brulé, G., and Fauquant, J. (1983). *Lait* **63,** 473–489.
Provencher, S. W., and Glöckner, J. (1981). *Biochemistry* **20,** 33–37.
Pyne, G. T., and McGann, T. C. A. (1960). *J. Dairy Res.* **27,** 9–17.
Pyne, G. T., and Ryan, J. J. (1932). *Sci. Proc. R. Dublin Soc.* **20,** 471–476.
Raap, J., Kerling, K. E. T., Vreeman, H. J., and Visser, S. (1983). *Arch. Biochem. Biophys.* **221,** 117–124.
Rau, D. C., Lee, B., and Parsegian, V. A. (1984). *Proc. Natl. Acad. Sci. U.S.A.* **81,** 2621–2625.
Reerink, H., and Overbeek, J. T. G. (1954). *Discuss. Faraday Soc.* **18,** 74–84.
Renugopalakrishnan, V., Horowitz, P. M., and Glimcher, M. J. (1985). *J. Biol. Chem.* **260,** 11406–11413.
Ribadeau Dumas, B., Grosclaude, F., and Mercier, J.-C. (1975). *Mod. Probl. Paediatr.* **15,** 46–62.
Richardson, B. C., Creamer, L. K., Pierce, K. N., and Munford, R. E. (1974). *J. Dairy Res.* **41,** 239–247.
Rollema, H. S., Vreeman, H. J., and Brinkhuis, J. A. (1984). *In* "Proceedings of the 22nd Congress Ampère" (K. A. Müller, R. Kind, and J. Roos, eds.), pp. 494–495. Ampère Committee, Zurich.

Rollema, H. S., Brinkhuis, J. A., and Vreeman, H. J. (1988). *Neth. Milk Dairy J.* **42,** 233–248.
Rose, D. (1968). *J. Dairy Sci.* **51,** 1897–1902.
Rose, D., and Colvin, J. R. (1966). *J. Dairy Sci.* **49,** 351–355.
Saito, Z. (1973). *Neth. Milk Dairy J.* **27,** 143–162.
Schmidt, D. G. (1980). *Neth. Milk Dairy J.* **34,** 42–64.
Schmidt, D. G. (1982). *In* "Developments in Dairy Chemistry" (P. F. Fox, ed.), pp. 61–86. Appl. Sci. Publ. Barking, U. K.
Schmidt, D. G., and Both, P. (1982). *Milchwissenschaft* **37,** 336–337.
Schmidt, D. G., and Buchheim, W. (1976). *Neth. Milk Dairy J.* **30,** 17–28.
Schmidt, D. G., and Koops, J. (1977). *Neth. Milk Dairy J.* **31,** 342–351.
Schmidt, D. G., and Poll, J. K. (1989). *Neth. Milk Dairy J.* **43,** 53–62.
Schmidt, D. G., van der Spek, C. A., Buchheim, W., and Hinz, A. (1974). *Milchwissenschaft* **29,** 455–459.
Schmidt, D. G., Koops, J., and Westerbeek, D. (1977). *Neth. Milk Dairy J.* **31,** 328–341.
Schmidt, D. G., Both, P., and Koops, J. (1979). *Neth. Milk Dairy J.* **33,** 40–48.
Scott-Blair, G. W., and Oosthuizen, J. C. (1961). *J. Dairy Res.* **28,** 165–173.
Singh, H., and Fox, P. F. (1985). *J. Dairy Res.* **52,** 65–76.
Slattery, C. W. (1975). *Biophys. Chem.* **3,** 83–89.
Slattery, C. W. (1977). *Biophys. Chem.* **6,** 59–64.
Slattery, C. W., and Evard, R. (1973). *Biochim. Biophys. Acta* **317,** 529–538.
Sleigh, R. W., Sculley, T. B., and Mackinlay, A. G. (1979). *J. Dairy Res.* **46,** 337–342.
Sleigh, R. W., Mackinlay, A. G., and Pope, J. M. (1983). *Biochim. Biophys. Acta* **742,** 175–183.
Snoeren, T. H. M., van Markwijk, B., and van Montfort, R. (1980). *Biochim. Biophys. Acta* **622,** 268–276.
Sood, S. M., Sidhu, K. S., and Dewan, R. K. (1976). *N. Z. J. Dairy Sci. Technol.* **11,** 79–82.
Sood, S. M., Gaind, D. K., and Dewan, R. K. (1979). *N. Z. J. Dairy Sci. Technol.* **14,** 32–34.
Sprang, S. R., Acharya, K. R., Goldsmith, E. J., Stuart, D. L., Varvill, K., Fletterick, R. J., Madsen, N. B., and Johnson, L. N. (1988). *Nature (London)* **336,** 215–221.
Stewart, A. F., Willis, I. M., and Mackinlay, A. G. (1984). *Nucleic Acids Res.* **12,** 3895–3907.
Stewart, A. F., Bonsing, J., Beattie, C. W., Shah, F., Willis, I. M., and Mackinlay, A. G. (1987). *Mol. Biol. Evol.* **4,** 231–241.
Stothart, P. H. (1989). *J. Mol. Biol.* **208,** 635–638.
Stothart, P. H., and Cebula, D. J. (1982). *J. Mol. Biol.* **160,** 391–395.
Sullivan, R. A., Fitzpatrick, M. M., and Stanton, E. K. (1959). *Nature (London)* **183,** 616–617.
Swaisgood, H. E. (1982). *In* "Developments in Dairy Chemistry" (P. F. Fox, ed.), pp. 1–59. Elsevier Appl. Sci., Barking, U. K.
Tai, M., and Kegeles, G. (1984). *Biophys. Chem.* **20,** 81–87.
Takase, K., Niki, R., and Arima, S. (1980). *Biochim. Biophys. Acta* **622,** 1–8.
Takeuchi, M., Tsuda, E., Yoshikawa, M., Sasaki, R., and Chiba, H. (1984). *Agric. Biol. Chem.* **48,** 2789–2797.
Takeuchi, M., Tsuda, E., Yoshikawa, M., Sasaki, R., and Chiba, H. (1985a). *Agric. Biol. Chem.* **49,** 2269–2276.

Takeuchi, M., Yoshikawa, M., Sasaki, R., and Chiba, H. (1985b). *Agric. Biol. Chem.* **49,** 1059–1069.
Taylor, J. E., and Husband, A. D. (1922). *J. Agric. Sci.* **12,** 111–124.
ter Horst, M. G. (1963). *Neth. Milk Dairy J.* **17,** 185–192.
Termonia, Y., and Meakin, P. (1985). *Phys. Rev. Lett.* **54,** 1083–1086.
Thompson, M. D., Dave, J. R., and Nakhasi, H. L. (1985). *DNA* **4,** 263–271.
Thompson, M. P., and Farrell, H. M., Jr. (1973). *Neth. Milk Dairy J.* **27,** 220–239.
Thompson, M. P., Gordon, W. G., Boswell, R. T., and Farrell, H. M., Jr. (1969). *J. Dairy Sci.* **52,** 1166–1172.
Threadgill, D. W., and Womack, J. E. (1990). *Nucleic Acids Res.* **18,** 6935–6942.
Thurn, A., Burchard, W., and Niki, R. (1987a). *Colloid Polymr. Sci.* **265,** 653–666.
Thurn, A., Burchard, W., and Niki, R. (1987b). *Colloid Polym. Sci.* **265,** 897–902.
van Bruggen, E. F. J., Booy, F. P., van Breemen, J. F. L., Brink, J., Keegstra, W. and Schmidt, D. G. (1986). *Proc. Int. Congr. Electron Microsc. Kyoto, 11th* pp. 2423–2424.
van Dongen, P. G. J., and Ernst, M. H. (1985). *Phys. Rev. Lett.* **54,** 1396–1399.
van Halbeek, H., Dorland, L., Vliegenthart, J. F. G., Fiat, A.-M., and Jollès, P. (1980). *Biochim. Biophys. Acta* **623,** 295–300.
van Hooydonk, A. C. M., and Walstra, P. (1987). *Neth. Milk Dairy J.* **41,** 19–47.
van Kemenade, M. J. J. M. (1988). Ph.D. Thesis, University of Utrecht, The Netherlands.
Virk, S. S., Kirk, K. J., and Shears, S. B. (1985). *Biochem. J.* **226,** 741–748.
Visser, S. (1981). *Neth. Milk Dairy J.* **35,** 65–88.
Visser, S., Schaier, R. W., and van Gorkom, M. (1979). *J. Dairy Res.* **46,** 333–335.
Visser, S., Slangen, C. J., and van Rooijen, P. J. (1987). *Biochem. J.* **244,** 553–558.
von Hippel, P. H., and Waugh, D. F. (1955). *J. Am. Chem. Soc.* **77,** 4311–4319.
von Smoluchowski, M. (1917). *Z. Phys. Chem.* **92,** 129–168.
Vreeman, H. J., Both, P., Brinkhuis, J. A., and van der Spek, C. (1977). *Biochim. Biophys. Acta* **491,** 93–103.
Vreeman, H. J., Brinkhuis, J. A., and van der Spek, C. A. (1981). *Biophys. Chem.* **14,** 185–193.
Vreeman, H. J., Visser, S., Slangen, C. J., and van Riel, J. A. M. (1986). *Biochem. J.* **240,** 87–97.
Vreeman, H. J., van Markwijk, B. W., and Both, P. (1989). *J. Dairy Res.* **56,** 463–470.
Wahlgren, N. M., Dejmek, P., and Drakenberg, T. (1990). *J. Dairy Res.* **57,** 355–364.
Walstra, P. (1979). *J. Dairy Res.* **46,** 317–323.
Walstra, P. (1990). *J. Dairy Sci.* **73,** 1965–1979.
Walstra, P., Bloomfield, V. A., Wei, G. J., and Jenness, R. (1981). *Biochim. Biophys. Acta* **669,** 258–259.
Walter, P., Gilmore, R., and Blöbel, G. (1984). *Cell (Cambridge, Mass.)* **38,** 5–8.
Watters, C. D. (1984). *Biochem. J.* **224,** 39–45.
Waugh, D. F. (1971). *In* "Milk Proteins" (H. A. Mackenzie, ed.), Vol. 2, pp. 3–85. Academic Press, New York.
Waugh, D. F., and Noble, R. W. (1965). *J. Am. Chem. Soc.* **87,** 2246–2257.
Waugh, D. F., and von Hippel, P. H. (1956). *J. Am. Chem. Soc.* **78,** 4576–4582.
Waugh, D. F., Slattery, C. W., and Creamer, L. K. (1971). *Biochemistry* **10,** 817–823.
West, D. W. (1981). *Biochim. Biophys. Acta* **673,** 374–386.
West, D. W., and Clegg, R. A. (1981). *Biochem. Soc. Trans.* **9,** 468–469.
White, J. C. D., and Davies, D. T. (1958). *J. Dairy Res.* **25,** 236–255.
Whitney, R. McL., Brunner, J. R., Ebner, K. E., Farrell, H. M., Jr., Josephson, R. V., Morr, C. V., and Swaisgood, H. E. (1976). *J. Dairy Sci.* **59,** 795–815.

Witten, T. A., Jr., and Sander, L. M. (1981). *Phys. Rev. Lett.* **47,** 1400–1403.
Wooding, F. B. P., and Morgan, G. (1978). *J. Ultrastruct. Res.* **63,** 323–333.
Yaguchi, M., Davies, D. T., and Kim, Y. K. (1968). *J. Dairy Sci.* **51,** 473–477.
Yamauchi, K., and Yoneda, Y. (1977). *Agric. Biol. Chem.* **41,** 2395–99.
Yamauchi, K., Yoneda, Y., Koga, Y., and Tsugo, T. (1969). *Agric. Biol. Chem.* **33,** 907–914.
Yoshikawa, M., Takahata, K., Sasaki, R., and Chiba, H. (1978). *Agric. Biol. Chem.* **42,** 1923–1926.
Yoshikawa, M., Sasaki, R., and Chiba, H. (1981). *Agric. Biol. Chem.* **45,** 909–914.
Yoshikawa, M., Takeuchi, M., Sasaki, R., and Chiba, H. (1982). *Agric. Biol. Chem.* **46,** 1043–1048.
Yoshimura, M., and Oka, T. (1989). *Gene* **78,** 267–275.
Zevaco, C., and Ribadeau Dumas, B. (1984). *Milchwissenschaft* **39,** 206–210.

PROTON NUCLEAR MAGNETIC RESONANCE STUDIES ON HEMOGLOBIN: COOPERATIVE INTERACTIONS AND PARTIALLY LIGATED INTERMEDIATES

By CHIEN HO

Department of Biological Sciences, Carnegie Mellon University, Pittsburgh, Pennsylvania 15213

I.	Introduction	154
II.	Experimental Procedures	166
	A. NMR Methodology	166
	B. NMR Techniques	185
	C. Sample Preparation	187
III.	Resonance Assignments: Subunit Interfaces and Heme Environments	192
	A. Strategies for Assignment of Proton Resonances of Hemoglobin	192
	B. Resonances from Subunit Interfaces	200
	C. Resonances from Heme Environments	202
IV.	^{1}H NMR Investigations of Ligand Binding to α and β Chains of Hemoglobin	214
	A. Ligand Binding to α and β Chains of Hb A	215
	B. Comparison between HbO$_2$ A and HbCO A	218
	C. Binding of O$_2$ and CO to Hemoglobin	219
	D. Binding of Alkyl Isocyanides to Hemoglobin	229
	E. Binding of Azide and Cyanide to Methemoglobin	236
V.	^{1}H NMR Investigations of Partially Oxygenated Species of Hemoglobin: Evidence for Nonconcerted Structural Changes during the Oxygenation Process	240
VI.	^{1}H NMR Investigations of Structures and Properties of Symmetric Valency Hybrid Hemoglobins: Models for Doubly Ligated Species	252
	A. M-Type Hemoglobins	253
	B. Synthetic Symmetric Valency Hybrid Hemoglobins	260
VII.	^{1}H NMR Investigations of Structures and Properties of Asymmetric Valency Hybrid Hemoglobins: Models for Singly and Doubly Ligated Species	261
	A. ^{1}H NMR Studies	261
	B. Equilibrium O$_2$ Binding Studies	267
VIII.	Influence of Salt Bridges on Tertiary and Quaternary Structures of Hemoglobin	273
IX.	Other Evidence for Existence of Ligation Intermediates of Hemoglobin	280
X.	X-Ray Crystallographic Investigations of Structural Characteristics of Partially Ligated Species of Hemoglobin	285

XI.	Possible Pathways for Heme–Heme Communication	291
XII.	Concluding Remarks	296
	References	303
	Note Added in Proof	312

I. Introduction

Hemoglobin (Hb) is the oxygen carrier of the blood. It carries O_2 from the lungs to the tissues and also carries carbon dioxide back to the lungs, where CO_2 can be expelled as a waste product. Human normal adult hemoglobin (Hb A) is a tetrameric protein molecule consisting of two α chains of 141 amino acid residues each and two β chains of 146 amino acid residues each. An iron [Fe(II)] protoporphyrin IX, known as the heme group, is contained in each α and β chain. Figure 1 gives a schematic drawing of a hemoglobin molecule. In the absence of oxygen, the four heme iron atoms in Hb A are in the high-spin ferrous state [Fe(II)] with four unpaired electrons each. In the presence of oxygen, each of the four heme iron atoms in Hb A can combine with an O_2 molecule and the four heme iron atoms are then converted to a low-spin, diamagnetic ferrous state. Under physiological conditions, the heme iron atoms of hemoglobin remain in the ferrous state. The oxygen dissociation curve for Hb exhibits sigmoidal behavior, with an overall association constant expression giving a greater than first-power dependence on the concentration of O_2. Thus, the oxygenation of Hb is a cooperative process, such that when one O_2 is bound, succeeding O_2 molecules are bound more readily. Hemoglobin is an allosteric protein, i.e., its functional properties are regulated by a number of metabolites other than its ligand, O_2, such as hydrogen ions, chlorides, carbon dioxide, and 2,3-diphosphoglycerate (2,3-DPG). [For detailed discussions on various aspects of Hb and allosteric proteins, see Antonini and Brunori (1971), Edsall (1972), Ho *et al.* (1982a), Dickerson and Geis (1983), Bunn and Forget (1986), and Perutz (1989, 1990)].

Research efforts on Hb during the past 60 years have been concentrated in two main areas, namely, the molecular basis for the cooperative oxygenation of Hb A and the molecular basis for the Bohr effect. The oxygen affinity of Hb depends on pH (German and Wyman, 1937). At a pH between 6.1 and 9.0, the O_2 dissociation curve is shifted to higher oxygen pressure with decreasing pH, i.e., Hb has a lower affinity for O_2 at lower pH values. This effect is known as the alkaline Bohr effect and is physiologically important in helping to unload O_2 from oxyhemoglobin (HbO$_2$) when muscle acidity indicates that more O_2 is needed for metabolic reactions. On

FIG. 1. A schematic representation of hemoglobin indicating a tetrameric molecule consisting of two α and two β chains, each containing a heme group. [From Dickerson and Geis (1969); illustration copyright by I. Geis].

the other hand, at a pH between 4.5 and 6.1, deoxy-Hb is a stronger acid than oxy-Hb, i.e., the so-called acid Bohr effect (German and Wyman, 1937). [For details on recent developments in the Bohr effect of Hb, refer to Shih and Perutz (1987), Shih *et al.* (1987), Ho and Russu (1987), Busch and Ho (1990), Busch *et al.* (1991), and references therein.] In spite of the fact that great progress has been made during the past decade in our understanding of the Bohr effect, we shall not discuss this important phenomenon in this review. The main emphasis of this article is on the molecular basis of the cooperative oxygenation of Hb A.

Hemoglobin has been a favorite protein molecule for designing and testing various models of the molecular basis of subunit interactions in multisubunit proteins. In the 1960s, two mechanisms were advanced to account for the sigmoidal nature of the oxygen binding curve for Hb A, one proposed by Monod *et al.* (1965) based on a concerted mechanism and the other proposed by Koshland *et al.* (1966) based on a sequential mechanism. The concerted mechanism postulates that there is an equilibrium between the low-affinity conformation (T form) and the high-affinity conformation (R form) and that this transition is a concerted process, i. e., an intermediate (or hybrid) conformation RT cannot exist. On the other hand, the sequential model does not assume the existence of an equilibrium between the T and R forms in the absence of ligand. The transition from the T form to the R form is induced by the binding of ligand, i.e., a sequential process. Thus, the hybrid species RT plays a very prominent role in the sequential model. Even though there are conceptual differences in these two mechanisms (in terms of the nature of conformational transitions induced by ligand binding), they both can account for the O_2 binding curve of Hb A quite satisfactorily. During the past three decades, extensive efforts by a large number of researchers have been devoted to attempting to understand the molecular basis for the oxygenation of hemoglobin. There is still a lack of consensus among scientists doing research on hemoglobin regarding the molecular mechanism for its cooperative oxygenation (for details, refer to Sections IX–XII).

A milestone in our understanding of the structural basis for the cooperative oxygenation of Hb came in the late 1960s when Perutz (1970) compared the atomic models, based on the X-ray crystallographic structural investigations, of human deoxyhemoglobin (deoxy-Hb A) and horse oxylike methemoglobin (met-Hb). As a result of this comparison, Perutz (1970) proposed a stereochemical mechanism for the cooperative oxygenation of Hb. In its original form, his model emphasizes the link between the cooperativity and the transition between two quaternary structures, i.e., two different arrangements of the four subunits, the deoxy quaternary structure, denoted by T, and the oxy quaternary structure, denoted by R. Figure 2A illustrates the quaternary motion of Hb in going from the deoxy to the oxy state. Figure 2B shows side views of oxy- and deoxy-Hb and indicates the rotation of the $\alpha_1\beta_1$ dimer by 15° relative to the $\alpha_2\beta_2$ dimer in going

---

FIG. 2. Quaternary structural transition in hemoglobin. (A) Subunit motion of hemoglobin in going from the deoxy (or T conformation) to the oxy (or R conformation) state; (B) side views of deoxy and oxy states of hemoglobin. In oxy-Hb, the $\alpha_1\beta_1$ dimer is rotated 15° relative to the $\alpha_2\beta_2$ dimer. [Adapted from Dickerson and Geis (1983); illustration copyright by I. Geis].

A

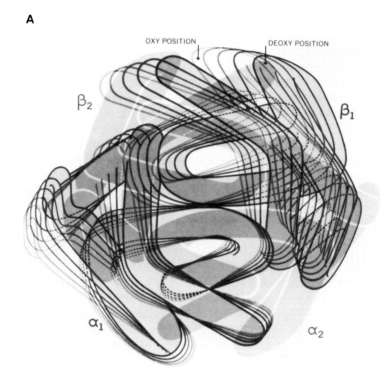

B

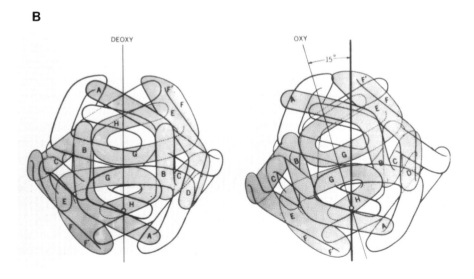

from the oxy to the deoxy form [for details, see Dickerson and Geis (1983)]. The $\alpha\beta$ contacts consist of two types: (1) the $\alpha_1\beta_1$ or $\alpha_2\beta_2$ contacts involve B, G, and H helices and GH corners, and are called the packing contacts, which hold the dimer together; and (2) the $\alpha_1\beta_2$ or $\alpha_2\beta_1$ contacts involve mainly C and G helices as well as the FG corner, and are called the sliding contacts, which undergo relative motion when there is a change in the ligation state of the heme. The letters (such as A or B) specify various helical segments in the α or β chain of Hb, and two-letter symbols (such as FG) signify the connection between two helical segments. The α chain and the β chain contain seven and eight helical segments, respectively, interrupted by nonhelical segments. Movement of the heme iron atoms and the sliding motion of the $\alpha_1\beta_2$ or $\alpha_2\beta_1$ subunit interface as well as the breaking of the intra- and intermolecular salt bridges and hydrogen bonds as a result of the ligation of the Hb molecule are among the central features of the stereochemical mechanism of Perutz for the cooperative oxygenation of Hb (Perutz, 1970; Dickerson and Geis, 1983). The mechanism of Perutz allows tertiary structural changes, i.e., changes in the conformation of the subunits, to take place each time a subunit is oxygenated, but a single concerted quaternary structural transition (i.e., $T \rightleftharpoons R$) is responsible for the cooperativity of the oxygenation process. Thus, the affinity for O_2 of an unligated subunit is not affected by the state of ligation of its neighbors within a given quaternary structure. The basic conceptual framework of the mechanism of Perutz shares many features of a two-state allosteric model, such as the one proposed by Monod *et al.* (1965).

Based on a comparison of the atomic coordinates of human deoxy-Hb A, horse met-Hb, and human carbonmonoxyhemoglobin (HbCO A), as well as computer graphics and least-squares fitting methods, Baldwin and Chothia (1979) have described the structural changes related to ligand binding and their implications for the cooperative oxygenation of Hb. From a detailed analysis of the various noncovalent interactions in the subunit interfaces ($\alpha_1\beta_1$, $\alpha_2\beta_2$, $\alpha_1\beta_2$, and $\alpha_2\beta_1$), Baldwin and Chothia (1979) have reinforced the early proposal by Perutz that there can be only two types of stable subunit interfaces, i.e., either the deoxy (T) or the ligated (R) type of quaternary structure, due to the dovetailed nature of the noncovalent interactions among various amino acid residues in the $\alpha_1\beta_2$ and $\alpha_2\beta_1$ interfaces. Figure 3 shows the $\alpha_1\beta_2$ subunit interface of deoxy-Hb A, which includes both the switch and flexible joint regions. Figure 4 is an enlarged representation of the interactions between FG corners and C helices showing various H bonds, salt bridges, and nonbonded

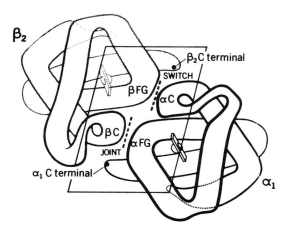

FIG. 3. The contact area in the $\alpha_1\beta_2$ subunit interface showing the switch and flexible joint regions of the hemoglobin molecule. [Adapted from Dickerson and Geis (1983); illustration copyright by I. Geis].

interactions characteristic of the deoxy quaternary structure as well as the amino acid residues located in both the distal and proximal sides of the heme pockets of α_1 and β_2 subunits. Figure 5A gives a close-up view of the differences in the helix and heme motion in the β chain, illustrating the conformational differences in Hisβ92(F8), Valβ98(FG5), and Tyrβ145(HC2), between the T and R states (Dickerson and Geis, 1983). Figure 5B illustrates ligation and strain in the heme environment: the unligated heme in deoxy-Hb (a), the ligated heme held in deoxy conformation before the T to R transition (b), and the relaxed, ligated heme after the T to R transition (c) (Dickerson and Geis, 1983). Figure 6 shows the packing of amino acid residues around the proximal histidyl residue (F8) of the β chain (Dickerson and Geis, 1983). Hisβ92(F8)is surrounded by a cage of hydrophobic amino acid residues, Leu(F4), Leu(F7), Leu(FG3), Val(FG5), and Leu(H19). Figure 7 illustrates the helix and the heme motion in the α chain of Hb during the transition between the T and R states (Dickerson and Geis, 1983). The X-ray crystallographic results as illustrated in Figs. 3–7 clearly suggest that there are close connections and interactions between the heme in the α_1 subunit and the heme in the β_2 subunit through the $\alpha_1\beta_2$ subunit contacts. These connections and contacts are important in understanding the molecular basis for the cooperative oxygenation of Hb (see Sections XI and XII).

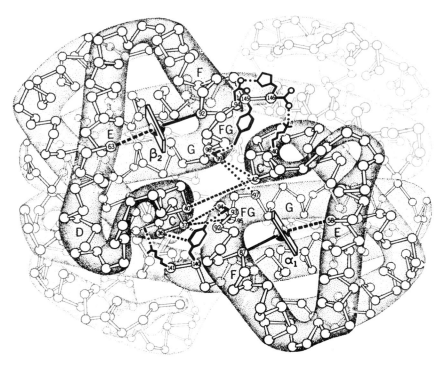

FIG. 4. An enlarged representation of the $\alpha_1\beta_2$ subunit interface illustrating extensive subunit interactions between FG corners and C helices and showing various hydrogen bonds (····), salt-bridges, and noncovalent interactions (----) characteristic of the deoxy quaternary structure of Hb. The connections of the proximal histidyl residues (position 87 in the α chain and position 92 in the β chain) to the heme iron atoms are indicated by a solid line. The connections of the distal histidyl residues (position 58 in the α chain and position 63 in the β chain) to the heme iron atoms are indicated by a dashed line. [Adapted from Dickerson and Geis (1983); illustration copyright by I. Geis].

Due to the highly cooperative nature of the oxygenation of Hb A, there are few partially oxygenated species [$Hb(O_2)_1$, $Hb(O_2)_2$, and $Hb(O_2)_3$] present during the oxygenation process, and thus the majority of the Hb species are either fully deoxy- or fully oxy-Hb. This fact, together with the lack of suitable techniques for investigating structural and functional properties of transient, partially ligated species, has led to the prevalence of two-state descriptions of the cooperative oxygenation process. Nevertheless, there are by now a number of reports of experimental results on Hb that are not consistent with a two-state or two-structure type of model for hemoglobin, suggesting the existence of intermediate forms. How much difference in the structure of a ligation intermediate from the fully deoxy (T-type) and the fully oxy (R-type) quaternary structures needs to be observed before one would classify it as a new quaternary structure? Bernal (1958) defined the quaternary structure of a protein molecule as the relative arrangement of subunits in a multisubunit protein. If one adopts the definition of Bernal, any stretching or weakening of intersubunit linkages would produce a new quaternary structure. However, this can become a semantic argument among researchers involved in hemoglobin studies (refer to Sections IX–XII).

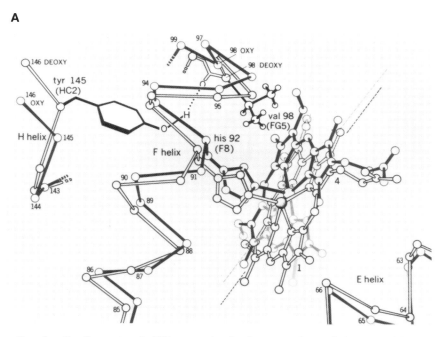

FIG. 5. Conformational differences in the heme pockets of deoxy and ligated forms of hemoglobin. (A) Helix and heme motion in the β chain of hemoglobin during the transition from deoxy (light line) to the carbonmonoxy (dark line) form of hemoglobin; (B) ligation and strain in the heme environment: (a) unligated heme in deoxy-Hb, with the proximal histidyl residue (F8) tilted, and the heme in contact with the side chain of Val(FG5); (b) ligated heme held in its deoxy conformation before the T → R transition [the steric strain between heme, proximal His(F8), ligand, and Val(FG5) is denoted by S]; and (c) relaxed, ligated heme after the T → R transition. The proximal His(F8) is perpendicular to the heme plane, and the heme is undomed. [From Dickerson and Geis (1983); illustration copyright by I. Geis].

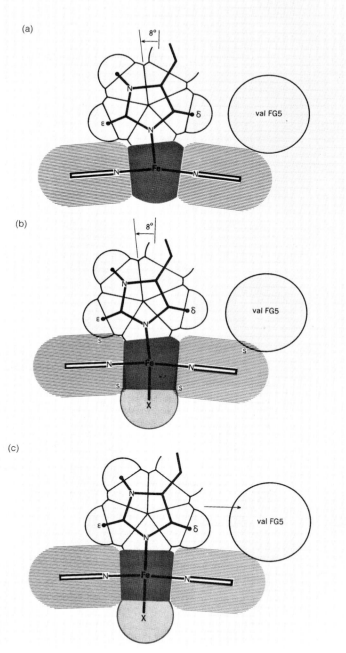

FIG. 5. (*continued*)

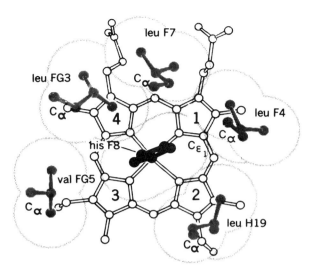

FIG. 6. Packing of the amino acid side chains around the proximal histidyl residue (F8) of the β chain of Hb. The proximal histidyl residue is surrounded by a cage of hydrophobic amino acid residues such as Leu(F4), Leu(F7), Leu(FG3), Val(FG5), and Leu(H19). [From Dickerson and Geis (1983); illustration copyright by I. Geis].

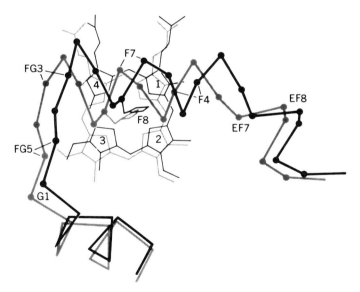

FIG. 7. Helix and heme motion in the α chain during the transition from the deoxy (heavy lines) to the carbonmonoxy (light lines) form of hemoglobin. [From Dickerson and Geis (1983); illustration copyright by I. Geis].

A major advance in our understanding of the molecular basis of 2,3-DPG in regulating the O_2 affinity of Hb came from the independent results of Chanutin and Curnish (1967) and Benesch and Benesch (1967). They found that 2,3-DPG can modify the O_2 binding to Hb A, i.e., it decreases the O_2 affinity of Hb A. Its concentration within red blood cells (~ 5 mM of packed cells) is approximately that of the Hb tetramer. At physiologic pH and ionic strength, 2,3-DPG is bound to hemoglobin in a 1:1 molar ratio with a much higher affinity for deoxy-Hb A than for oxy-Hb A (the dissociation constant K_D for deoxy-Hb A is $\sim 2.5 \times 10^{-5}$ M versus $\sim 3 \times 10^{-3}$ M for oxy-Hb A) (Benesch and Benesch, 1967; Imai, 1982). X-ray diffraction results by Arnone (1972) indicate that 2,3-DPG binds to the central cavity between the two β chains, i.e., two each of the N-terminal amino group of Valβ1(NA1), Hisβ2(NA2), Lysβ82(EF6), and Hisβ143(H21)0. ^{1}H and ^{31}P nuclear magnetic resonance (NMR) studies suggest that 2,3-DPG is likely to bind to the same amino acid residues in oxy-Hb A as in deoxy-Hb A (Gupta *et al.*, 1979; Russu *et al.*, 1990). Inositol hexaphosphate (IHP), which occupies the same binding sites as those of 2,3-DPG in deoxy-Hb A (Arnone and Perutz, 1974), is an even more powerful allosteric effector for hemoglobin than 2,3-DPG. Its binding constant to deoxy-Hb A ($K_D \sim 4 \times 10^{-8}$ M) is also much larger than that to oxy-Hb A, the latter being comparable to that of 2,3-DPG for deoxy-Hb A (Benesch *et al.*, 1977; Imai, 1982).

In this review, we emphasize our proton nuclear magnetic resonance investigations of human normal, abnormal, and modified hemoglobins. We correlate our ^{1}H NMR results with published data on the energetics of the oxygenation of Hb and on the structures of various forms of Hb in order to gain new insights into the molecular basis for the cooperative oxygenation of Hb. By using appropriate mutant and modified Hbs, we have assigned a number of proton resonances to specific amino acid residues of Hb A. Several of these resonances serve as excellent tertiary and quaternary structural markers in both deoxy and oxy forms of Hb A. ^{1}H NMR techniques have allowed us to monitor the binding of ligands to the α and β chains of an intact tetrameric Hb molecule and the ligand-induced structural changes (both tertiary and quaternary) of Hb A. We have reached the following major conclusions: (1) The α and β chains of Hb A are nonequivalent, both structurally and functionally. The relative ligand affinities of the α and β chains of Hb A depend not only on the nature of heme ligands but also on the nature of allosteric effectors. For example, the α and β chains have equal affinity for O_2 in the absence of organic phosphates. However, in the presence of 2,3-DPG or IHP, the α chains have a higher affinity for O_2 than do the β chains. (2) Structural changes that occur in Hb on oxygenation are not concerted. Ligation of one subunit can affect the structure of unligated subunits within a tetrameric Hb molecule, thus altering the ligand affinity of the unligated subunits. Some cooperativity must be present within the deoxy quaternary structural state during the oxygenation process. (3) Strong evidence exists that the structure of the singly ligated Hb species is likely to be different from that of the fully deoxy- or oxy-Hb (especially in the $\alpha_1\beta_2$ subunit interface). These results indicate that there are more than two quaternary structures during the transition from the deoxy to the oxy state and that two-structure allosteric models do not adequately describe the cooperative oxygenation of Hb A. These results are more consistent with a sequential (or induced-fit) description for the oxygenation of hemoglobin. Experimental evidence to support these conclusions is given in this article. This review is not intended to be comprehensive. It gives emphasis to our ^{1}H NMR results on the structural changes associated with the cooperative oxygenation of Hb A and to the relation between our work and other relevant results published in the literature.

In many aspects, hemoglobin remains an excellent model for testing specific hypotheses regarding the relationship between structure and function in a protein molecule. With the recent advances in protein engineering by recombinant DNA methodology, one can, in principle, design a protein to desired specifications, provided that there is a cloned gene available that can be expressed in a suitable expression system. α-Globin and β-globin genes have been expressed in several expression systems (Triesman *et al.*, 1983; Nagai and Thøgersen, 1984; Chada *et al.*, 1985; Grosveld *et al.*, 1987; Ryan *et al.*, 1989; Hanscombe *et al.*, 1989; Hoffman *et al.*, 1990; Wagenbach *et al.*, 1991; Groebe *et al.*, 1990, 1992). Recombinant Hb A and a number of recombinant mutant Hbs have been produced and their structures and properties have been investigated (for example, see Nagai *et al.*, 1985, 1987; Olson *et al.*, 1988; Mathews *et al.*, 1989, 1991; Tame *et al.*, 1991; Imai *et al.*, 1991). In spite of tremendous amounts of effort having been devoted to the Hb problem, we do not fully understand this remarkable protein molecule. There are still many exciting discoveries waiting for protein chemists. The recent discovery of a new quaternary structure based on X-ray crystallographic analysis of a mutant hemoglobin HbCO Ypsilanti (β99Asp → Tyr), reported by Smith *et al.* (1991), kinetic and thermodynamic investigations of the subunit assembly leading to a molecular code for cooperative switching in hemoglobin (Smith *et al.*, 1987; Daughtery *et al.*, 1991; Ackers *et al.*, 1992), the ability to use recombinant DNA methodology to design desired mutant Hbs for structural and functional studies, and new insights into the nature of the Bohr effect (Busch and Ho, 1990; Busch *et al.*, 1991) are excellent examples of new advances relating structure to function in hemoglobin.

II. Experimental Procedures

A. *NMR Methodology*

In many aspects, one-dimensional proton nuclear magnetic resonance spectroscopy is an excellent technique for investigating the structure–function relationship in hemoglobin. Due to the presence of the unpaired electrons in the high-spin ferrous atoms in deoxy-Hb and to the presence of the highly conjugated porphyrins in the Hb molecule, the proton chemical shifts of various Hb derivatives cover a wide range. Resonances vary from about 20 ppm upfield from H_2O to about 90 ppm downfield from H_2O depending on the spin state of the iron atoms and the nature of ligands attached to the heme groups. Figures 8A–K give a summary of the proton resonances of the unligated and various ligated forms of Hb A and the effects of magnetic field on the proton resonances. Figures 8A and 8B give the entire spectral range covered for oxy- and deoxy-Hb in H_2O, respectively. The unusually large spread of proton chemical shifts (~100 ppm) for deoxy-Hb A provides the selectivity and the resolution necessary to investigate specific regions of the Hb molecule. The ^1H NMR spectrum of a Hb molecule can be divided into the following spectral regions, which have been used to monitor spectral (or structural) changes associated with the ligation of Hb A: (1) Resonances in the region +50 to +80 ppm downfield from H_2O arise from the $N_\delta H$-exchangeable protons of the proximal histidyl residues of the α and β chains of deoxy-Hb A (Fig. 8C). These resonances have been shifted more than 50 ppm downfield from their normal diamagnetic resonance regions due to the hyperfine interactions between the unpaired electrons of the high-spin ferrous iron atoms of the heme group and the $N_\delta H$ protons of the proximal histidyl residues of Hb A. They are markers for the proximal histidyl residues of the α and β chains of deoxy-Hb A and can be used to monitor the binding of O_2 to the α and β hemes of Hb A. (2) Resonances in the region +10 to +90 ppm downfield from HDO arise from the protons of the heme groups and their nearby amino acid residues due to the hyperfine interactions between these protons and unpaired electrons of the high-spin ferric heme iron [Fe(III)] atoms of met-Hb A (Fig. 8D). (3) Resonances in the region +5 to +25 ppm downfield from HDO arise from the protons of the heme groups and their nearby amino acid residues due to hyperfine interactions of these protons and the unpaired electrons of the low-spin ferric

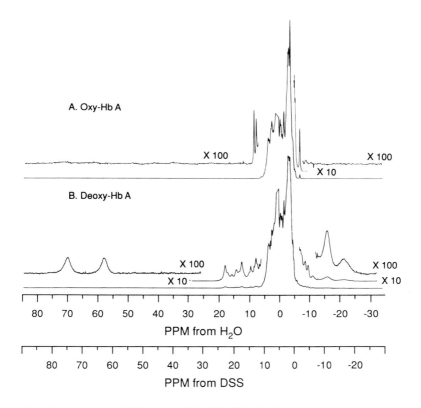

FIG. 8. A summary of 300- and 500-MHz ^{1}H NMR spectra of 1.4 mM human normal adult hemoglobin in 0.1 M phosphate at pH 7.1 in H$_2$O or D$_2$O at 29°C as a function of experimental conditions. (A) 300-MHz ^{1}H NMR spectrum of oxy-Hb A in H$_2$O over the entire spectral range from −30 to +90 ppm (×100 means that the spectrum is expanded 100 times the original spectrum). (B) 300-MHz ^{1}H NMR spectrum of deoxy-Hb A in H$_2$O over the entire spectral range from −30 to +90 ppm (×10 and ×100 mean that the spectrum is expanded 10 times or 100 times the original respective spectrum). (C) Hyperfine-shifted exchangeable proton reso-

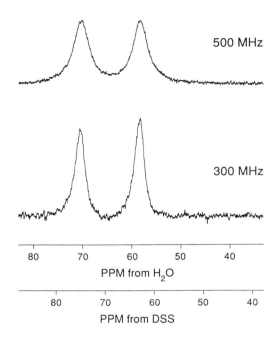

C. Deoxy-Hb A : Hyperfine-Shifted, Exchangeable Proton Resonances, N$_\delta$H, of Proximal Histidyl Residues

nances, N$_\delta$H, of the proximal histidyl residues (F8) of the α and β chains of deoxy-Hb in H$_2$O over the spectral region from +50 to +80 ppm from H$_2$O. (D) High-spin ferric hyperfine-shifted proton resonances of met-Hb A in D$_2$O over the spectral range from +20 to +90 ppm from HDO. (E) Low-spin ferric hyperfine-shifted proton resonances of cyanomethemoglobin and azidomethemoglobin in D$_2$O over the spectral region from +5 to +30 ppm from HDO. (F) High-spin ferrous hyperfine-shifted proton resonances and exchangeable proton resonances of deoxy-Hb A in H$_2$O (a) over the spectral region from +6 to +22 ppm from H$_2$O, and high-spin ferrous hyperfine-shifted proton resonances of deoxy-Hb A in D$_2$O (b) over the spectral region from +6 to +22 ppm from HDO. (G) Exchangeable proton resonances of oxy-Hb A in H$_2$O over the spectral region from +5 to +8.5 ppm from H$_2$O. (H) Aromatic proton resonances of deoxy- and oxy-Hb A in D$_2$O over the spectral region from +2 to +6 ppm from HDO. (I) The aliphatic proton resonances

D. Met-Hb A : High-Spin Ferric Hyperfine-Shifted Proton Resonances

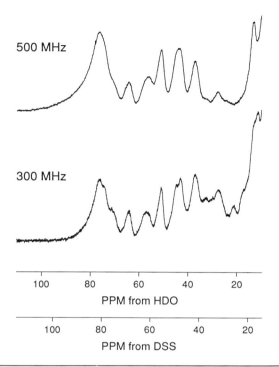

of deoxy- and oxy-Hb A in H_2O over the spectral range from -1.0 to -5.0 ppm from H_2O. (J) The ring-current-shifted proton resonances of oxy-Hb A in H_2O over the spectral range from -5 to -8 ppm from H_2O. (K) Ring-current-shifted and hyperfine-shifted proton resonances of deoxy-Hb A in H_2O over the spectral region from -6.0 to -20 ppm from H_2O ($\times 5$ means that the spectrum is expanded five times the original spectrum). The ^1H NMR spectra as shown in Figs. 8A to 8K are referred to two sets of chemical shift scale: one is the proton chemical shift of H_2O or HDO and the other is the proton resonance of the methyl group of the sodium salt of 2,2-dimethyl-2-silapentane 5-sulfonate (DSS), which is $+4.73$ ppm upfield from that of H_2O (or HDO) at 29°C. It should be noted that the linewidths of some of the proton resonances of Hb A depend on the magnetic field. The mechanisms giving rise to this field-dependent phenomenon are discussed in the text. (The ^1H NMR spectra are the unpublished results of V. Simplaceanu, N. T. Ho, and C. Ho.)

E. Low-Spin Ferric Hyperfine-Shifted Proton Resonances

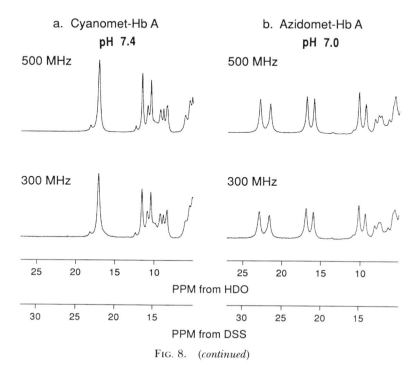

FIG. 8. (continued)

heme iron atoms of met-HbCN or met-HbN$_3$ (Fig. 8E, a and b). (4) Resonances in the region +6 to +22 ppm downfield from H$_2$O or D$_2$O of deoxy-Hb A are due to two sources (Fig. 8F). First, they arise from the protons on the heme groups and their nearby amino acid residues due to the hyperfine interactions between these protons and unpaired electrons of Fe(II) in the heme iron atoms. Second, they arise from the exchangeable protons due to intra- and intermolecular hydrogen bonds in Hb A. Figure 8F (a) shows the proton resonances of deoxy-Hb A in H$_2$O over this spectral range. Both hyperfine-shifted (hfs) and exchangeable proton resonances are present. Figure 8F (b) shows the proton resonances of deoxy-Hb A in D$_2$O. In this case, the exchangeable protons have been exchanged for deuterons, which are not visible in ^{1}H NMR spectroscopy, and only the high-spin ferrous hyperfine-shifted resonances are present. Exchangeable and hfs proton resonances are excellent tertiary and quaternary struc-

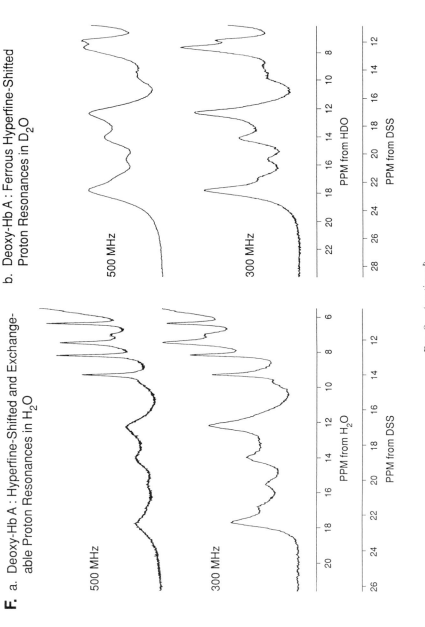

FIG. 8. (continued)

G. Oxy-Hb A : Exchangeable Proton Resonances

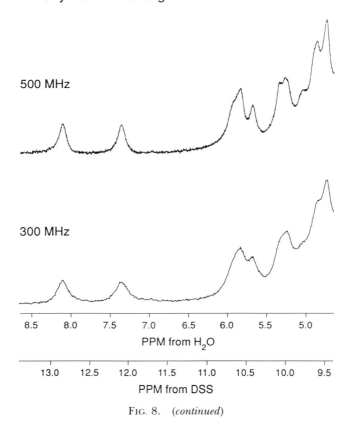

FIG. 8. (continued)

tural markers of deoxy-Hb. (5) Resonances in the region +5 to +8.5 ppm from H_2O arise from the exchangeable proton resonances of oxy-Hb A (Fig. 8G). They are excellent markers for the subunit interfaces and oxy quaternary structure. (6) Resonances in the region +2 to +6 ppm downfield from HDO arise from the protons of aromatic amino acid residues (including the C-2 and C-4 protons of the histidyl residues) of oxy- and deoxy-Hb A (Fig. 8H). The C-2 proton resonances have been used to investigate the molecular basis of the Bohr effect of Hb A (for example, see Ho and Russu, 1987;

H. Aromatic Proton Resonances

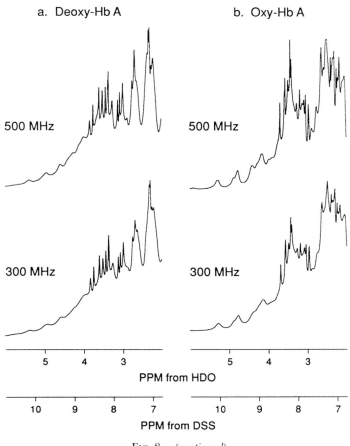

FIG. 8. (*continued*)

Busch and Ho, 1990; Busch *et al.*, 1991, and the references therein). (7) Resonances in the region −1.0 to −5.0 ppm upfield from HDO (Fig. 8I) arise from the protons of aliphatic amino acid residues of oxy- and deoxy-Hb A. This is a very crowded spectral region due to the large number of CH_2 and CH_3 groups in the amino acid residues. (8) Resonances in the region −5 to −8 ppm from HDO arise from the ring-current-shifted protons due to those protons located above or below the aromatic amino acid residues and the porphyrins of oxy-Hb A (Fig. 8J). (9) Resonances in the region −6 to −20 ppm

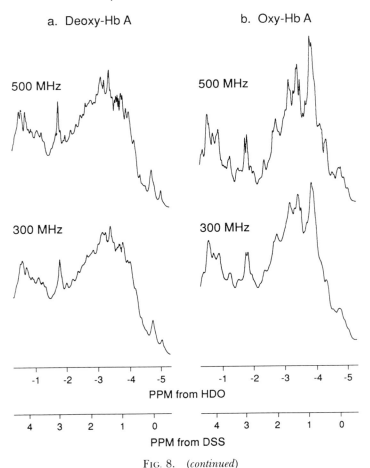

FIG. 8. (continued)

from H$_2$O arise from ring-current-shifted protons and the protons on the heme groups and their nearby amino acid residues due to the hyperfine interactions between these protons and the unpaired electrons of the high-spin ferrous heme iron atoms of deoxy-Hb A (Fig. 8K).

^{1}H NMR spectroscopy has played an important role in our current understanding of the structural changes associated with the cooperative oxygenation of Hb and of the structural features of the partially ligated Hb species. In this section we give a brief summary of

J. Oxy-Hb A : Ring-Current Shifted Proton Resonances

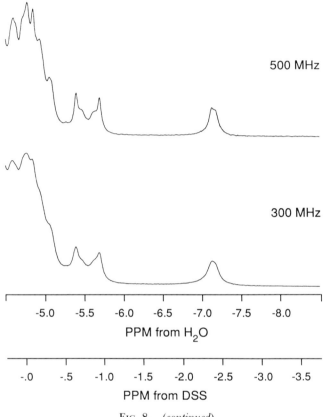

FIG. 8. (*continued*)

the theoretical background needed for the application of ^{1}H NMR spectroscopy to the Hb problem and of the NMR techniques that have been used to investigate Hb [for reviews on this subject, see Ho *et al.* (1975, 1978), Shulman *et al.* (1975), and Ho and Russu (1981, 1985).].

A resonance in NMR spectroscopy is generally characterized by five parameters, the chemical shift (δ), the intensity or the area of the resonance (proportional to concentration), the multiplet structure [related to the spin–spin coupling constant (J)], and two relaxation

K. Deoxy-Hb A : Ring-Current Shifted and
Hyperfine-Shifted Proton Resonances

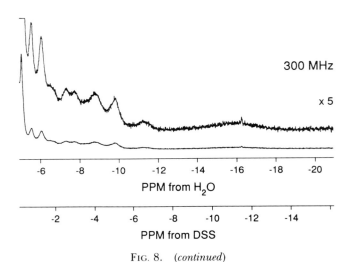

FIG. 8. (*continued*)

times, i.e., the spin–lattice (or longitudinal) relaxation time (T_1) and the spin–spin (or tranverse) relaxation time (T_2). In this section we shall summarize briefly how these parameters can be used to derive information about the Hb molecule under different experimental conditions [for a general description of NMR, refer to Pople *et al.* (1959), Carrington and McLachlan (1967), Becker (1980), and Slichter (1990).]

1. Hyperfine-Shifted Resonances

The origin of chemical shift is the screening of the applied magnetic field produced by the surrounding electrons of a nucleus. The entire proton chemical shift range for HbO_2 (a diamagnetic protein molecule) is about 16 ppm (Fig. 8A), whereas that for deoxy-Hb A (a paramagnetic protein molecule) is about 100 ppm (Fig. 8B). The main physical mechanisms responsible for the large spread in the proton chemical shifts in deoxy-, met-, cyano-, and azido-Hb (paramagnetic forms of Hb) are the hyperfine interactions between the protons of the heme groups as well as between the protons of the amino acid residues situated in the vicinity of the heme groups and the unpaired electrons of the iron atoms in the hemes of these Hbs. The magnitude and direction of these chemical shifts depend on the spin state of the heme iron atom and the nature of the ligand attached to these heme iron atoms. There are two general mechanisms giving rise to the hyperfine-shifted resonance, i.e., contact shifts (δ_c; also known as Fermi contact shifts) and pseudocontact shifts (δ_{pc}).

The contact (or Fermi) shifts arise from the delocalization of unpaired electrons of the paramagnetic metal atoms to the resonating protons through chemical bonds or hyperconjugation. In the case of Hb, the contact shifts are expected to influence the resonances of those protons that have direct contact with the paramagnetic iron atoms, such as those of the porphyrins and of the proximal histidyl residues. According to Bloembergen (1957), the magnitude of a contact shift (relative to the corresponding position in a diamagnetic analog) is given by Eq. (1),

$$\delta_c = -\left(\frac{A}{\hbar}\right)\frac{g\beta S(S+1)}{3kT\gamma_H} \qquad (1)$$

where $A/\hbar$ is the hyperfine coupling constant (rad sec^{-1}), $\hbar$ is Planck's constant h divided by 2π, g is the Landé factor, β is the Bohr magneton, S is the total electronic spin, k is the Boltzmann constant, T is the absolute temperature, and γ_H is the proton gyromagnetic ratio (rad sec^{-1} gauss^{-1}). It should be noted that Eq. (1) is valid for systems in which the anisotropy of the g tensor can be neglected and only the ground electronic state is populated. [For more general expressions for the contact shifts, refer to Jesson (1967) and Kurland and McGarvey (1970).] In general, the contact shifts are very sensitive to the details of the distribution of the unpaired electrons from the iron atoms to the attached protons because the hyperfine coupling constant, $A/\hbar$, is directly related to the density of the unpaired electrons at the resonating proton (McConnell, 1956; Shulman et al., 1971).

The pseudocontact shifts arise from the dipolar interaction between the magnetic moment of the unpaired electrons in the paramagnetic iron atoms and the resonating protons. The general form for the magnitude of the pseudocontact shifts in transition metal complexes can be given as follows:

$$\delta_{pc} = \frac{2\beta S(S+1)}{3kT} F\left(\tilde{g}\frac{\Omega,\theta}{r^3}\right) \qquad (2)$$

where F is a function of the anisotropy of the g tensor and the geometric factors Ω, θ, and r. The other symbols have the same meaning as those given in Eq. (1). [For details on pseudocontact shifts, refer to McConnell and Robertson (1958), Jesson (1967), Kurland and McGarvey (1970), and Wüthrich (1976).] The pseudocontact interactions are expected to affect those protons in the porphyrin and in the nearby amino acid residues that are close to the heme iron atom.

There are two common features of these two types of hyperfine interactions. First, both δ_c and δ_{pc} are inversely dependent on temperature. Second, they both depend on the total electronic spin S of the iron atom as shown in Eqs. (1) and (2). The magnitude of the hfs resonances depends on the S value of each Hb derivative. For example, met-Hb is a high-spin ferric complex with five unpaired electrons per heme, deoxy-Hb is a high-spin ferrous complex with four unpaired electrons per heme, and both cyanomet- and azidomet-Hb are low-spin ferric complexes, each with one unpaired electron per heme. The hyperfine interactions can shift resonances either upfield or downfield from their diamagnetic counterparts. It should be noted that both HbO_2 and HbCO are low-spin ferrous complexes that are diamagnetic systems ($S = 0$) and they will not give rise to hyperfine interactions.

2. Ring-Current-Shifted Resonances

A secondary magnetic field can be generated when an aromatic ring-containing molecule is oriented perpendicular to the applied magnetic field, H_0. The ring-current effect affects the protons that are situated close to ringlike conjugated structures such as porphyrin and aromatic amino acid residues in proteins. When an aromatic ring is placed in a magnetic field, the delocalized π electrons of the conjugated system set up an induced current. This induced current produces a small magnetic field that can either add to (for protons situated on the periphery but exterior to the plane of the ring) or subtract from (for protons located directly above or below the plane of the ring) the applied magnetic field. The proton resonances are shifted either to lower or higher fields, respectively. This ring-current effect can be explained qualitatively using the free-electron mode of Pauling (1936). Several theories for the ring-current effect have been proposed (for example, see Pople, 1956; Johnson and Bovey, 1958; Haigh and Mallion, 1971).

The ring-current-shifted proton resonances are very sensitive to the geometric relationship between the ring structure and the neighboring protons. The magnitude of the ring-current-induced chemical shifts is directly proportional to the area of the ring and the number of π electrons. The ring-current-induced shifts produced by a porphyrin are larger than those provided by aromatic amino acids. In addition, the magnetic field of these shifts decreases rapidly on increasing the vertical distance of the proton of interest from the ring as well as on increasing the distance of the proton from the symmetry axis perpendicular to the center of the ring. Thus, the ring-current-shifted proton resonances are sensitive probes of tertiary structure for those protons in the vicinity of porphyrins and aromatic amino acid residues in proteins (McDonald and Phillips, 1967; Sternlicht and Wilson, 1967). Several models have been developed to correlate the ring-current-shifted proton resonances with the positions of the protons relative to the heme ring (Perkins, 1980). In the absence of any temperature-induced conformational changes in proteins, the ring-current-shifted proton resonances are independent of temperature. Hence, this property can be used to distinguish ring-current-shifted proton resonances from hfs proton resonances in Hb.

3. Relaxation Processes

There are two kinds of relaxation processes, spin–lattice (or longitudinal) relaxation and spin–spin (or transverse) relaxation. T_1^{-1} characterizes spin–lattice relaxation and is a measure of the rate at which the spins return to equilibrium along the direction of the applied magnetic field after being perturbed in an NMR experiment. Spin–lattice relaxation occurs as a result of the interaction of nuclear spin dipoles with random, fluctuating magnetic fields caused by the motion of surrounding dipoles in the lattice. These dipoles need to have components with the same frequency as the resonance frequency. Spin–spin relaxation is caused by random magnetic fields from the neighboring nuclei in the sample that are not fluctuating and is characterized by T_2^{-1}. The sources of the random magnetic fields giving rise to T_1 relaxation will also lead to T_2 relaxation. Other relaxation mechanisms may contribute to T_2 but do not contribute to T_1. The T_2^{-1} is related to the linewidth of the resonance by $T_2^{-1} = \pi \Delta \nu$, where $\Delta \nu$ [in hertz (Hz)] is the width of the resonance at half-height.

For a paramagnetic protein molecule such as deoxy-Hb, the T_1^{-1} and T_2^{-1} values in the absence of chemical exchange have been calculated as the sum of the dipole–dipole interactions $\left(T_{1d}^{-1} \text{ and } T_{2d}^{-1}\right)$ and the paramagnetic contribution from the unpaired electrons of the heme iron atom $\left(T_{1p}^{-1} \text{ and } T_{2p}^{-1}\right)$. According to Solomon (1955), the dipole–dipole relaxation between two nonequivalent spins is given as follows:

$$T_{1d}^{-1} = \gamma^4 \hbar^2 \left(\frac{1}{16}\right) \sum [J(\omega_i - \omega_j) + 3J(\omega_i) + 6J(\omega_i + \omega_j)] \quad (3)$$

$$T_{2d}^{-1} = \gamma^4 \hbar^2 \left(\frac{1}{32}\right) \sum [5J(\omega_i - \omega_j) + 9J(\omega_i) + 6J(\omega_i + \omega_j)] \quad (4)$$

where $\omega_{i,j}$ are the Larmor frequencies of the interacting spins, γ is the gyromagnetic ratio, and $J(\omega)$ is the spectral density function given by the following expression:

$$J(\omega) = \left(\frac{8}{5r_{ij}^6}\right)\left(\frac{\tau_c}{1 + \omega^2 \tau_c^2}\right) \quad (5)$$

where r_{ij} is the distance between the interacting atoms and τ_c is the rotational correlation time for the motion of the particular atom being studied.

The paramagnetic contribution to the spin–lattice relaxation (for example, the C-2 proton of a histidyl residue of Hb) from the electronic spin of the heme iron is given by

$$T_{1p}^{-1} = \left(\frac{2}{15}\right)\nabla^2 S(S+1)\left[\frac{3T_{1e}}{1 + \omega_H^2 T_{1e}^2} + \frac{7T_{1e}}{1 + \omega_e^2 T_{1e}^2}\right] \quad (6)$$

where $\nabla = \gamma_H \beta S/r^3$, γ_H is the proton gyromagnetic ratio, r is the distance between the proton and the paramagnetic heme iron, T_{1e} is the longitudinal relaxation time of the electronic spin, and ω_H and ω_e are the proton and electron Larmor frequencies, respectively [for details, see Abragam (1961)]. In Eq. (6), the Curie spin contribution has been neglected because this contribution is much smaller than the standard paramagnetic term (Guéron, 1975). Deoxy-Hb provides a good model for investigating the influence of cross-relaxation on proton longitudinal relaxation in proteins at the slow motion limit and in the presence of paramagnetic centers. For the His C-2 protons of surface histidyl residues, the cross-relaxation resulting from the interresidue dipolar interaction makes an important contribution to the longitudinal relaxation. [For a detailed discussion of various factors affecting the T_1^{-1} values of the histidyl residues in Hb, see Russu and Ho (1982).]

The paramagnetic contribution to the spin–spin relaxation is given by

$$T_{2p}^{-1} = (7/15) \nabla^2 S(S+1) T_{1e} + (4/5) \nabla^2 S_c^2 \tau_c \qquad (7)$$

with $S_c = g\beta S(S+1) H_0/3kT$, where H_0 is the applied magnetic field. The first term in Eq. (7) corresponds to the standard paramagnetic contribution. The second term gives the Curie contribution [for details, see Guéron (1975) and Johnson et al. (1977)]. The second term in Eq. (7) gives the linewidth of the hfs proton resonances a strong dependence on the applied magnetic field (Johnson et al., 1977). Figure 9 shows that the linewidth ($\Delta \nu_{1/2}$) of the hfs proton resonances of deoxy-Hb A is proportional to the square of the resonance frequency (ν_0).

The total relaxation rates (in the absence of spin diffusion) can be written as

$$T_1^{-1} = T_{1d}^{-1} + T_{1p}^{-1} \qquad (8)$$

$$T_2^{-1} = T_{2d}^{-1} + T_{2p}^{-1} \qquad (9)$$

The resonance frequencies of the C-2 protons of histidyl residues of deoxy-Hb A and HbCO A depend on the pH of the solution (Russu et al., 1980, 1982) because the observed proton chemical shifts are a weighted average of the chemical shifts of the protonated and unprotonated states of histidyl residues in Hb. Due to this exchange,

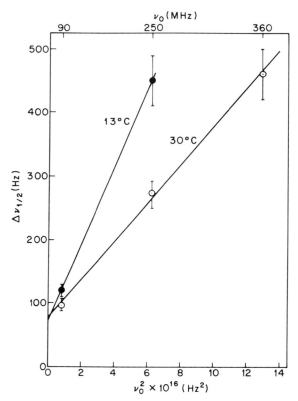

FIG. 9. The linewidth ($\Delta\nu_{1/2}$) of a hyperfine-shifted proton resonance of deoxy-Hb A as a function of the resonance frequency (ν_0) and of temperature. [From Johnson et al. (1977)].

the apparent (or observed) spin–spin relaxation rate values of His C-2 proton resonances of Hb at pH 7 vary as a function of the square of the resonance frequency. Figure 10 illustrates this point [for details, see Madrid et al. (1990)]. The expression for the apparent spin–spin relaxation rate $\left(T_{2a}^{-1}\right)$ is given by T_2^{-1} plus the chemical exchange contribution (in the fast exchange limit):

$$T_{2a}^{-1} = T_2^{-1} + f_+ f_0 \, (2\pi\Delta\delta)^2 \, \tau_e \qquad (10)$$

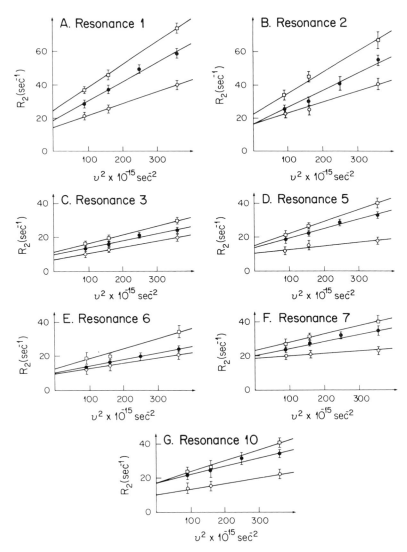

FIG. 10. Spin–spin relaxation rates (R_2) of C-2 protons of histidyl residues of 10% deoxy-Hb S at pH 7.0 as a function of the square of the resonance frequency, ν (MHz), and of temperature (□, 17°C; ●, 25°C; ○, 37°C). Each resonance line is shown separately. Similar results were obtained for Hb S and Hb A at all the concentrations investigated (10, 14, and 16%). [From Madrid et al. (1990)].

where T_{2a}^{-1} is the apparent spin–spin relaxation rate in the presence of chemical exchange, T_2^{-1} is the spin–spin relaxation rate in the absence of chemical exchange, f_+ and f_0 are the fractional populations of the ionized and neutral states of the nitrogens of a histidyl imidazole residue, $\Delta\delta = \delta_+ - \delta_0$, δ_+ and δ_0 are the corresponding proton chemical shifts of the His C-2, and τ_e is the lifetime between exchanges. The difference in chemical shift ($\Delta\delta$) is proportional to the applied magnetic field. Hence, the chemical exchange contribution to the T_{2a}^{-1} increases with the square of the resonance frequency (Allerhand and Gutowsky, 1964, 1965).

4. Spin–Spin Coupling

High-resolution NMR spectra often exhibit multiplet structures that arise from weak interactions between magnetic nuclei. These interactions are communicated and transmitted between the nuclei by the electrons in a chemical bond. The size of this interaction is defined by the spin–spin coupling constant (J), which is expressed in Hz. The magnitude of J depends on both the nature of the bond and the number of bonds involved. There is also a relationship between the bond angle and the observed value of J (Karplus, 1959, 1963). The coupling pattern and its J value are a good indication of the nature of the chemical group involved. Proton–proton coupling constants are normally ≤10 Hz (Pople et al., 1959; Becker, 1980). Two-dimensional J-correlated spectroscopy (COSY) has been used successfully to identify various amino acid residues in small protein molecules with molecular weights less than 15,000 (for example, see Wüthrich, 1986). For proteins with molecular weights greater than 15,000, the proton linewidth becomes comparable, or in some cases exceeds, the ^{1}H–^{1}H coupling constants. In such cases, spin–spin multiplet structure cannot be observed (see Wüthrich, 1986).

5. Nuclear Overhauser Effect

The nuclear Overhauser effect (NOE) is a consequence of the modulation of the dipole–dipole interactions (through space) between different nuclei and is correlated with the inverse sixth power of the internuclear distance. Experimentally, the NOE is the fractional change in intensity of one resonance when another resonance is irradiated in a double-irradiation experiment. The NOE phenomenon is intimately related to spin relaxation. The NOE varies as a function of the product of the Larmor frequency, ω_0, and the rotational correlation time, τ_c. In small molecules, τ_c is short relative to ω_0^{-1}. In this "extreme" motional narrowing situation, the frequency covered by the rotational motion of the spin system includes ω_0 and $2\omega_0$, which enable dissipative transitions between spin states. Two-dimensional nuclear Overhauser and exchange spectroscopy (NOESY) has been used to provide interproton distances in small protein molecules with molecular weights less than 15,000 (for example, see Wüthrich, 1986). In large molecules, τ_c is long relative to ω_0^{-1} and the frequencies of the rotational motions are too low to allow efficient coupling with the nuclear spin transitions. Thus, energy-conserving transitions of the type $\alpha_i\beta_j \rightarrow \beta_i\alpha_j$ (cross-relaxation) are favored [for details, see Noggle and Schirmer (1971) and Wüthrich (1986)].

6. *Multidimensional NMR Spectroscopy*

Recent advances in multidimensional NMR spectroscopy offer unique opportunities for making ^{1}H resonance assignments and for determining the three-dimensional (3D) structures of proteins with molecular weights up to approximately 15,000 [for reviews on this topic, see Wüthrich (1986) and Bax (1989)]. For proteins of molecular weights 15,000–40,000, newer 3D and 4D NMR techniques, in combination with isotopic labeling, offer hope for detailed solution structures (see Bax, 1989; Clore and Gronenborn, 1991; Tjandra *et al.*, 1992, and references therein). For larger proteins such as Hb, the problems of overlapping cross-peaks and ambiguous connectivities in 2D ^{1}H NMR spectra pose severe difficulties in making precise resonance identification. Hence, 2D ^{1}H NMR spectroscopy has so far had only limited applications in the case of Hb. Nevertheless, some 2D ^{1}H NMR experiments have been carried out and new methods may be developed for making multinuclei, multidimensional NMR useful for proteins the size of Hb (especially if cloned genes can be expressed in suitable expression systems for incorporation of isotopic labels).

B. *NMR Techniques*

Any modern Fourier transform NMR spectrometer manufactured in the 1980s by major instrument companies is capable of performing various types of ^{1}H NMR experiments needed for studies of hemoglobin. With a modern 7.0-Tesla high-resolution NMR spectrometer operating at 300 MHz for ^{1}H, a satisfactory ^{1}H NMR spectrum (with a signal-to-noise ratio of ~20 or better) of 0.3–0.5 ml Hb in millimolar concentration contained in a 5-mm sample tube can be obtained in a few minutes.

^{1}H NMR studies of proteins are carried out either in deuterated or in normal aqueous media. When proteins are dissolved in H_2O for NMR studies, the concentration of protons is about 110 M whereas that of the protein is around 1 mM. Thus, there is a severe dynamic range problem. A traditional way to overcome the dynamic range problem is to dissolve proteins in D_2O rather than in H_2O. Solvent-accessible exchangeable protons are then replaced by deuterons. Hence, these protons, which have been replaced by deuterons, will not be observable in a ^1H NMR spectrum. If one would like to detect these exchangeable proton resonances in the presence of H_2O, solvent suppression techniques to reduce the dynamic range problem are needed. We have found that the jump-and-return pulse sequence developed by Plateau *et al.* (1983) is quite useful for the spectral range from 8 ppm upfield from H_2O to 22 ppm downfield from H_2O (see Figs. 8F (a) and 8J). For the very low-field hyperfine-shifted exchangeable proton resonances (50 to 90 ppm downfield from H_2O), the $1\bar{2}1$ soft pulse sequence developed in our laboratory (Yao *et al.*, 1986) is quite acceptable. When the Hb samples are dissolved in D_2O media, standard pulse sequences available in a modern Fourier transform NMR spectrometer are capable of providing excellent ^1H NMR spectra of Hb samples. Hochmann and Kellerhals (1980) described a pulse sequence in which one selects for observation the fastest relaxing resonances and suppresses the slow relaxing ones (i.e., the diamagnetic proton resonances). This technique is quite useful for observing the hfs proton resonances of Hb. [For details on various ^1H NMR techniques used to obtain Hb spectra, refer to our publications and those of other investigators mentioned in this article.]

What is the optimal magnetic field strength for ^1H NMR studies of Hb? For hfs proton resonances of deoxy-Hb and met-Hb, the linewidths of these resonances increase with the square of the resonance frequency [because of the paramagnetic contribution to the spin–spin relaxation as described in Eq. (7)], whereas the resolution between the resonances only increases linearly with frequency (see Fig. 9) (Johnson *et al.*, 1977). Thus, there is a magnetic field at which both optimal sensitivity and optimal resolution are obtained. For the hfs proton resonances of deoxy-Hb A, the optimal magnetic field appears to be around 7.0 Tesla (or 300 MHz). For the diamagnetic proton resonances (such as exchangeable and ring-current-shifted proton resonances), it appears that a 500- or 600-MHz instrument gives better sensitivity and resolution than an instrument operating at 300 MHz (see Figs. 8G and 8J). On the other hand, for the C-2 proton resonances of histidyl residues, due to the chemical exchange contribution to the spin–spin relaxation, the linewidths of the His C-2 proton resonances increase with the square of the resonance frequency [see Eq. (10) and Fig. 10] (Madrid *et al.*, 1990). There is no improvement in resolution in going from 300 to 600 MHz for these histidyl resonances of Hb A (Fig. 8H). Thus, the His C-2 proton resonances of Hb can be readily obtained in a 300-MHz NMR spectrometer.

A commonly used proton chemical shift standard in ^{1}H NMR studies of proteins in aqueous solution is the proton resonance of the methyl group of the sodium salt of 2,2-dimethyl-2-silapentane 5-sulfonate (DSS). This resonance is +4.73 ppm upfield from the proton resonance of water at 29°C. Because all our Hb samples contain H_2O (in the case of the samples in D_2O, there is residual HDO present), we have found that it is quite useful to use the proton resonance of H_2O in each sample as the internal reference. The ^{1}H chemical shift of H_2O varies with temperature. As long as we know its variation as a function of temperature, we can always refer the ^{1}H chemical shift of H_2O to that of DSS. In our ^{1}H NMR studies, we have used both H_2O and DSS as proton chemical shift references. At the recommendation of the International Union of Pure and Applied Chemistry (No. 38, August, 1974), the chemical shift scale has been defined as positive in the region of the ^{1}H NMR spectrum at a lower field than the resonance of a standard. The chemical shifts that are upfield from that of a standard carry a negative sign. It should be mentioned that in some of our earlier publications, we used the negative sign to indicate a chemical shift that was downfield from that of the reference.

C. Sample Preparation

Hb A is normally isolated from human red blood cells obtained from the local blood bank. Human abnormal hemoglobins are isolated from blood samples of individuals whose blood contains such hemoglobins, provided to us by colleagues around the world. The samples can be prepared in either H_2O or D_2O media. Converting Hb samples from H_2O to D_2O can readily be carried out by repeated dilution with D_2O and subsequent concentration through an ultrafiltration membrane or dialysis tubing. For Hb samples in D_2O, the isotope effect on the glass electrode can be corrected by the following expression (Glasoe and Long, 1960): pD = pH (meter reading) + 0.40.

Deoxy-, oxy-, or carbon monoxy-Hb samples are prepared by standard procedures and then transferred under N_2, O_2, or CO pressure,

TABLE I
Partial List of Human Mutant and Modified Hemoglobins Used in Research

Hemoglobin	Properties[a]
Hb A ($\alpha_2\beta_2$)	Human normal adult hemoglobin
Deoxy-Hb A	Deoxyhemoglobin; the heme iron atoms are in the high-spin ferrous state with four unpaired electrons per heme iron
Oxy-Hb A (or HbO$_2$ A)	Oxyhemoglobin; the heme iron atoms are in the low-spin ferrous, diamagnetic state
Carbon monoxy-Hb A (or HbCO A)	Carbon monoxyhemoglobin; the heme iron atoms are in the low-spin ferrous, diamagnetic state
Met-Hb A	Methemoglobin; the heme iron atoms are in the high-spin ferric state with five unpaired electrons per heme iron
Met-HbCN ($\alpha^{+CN}\beta^{+CN}$)$_2$	Cyanomethemoglobin; the heme iron atoms are in the low-spin ferric state with one unpaired electron per heme iron
Met-HbN$_3$	Azidomethemoglobin; the heme iron atoms are in the low-spin ferric state with one unpaired electron per heme iron
Hb F ($\alpha_2\gamma_2$)	Human fetal hemoglobin
Hb C [β6(A3)Glu → Lys]	Mutation located in the amino-terminal region of the β chain
Hb Chesapeake [α92(FG4)Arg → Leu]	High-affinity mutant with the mutation located in the $\alpha_1\beta_2$ subunit interface
Hb Deer Lodge [β2(NA2)His → Arg]	Mutation located in the amino-terminal region of the β chain
Hb J Capetown [α92(FG4)Arg → Gln]	High-affinity mutant with the mutation located in the $\alpha_1\beta_2$ subunit interface
Hb Kempsey [β99(G1)Asp → Asn]	High-affinity mutant with the mutation located in the $\alpha_1\beta_2$ subunit interface
Hb M Boston [α58(E7)His → Tyr]	Mutation in the distal heme pocket of the α chain; the heme iron atoms in the α chains are in the high-spin ferric state, i.e., ($\alpha_2^+\beta_2$), with five unpaired electrons per heme iron in the α chains and with four unpaired electrons per heme iron in the β chains; a naturally occurring valency hybrid hemoglobin

TABLE I (*Continued*)

Hemoglobin	Properties[a]
Hb M Milwaukee [β67(E11)Val → Glu]	Mutation in the distal heme pocket of the β chain; the heme iron atoms in the β chains are in the high-spin ferric state, i.e., ($\alpha_2\beta_2^+$), with five unpaired electrons per heme iron in the β chains and four unpaired electrons per heme iron in the α chains; a naturally occurring valency hybrid hemoglobin
Hb Sydney [β67(E11)Val → Ala]	Mutation located in the heme pocket of the β chain
Hb Yakima [β99(G1)Asp → His]	High-affinity mutant with the mutation located in the $\alpha_1\beta_2$ subunit interface
Hb Zürich [β63(E7)His → Arg]	Mutation located in the heme pocket of the β chain
Hb$(\alpha\beta)_A(\alpha\beta)_C$XL	Cross-linked hemoglobin; the subscript A or C denotes that the αβ dimer in parentheses comes from either Hb A or Hb C, respectively; XL symbolizes a cross-linked hemoglobin prepared by reaction with a bifunctional reagent, bis(3,5-dibromosalicyl) fumarate, which cross-links the ε-amino group of the lysyl residue at position 82 of the two β chains
Hb$(\alpha^{+CN}\beta)_A(\alpha\beta)_C$XL	Cross-linked mixed-valency hybrid hemoglobin with one α chain in the ligated state (i.e., α^{+CN}), serving as a model for a singly ligated species
Hb$(\alpha\beta^{+CN})_A(\alpha\beta)_C$XL	Cross-linked mixed-valency hybrid hemoglobin with one β chain in the ligated state (i.e., β^{+CN}), serving as a model for a singly ligated species
Hb$(\alpha^{+CN}\beta^{+CN})_A(\alpha\beta)_C$XL	Cross-linked mixed-valency hybrid hemoglobin with one α chain and one β chain ligated (i.e., as an $\alpha_1\beta_1$ dimer), serving as a model for a doubly ligated species
Hb$(\alpha\beta^{+CN})_A(\alpha^{+CN}\beta)_C$XL	Cross-linked mixed-valency hybrid hemoglobin with one α chain and one β chain ligated (i.e., as an $\alpha_1\beta_2$ dimer), serving as a model for a doubly ligated species
Des-His(β146)-Hb A	The carboxy-terminal amino acids His(β146) are removed from both β chains in Hb A

(*continued*)

TABLE I (Continued)

Hemoglobin	Properties[a]
Des-Arg(α141)-Hb A	The carboxyl-terminal amino acids Arg(α141) are removed from both α chains in Hb A
Des-Arg(α141)-Tyr(α140)-Hb A	The carboxyl-terminal amino acids Arg(α141) and Tyr(α140) are removed from both α chains in Hb A
NES-Hb A	Both sulfhydryl groups of Cys(β93) in Hb A reacted with N-ethylmaleimide
NES-Des-Arg(α141)-Hb A	Both sulfhydryl groups of Cys(β93) in Des-Arg(α141)-Hb A reacted with N-ethylmaleimide
[α(Des-Arg)β]$_A$[$\alpha\beta$]$_C$XL	Cross-linked asymmetrically modified hemoglobin
[α(Des-Arg-Tyr)β]$_A$[$\alpha\beta$]$_C$XL	Cross-linked asymmetrically modified hemoglobin
[α(Des-Arg)β(NES)]$_A$[$\alpha\beta$]$_C$XL	Cross-linked asymmetrically modified hemoglobin
[α(Des-Arg)β]$_A$[$\alpha\beta$(NES)]$_C$XL	Cross-linked asymmetrically modified hemoglobin

[a]For details, see text.

respectively, to NMR sample tubes previously flushed with the corresponding gas. The preparation of deoxy-Hb A samples for ^{1}H NMR studies is described by Lindstrom and Ho (1972). In preparing deoxy-Hb, it is preferable to carry out the deoxygenation of the HbO$_2$ sample by prolonged flushing with N$_2$ gas at 4°C rather than using sodium dithionite. This reducing agent can affect the ^{1}H NMR spectra by binding to Hb (Ferrige et al., 1979) and by forming peroxides (Dalziel and O'Brien, 1957a,b).

Hb samples at partial O$_2$ saturations can be prepared by mixing approximate amounts of oxy- and deoxy-Hb solutions. The percentage oxygenation of the sample can be measured directly in a specially constructed NMR sample tube by monitoring the optical densities at 540, 560, and 577 nm (Huang and Redfield, 1976). Another method for determining the percentage oxygenation is to calculate it from the mixed volumes of oxy- and deoxy-Hb solutions, corrected for dissolved oxygen (Viggiano et al., 1979). Significant amounts of met-Hb can be formed (especially from high-affinity mutant or modified Hbs) during the course of ^{1}H NMR measurements of HbO$_2$ or Hb samples at partial O$_2$ saturation. The met-Hb reductase system of Hayashi *et al.* (1973) can be added to the Hb samples to prevent or to greatly reduce the formation of met-Hb. This reductase system does not appear to affect either the oxygenation properties or the ^{1}H NMR spectra of Hb (Hayashi *et al.*, 1973; Wiechelman *et al.*, 1974).

Table I gives a partial list of human mutant and modified hemoglobins that have been investigated in our research. Figure 11 shows the location of the mutation in some of the human mutant Hbs used in our research.

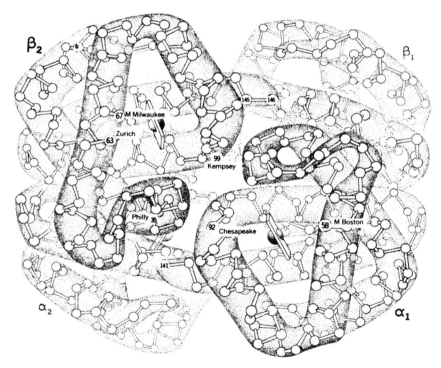

FIG. 11. Locations of selected human mutant hemoglobins used in our ^{1}H NMR studies [Adapted from Dickerson and Geis (1983); illustration copyright by I. Geis].

III. Resonance Assignments: Subunit Interfaces and Heme Environments

A. *Strategies for Assignment of Proton Resonances of Hemoglobin*

The first step in any NMR investigation of the structural and dynamic properties of a protein molecule is the assignment of resonances to specific amino acid residues. For a molecule the size of Hb, one of the most widely used methods to assign proton resonances to specific amino acid residues is to compare the ^{1}H NMR spectra of normal and appropriate mutant or chemically modified Hb molecules. This method is exemplified in Fig. 12, where the aromatic proton resonance region of the ^{1}H NMR spectrum of deoxy-Hb A is compared to those of mutant Hb Deer Lodge (β2His → Arg) and of enzymatically modified des-His(β146)-Hb (β146His deleted). As a result of the replacement by arginine, a nonaromatic amino acid, of the histidyl residue at position β2, two resonances at +3.32 and +2.30 ppm from HDO are missing from the ^{1}H NMR spectrum of deoxy-Hb Deer Lodge as compared to that of deoxy-Hb A, both in 0.1 M Bis–Tris buffer in D_2O at pH 6.30. There are no other observable differences between these two spectra. Thus, these two resonances can be assigned to the two protons of β2His of deoxy-Hb A, the one at +3.32 ppm to the C-2 proton and the one at +2.30 ppm to the C-4 proton (Russu *et al.*, 1982). Similarly, by comparing the ^{1}H NMR spectrum of deoxy-Hb A to that of deoxy-des His(β146)-Hb, both in 0.1 M Bis–Tris in D_2O at pH 6.30, the resonances at +3.80 and +2.76 ppm from HDO can be assigned to the C-2 and C-4 protons of β146His of deoxy-Hb A, respectively.

The potential of the method of using mutant and chemically modified Hbs to assign proton resonances of Hb is greatly enhanced when correlated with both intensity measurements and calculations from the crystal structure of Hb as determined by X-ray diffraction. A representative example of this approach is E11Val in HbCO A. Based on intensity measurements on a mixture of HbCO A and ferrocytochrome c, Lindstrom *et al.* (1972b) have shown for the ring-current-shifted proton resonances at −5.86, −6.48, and −6.58 ppm from HDO that each arises from one CH_3 per $\alpha\beta$ dimer. In the spectrum of a mixture of HbCO A and HbCO Sydney [β67(E11)Val → Ala], the resonances at ∼ −5.9 and ∼ −6.6 ppm from HDO have decreased intensities as compared to the corresponding ones in HbCO A (Fig. 13). These findings suggest that the resonances at −5.86 and −6.58

PROTON NUCLEAR MAGNETIC RESONANCE STUDIES

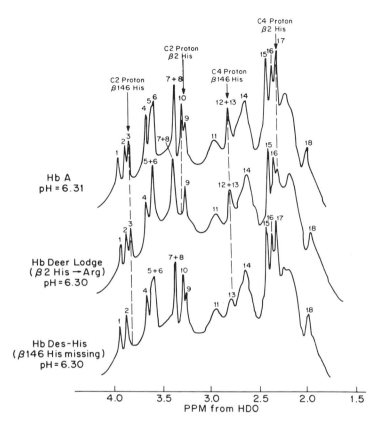

FIG. 12. Assignments of the proton resonances of the C-2 protons of histidyl residues at β2 and β146 of human normal adult hemoglobin: 250-MHz aromatic proton resonances of deoxy-Hb A, deoxy-Hb Deer Lodge (β2His → Arg), and deoxy-des-His-(β146-deleted)-Hb A in 0.1 M Bis–Tris in D$_2$O at pH 6.3 and 27°C. The resonance lines are labeled according to the notation of Russu *et al.* (1982). [From Ho and Russu (1981)].

ppm from HDO may originate from the γ_1- and γ_2-methyl groups of β67E11Val. Further work has confirmed the assignment of the -6.58-ppm resonance to the γ_2-methyl group of β67E11Val, but not the assignment of the -5.86-ppm resonance (see Section III,C,2).

This lack of confirmation of an apparent assignment points out that the method of using mutant or chemically modified Hbs to assign resonances as described above can be used correctly only if the single

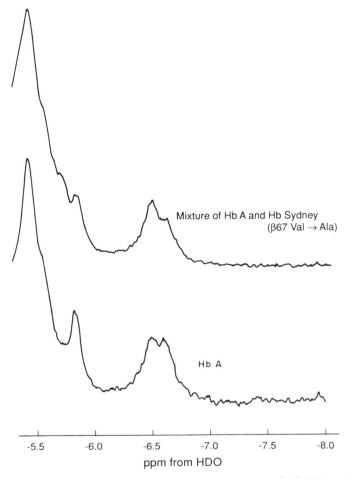

FIG. 13. 250-MHz ring-current-shifted proton resonances of HbCO A and HbCO Sydney (β97Val → Ala). [Adapted from Lindstrom et al. (1972b)].

mutation or chemical modification does not produce significant conformational alterations in other regions of the protein molecule. Such effects can result in changes in the ^{1}H NMR spectrum, making spectral assignments ambiguous. This limitation may be overcome by the following two approaches. First, one could use several mutant or chemically modified Hbs with a change in the same amino acid residue (for example, see Lindstrom et al., 1972b; Fung and Ho, 1975; Viggiano et al., 1978). Second, as in the case with the β67E11Val

TABLE II
Assignments of Proton Resonances of Human Normal Adult Hemoglobin[a]

Resonance position (ppm from H_2O or HDO)	Hb derivative	Experimental conditions	Assignment	Ref.
+71.0	Deoxy-Hb	0.1 M Bis–Tris in H_2O, pH 6.7, 27°C	β63(E7)His $N_\delta H$ proton, exchangeable hfs	b
+58.5	Deoxy-Hb	0.1 M Bis–Tris in H_2O, pH 6.7, 27°C	α58(E7)His $N_\delta H$ proton, exchangeable hfs	b
+17.5	Deoxy-Hb	0.1 M Bis–Tris, pH 6.7, 27°C	β chain, hfs	b,c,d,e
+16.8	Deoxy-Hb	0.1 M Bis–Tris, pH 6.7, 27°C	β chain, hfs	b,e
+15.6	Deoxy-Hb	0.1 M Bis–Tris, pH 6.7, 27°C	α chain, hfs	b,e
+14.2	Deoxy-Hb	0.1 M Bis–Tris, pH 6.7, 27°C	β chain, hfs	b,e
+13.2	Deoxy-Hb	0.1 M Bis–Tris, pH 6.7, 27°C	α chain, hfs	b,e
+12.1	Deoxy-Hb	0.1 M Bis–Tris, pH 6.7, 27°C	α chain, hfs	b,c,d,e
+9.5	Deoxy-Hb	0.1 M Bis–Tris, pH 6.7, 27°C	β chain, hfs	b
+9.4	Deoxy-Hb	0.1 M Bis–Tris in H_2O, pH 6.6, 27°C	H bond between α42Tyr and β99Asp, $α_1β_2$ interface	f,g
+8.2	Deoxy-Hb	0.1 M phosphate in H_2O, pH 7.0, 29°C	H bond between α126Asp and β35Tyr, $α_1β_1$ interface	g,h
+8.1	HbO_2 and HbCO	0.1 M phosphate in H_2O, pH 7.0, 29°C	H bond between α126Asp and β35Tyr, $α_1β_1$ interface	g
+7.8	Deoxy-Hb	0.1 M Bis–Tris, pH 6.7, 27°C	α chain, hfs	b,c,d
+7.5	Deoxy-Hb	0.1 M phosphate in H_2O, pH 7.0, 29°C	H bond between α103His and β108Asp, $α_1β_1$ interface	g
+7.4	HbO_2 and HbCO	0.1 M phosphate in H_2O, pH 7.0, 29°C	H bond between α103His and β108Asp, $α_1β_1$ interface	g
+7.2	Deoxy-Hb	0.1 M Bis–Tris, pH 6.7, 27°C	β chain, hfs	b

(continued)

TABLE II (*Continued*)

Resonance position (ppm from H_2O or HDO)	Hb derivative	Experimental conditions	Assignment	Ref.
+6.4	Deoxy-Hb	0.1 M Bis–Tris in H_2O, pH 6.8, 27°C	H bond between α94Asp and β37Trp	f,i
+5.9	HbO_2	0.1 M Bis–Tris in H_2O, pH 6.6, 27°C	H bond between β94Asp and β102Asn, $\alpha_1\beta_2$ interface	f
+5.5	HbCO	0.1 M Bis–Tris in H_2O, pH 6.6, 27°C	H bond between α94Asp and β102Asn, $\alpha_1\beta_2$ interface	f
+5.69	HbCO	0.1 M phosphate, pH 5.6, 37°C	Meso proton γ of the heme of α chain	j
+5.59	HbCO	0.1 M phosphate, pH 5.6, 37°C	Meso proton γ of the heme of β chain	j
+5.42	HbCO	0.1 M phosphate, pH 5.6, 37°C	Meso proton α of the heme of β chain	j
+5.09	HbCO	0.1 M phosphate, pH 5.6, 37°C	Meso proton δ of the heme of α and β chains	j
+4.97	HbCO	0.1 M phosphate, pH 7.2, 29°C	Meso proton δ of the hemes of α and β chains	k
+4.95	HbCO	0. M phosphate, pH 5.6, 37°C	Meso proton α of the heme of α chain	j
+4.76	HbO_2	0.1 M phosphate, pH 7.2, 29°C	Meso proton δ of hemes of α and β chains	k
+4.73	HbCO	0.1 M phosphate, pH 5.6, 37°C	Meso proton β of the heme of β chain	j
+4.71	HbCO	0.1 M phosphate, pH 5.6, 37°C	Meso proton β of the heme of α chain	j
+3.95	HbCO	0.1 M Bis–Tris, pH 6.3, 27°C	β97His, C-2 proton	l,m
+3.80	Deoxy-Hb	0.1 M Bis–Tris, pH 6.3, 27°C, or 0.1 M phosphate + 0.2 M NaCl, pH 6.2, 30°C	β146His, C-2 proton	n,o

TABLE II (Continued)

Resonance position (ppm from H_2O or HDO)	Hb derivative	Experimental conditions	Assignment	Ref.
+3.68	HbCO	0.2 M phosphate + 0.2 M NaCl, pH 6.2, 27°C	β2His, C-2 proton	p
+3.60	Deoxy-Hb	0.1 M HEPES, pH 7.5, 29°C	β146His, C-2 proton	q
+3.47	HbCO	0.2 M phosphate + 0.1 M HEPES, pH 6.7, 29°C	β146His, C-2 proton	q
+3.40	HbCO	0.1 M Bis–Tris, pH 6.3, 29°C	β2His, C-2 proton	r
+3.38	HbCO	0.1 M HEPES + 0.2 M phosphate, pH 6.7, 29°C	β2His, C-2 proton	r
+3.32	Deoxy-Hb	0.1 M Bis–Tris, pH 6.3, 27°C	β2His, C-2 proton	r
+3.31	HbCO	0.1 M Bis–Tris, pH 6.3, 29°C	β146His, C-2 proton	q
+3.27	HbCO	0.2 M phosphate + 0.2 M NaCl, pH 7.1, 29°C	β146His, C-2 proton	o
+3.20	HbCO	0.1 M Bis–Tris, pH 6.6, 27°C	β2His, C-2 proton	p
+3.23 ⎫ +3.17 ⎭	HbCO	0.1 M Bis–Tris, pH 6.3, 29°C	β116His or β117His, C-2 proton	q,s
+3.21	HbCO	0.2 M phosphate + 0.2 M NaCl, pH 7.1, 29°C	β2His, C-2 proton	r
+3.18 ⎫ +3.09 ⎭	Deoxy-Hb	0.1 M HEPES, pH 7.0, 29°C	β116His or β117His, C-2 proton	q,s
+3.13 ⎫ +3.02 ⎭	HbCO	0.1 M HEPES, pH 6.8, 29°C	β116His or β117His, C-2 proton	q,s
+3.11	HbCO	0.1 M HEPES, pH 6.8, 29°C	β2His, C-2 proton	r
+3.05	HbCO	0.1 M HEPES, pH 6.8, 29°C	β146His, C-2 proton	q
+2.93	Deoxy-Hb	0.1 M HEPES, pH 7.5, 29°C	β2His, C-2 proton	r

(continued)

TABLE II (Continued)

Resonance position (ppm from H_2O or HDO)	Hb derivative	Experimental conditions	Assignment	Ref.
+2.93 +2.90	HbCO	0.2 M phosphate + 0.2 M NaCl, pH 7.1, 29°C	β116His or β117His, C-2 proton	q,s
−0.96	HbCO	0.1 M phosphate, pH 5.6, 37°C	α62Val(E11), α-CH	j
−1.06	HbCO	0.1 M phosphate, pH 5.6, 37°C	3-CH_3 of the heme of β chain; 8-CH_3 of the heme of α chain; β67Val(E11), α-CH	j
−1.09	HbCO	0.1 M phosphate, pH 5.6, 37°C	8-CH_3 of the heme of β chain	j
−1.14	HbCO	0.1 M phosphate, pH 5.6, 37°C	1-CH_3 of the heme of α chain	j
−1.21 −1.56	HbCO	0.1 M phosphate, pH 7.2, 29°C	1- and 8-CH_3 of the hemes of α and β chains	k
−1.27 −1.54	HbO_2	0.1 M phosphate, pH 7.2, 29°C	1- and 8-CH_3 of the hemes of α and β chains	k
−1.38	HbCO	0.1 M phosphate, pH 5.6, 37°C	1-CH_3 of the heme of β chain	j
−2.00	HbCO	0.1 M phosphate, pH 5.6, 37°C	5-CH_3 of the heme of α chain	j
−3.15	HbCO	0.1 M phosphate, pH 5.6, 37°C	β67Val(E11), β-CH	j
−3.28	HbCO	0.1 M phosphate, pH 7.2, 29°C	β67Val(E11), β-CH	k
−3.35	HbCO	0.1 M phosphate, pH 5.6, 37°C	α62Val(E11), β-CH	j
−3.51	HbCO	0.1 M phosphate, pH 7.2, 29°C	α62Val(E11), β-CH	k
−4.47	HbCO	0.1 M phosphate, pH 5.6, 37°C	Val(E11), γ-CH_3 of α and β chains	j

TABLE II (Continued)

Resonance position (ppm from H_2O or HDO)	Hb derivative	Experimental conditions	Assignment	Ref.
−4.54	HbCO	0.1 M phosphate, pH 7.2, 29°C	β67Val(E11), γ_1-CH_3	k
−4.60	HbCO	0.1 M phosphate, pH 7.2, 29°C	α62Val(E11), γ_1-CH_3	k
−4.80	HbO_2	0.1 M phosphate, pH 7.2, 29°C	Val(E11), γ_1-CH_3 of α and β chains	k
−6.32	HbCO	0.1 M phosphate, pH 5.6, 37°C	α62Val(E11), γ-CH_3	j
−6.46	HbCO	0.1 M phosphate, pH 5.6, 37°C	β67Val(E11), γ-CH_3	j
−6.51	HbCO	0.1 M phosphate, pH 7.2, 29°C	Val(E11), γ_2-CH_3 of α and β chains	k
−7.11	HbO_2	0.1 M phosphate, pH 7.2, 29°C	Val(E11), γ_2-CH_3 of α and β chains	k

[a]The proton chemical shift of H_2O is 4.73 ppm downfield from that of the methyl group of 2,2-dimethyl-2-silapentane 5-sulfonate (DSS) at 29°C. [Modified from Ho and Russu (1981)]. The particular resonance can be observed in both H_2O and D_2O, unless specifically stated otherwise.
[b]Takahashi et al. (1980).
[c]Davis et al. (1971).
[d]Lindstrom et al. (1972a).
[e]Ho et al. (1982b).
[f]Fung and Ho (1975).
[g]Russu et al. (1987).
[h]Asakura et al. (1976).
[i]Ishimori et al. (1992). For further discussion, see Note Added in Proof.
[j]Craescu and Mispelter (1989).
[k]Dalvit and Ho (1985).
[l]Russu et al. (1982).
[m]Ho and Russu (1987).
[n]Russu et al. (1980).
[o]Kilmartin et al. (1973).
[p]Fung et al. (1975).
[q]Busch et al. (1991).
[r]Russu et al. (1989).
[s]Russu et al. (1984).

assignment, another approach is to use an independent technique to confirm the resonance assignments made by mutant or chemically modified Hbs. We have confirmed a few resonance assignments and have also assigned a number of additional resonances using one-dimensional NOE and two-dimensional NMR techniques. These methods also have pitfalls and limitations (see below), but can be very useful when used with care.

Table II gives a summary of the present state of the assignment of proton resonances of Hb A. A number of proton resonances for the isolated α and β chains in the CO form (especially the resonances of the porphyrin) have been assigned by us (Dalvit and Ho, 1985) and others (Dalvit and Wright, 1987; Schaeffer *et al.*, 1988; Craescu and Mispelter, 1988), but these are not included in Table II. Craescu and co-workers (Craescu and Mispelter, 1989; Craescu *et al.*, 1990) have combined 2D NMR techniques (COSY and NOESY) and X-ray structural information to make resonance assignments of HbCO A in 0.1 M phosphate in D_2O at pH 5.6 and 37°C. They have assigned a number of proton resonances of the heme groups and of several amino residues near the heme groups of HbCO A and Hb Saint Mandé (β102Asn → Tyr) based on their two-dimensional NMR results obtained at 400 MHz. We, too, have obtained ^1H COSY and NOESY spectra of HbCO A as a function of pH and temperature at 300 and 500 MHz (V. Simplaceanu, T.-J. Huang, M. Madrid, N. T. Ho, and C. Ho, unpublished results). However, one needs to be cautious in making specific spectral assignments of various amino acid residues of HbCO A until one is sure that spectral artifacts have been eliminated. Thus, we have decided not to include in Table II many of the spectral assignments of HbCO A based on two-dimensional NMR techniques.

B. *Resonances from Subunit Interfaces*

Exchangeable proton resonances of special interest for understanding the quaternary structural transitions in Hb are those of amino acid residues that are located at the subunit interfaces of the Hb tetramer and are involved in the hydrogen-bonding interactions responsible for the quaternary structures of deoxy-Hb and ligated Hb. Several of these amino acid residues give rise to exchangeable proton resonances over the spectral region +5.0 to +10.0 ppm downfield from the water proton resonance (Fung and Ho, 1975). Figure 14 shows the exchangeable proton resonances of Hb A in 0.1 M phosphate in H_2O at pH 7.0 and 29°C in deoxy, oxy, and CO forms. The resonances at +6.4, +7.5, +8.2, and +9.4 ppm are

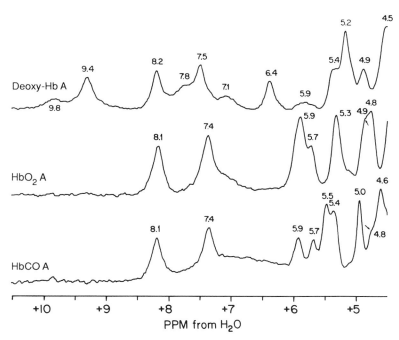

FIG. 14. 300-MHz ^{1}H NMR spectra of deoxy-Hb A, HbO$_2$ A, and HbCo A in 0.1 M phosphate in H$_2$O at pH 7.0 and 29°C: effects of ligation on the exchangeable proton resonances. [From Russu et al. (1987)].

absent in the spectra of isolated α and β chains, suggesting that they are specific markers for the quaternary structure of the Hb tetramer (Russu et al., 1987). The +6.4- and +9.4-ppm resonances disappear on the binding of ligand to deoxy-Hb A (Fung and Ho, 1975). The resonances at +8.2 ppm and +7.5 ppm are shifted by only about 0.1 ppm on ligand binding (Russu et al., 1987). Using appropriate mutant Hbs, the resonances at +6.4- +8.2, and +9.4 ppm have been assigned by our laboratory to specific amino acid residues. The +9.4-ppm resonance has been assigned to the intersubunit hydrogen bond between α42(C7)Tyr and β99(G1)Asp (an important deoxy quaternary feature in the α$_1$β$_2$ subunit interface), the +8.2-ppm resonance has been assigned to the intersubunit hydrogen bond between α126(H9)Asp and β35(C1)Tyr in the α$_1$β$_1$ subunit interface, and, with less certainty, the +6.4-ppm resonance has been tentatively assigned to the intrasubunit hydrogen bond between the hydroxy group of β145(HC2)Tyr and the carboxyl group of β98(FG5)Val (an important feature in the deoxy tertiary structure) [for details, see Fung and Ho (1975), Asakura *et al.* (1976), and Viggiano *et al.* (1978)]. Very recently, Ishimori *et al.* (1992) had proposed that the +6.4-ppm resonance arises from the intersubunit hydrogen bond between α94Asp and β37Trp in the $\alpha_1\beta_2$ subunit interface of the deoxy quaternary structure. However, this assignment needs further investigation. For details, see Note Added in Proof.

By using the initial NOE buildup rates on the exchangeable proton resonances at +7.5, +8.2, and +9.4 ppm from H_2O, we have observed specific NOEs for each of these exchangeable proton resonances in deoxy-Hb A (Fig. 15). To gain insight into the molecular origin of the observed NOEs, we have attempted to correlate the patterns of the NOEs to the predictions made based on the X-ray crystal structures of deoxy-Hb A and HbCO A (Fermi, 1975; Baldwin, 1980; Fermi and Perutz, 1981) as well as based on the assignments previously proposed by this laboratory for the +9.4- and +8.2-ppm resonances (Fung and Ho, 1975; Asakura *et al.*, 1976). From these studies, we have confirmed the assignment of the exchangeable proton resonances at +9.4 and +8.2 ppm and have assigned the exchangeable proton resonance at +7.5 ppm to the intermolecular hydrogen bond between α103His and β108Asn in the $\alpha_1\beta_1$ subunit interface (Russu *et al.*, 1987).

C Resonances from Heme Environments

1. Heme Pockets in Unligated Form

Considerable effort has been devoted to the assignment of hfs proton resonances (i.e., those resonances coming from protons on or near the heme groups) to the α and β chains of Hb A in the deoxy form. This represents the first step when using ^{1}H NMR to investigate the binding of ligands to the α and β chains of deoxy-Hb A and the effect of ligation of one subunit on the conformation of the neighboring unligated subunits. There are two spectral regions that are of interest, namely, from +6 to +20 ppm downfield from H_2O or HDO and from +50 to +80 ppm downfield from H_2O. In Fig. 16, the hfs resonances of deoxy-Hb A in the region +6 to +20 ppm downfield from HDO are compared to those of the naturally occurring valency hybrid hemoglobins, Hb M Boston (α58E7His → Tyr) and Hb M Milwaukee (β67E11Val → Glu) in the deoxy form. In Hb M Boston ($\alpha_2^+\beta_2$), the iron atoms of the α chains are in the ferric state, whereas those in the β chains are in the ferrous state. On the other hand, the iron atoms of the β chains in Hb M Milwaukee ($\alpha_2\beta_2^+$) are in the ferric state, whereas those in the α chains are in

A. Control Spectrum

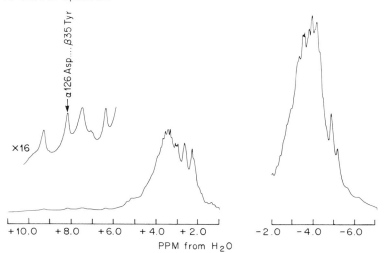

B. NOE on the Resonance at 8.2 ppm

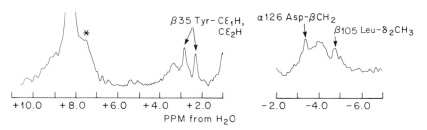

C. NOE on the Resonance at 7.5 ppm

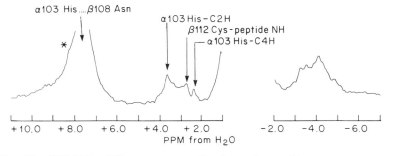

FIG. 15. 300-MHz difference spectra for the exchangeable proton resonances of deoxy-Hb A in 0.1 M phosphate in H_2O at pH 7.0 and 29°C. The irradiation time at +8.2 or +7.5 ppm was 50 msec. The asterisk denotes off-resonance spillage. [From Russu *et al.* (1987)].

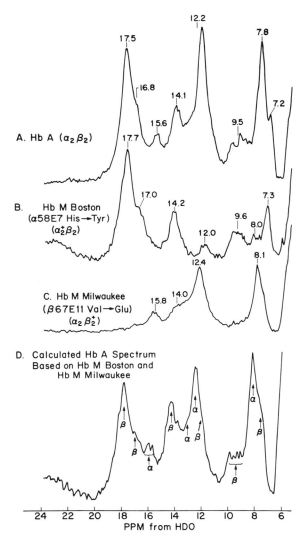

FIG. 16. 250-MHz ferrous hyperfine-shifted proton resonances of deoxyhemoglobins in 0.1 M Bis–Tris in D_2O at pH 6.6 and 27 °C. (A) Hb A; (B) Hb M Boston (α58His → Tyr, $\alpha_2^+ \beta_2$); (C) Hb M Milwaukee (β67Val → Glu, $\alpha_2 \beta_2^+$); (D) calculated Hb A spectrum obtained by the spectral sum of Hb M Boston and Hb M Milwaukee. [From Takahashi et al. (1980)].

the ferrous state. Because the hyperfine interactions are strongly dependent on the spin state of the iron atoms, the hfs resonances of the ferric α chains of Hb M Boston are shifted from the spectral positions of the ferrous α chains of deoxy-Hb A. Hence, the hfs resonances at +7.8, +12.2, and +15.6 ppm from HDO in the spectrum of deoxy-Hb A can be assigned to the protons belonging to the

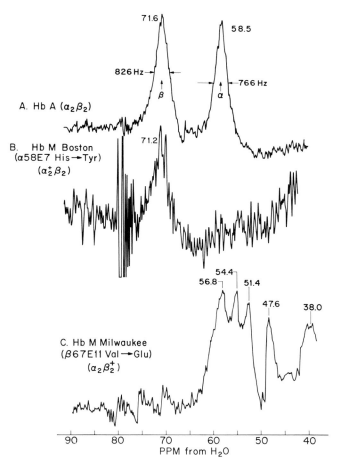

FIG. 17. 250-MHz ^{1}H NMR spectra of deoxyhemoglobins in 0.1 M Bis–Tris in H$_2$O at pH 6.7 and 27°C. (A) Hb A; (B) Hb M Boston (α58His → Tyr); (C) Hb M Milwaukee (β67Val → Glu). The beat appearing at ~80 ppm in spectrum B was generated by the 20-kHz time-sharing operation of the NMR spectrometer and was not a signal. [From Takahashi *et al.* (1980)].

α chains. Similarly, the hfs resonances missing from the ^{1}H NMR spectrum of Hb M Milwaukee (i.e., the resonances at +7.2, +9.5, +14.1, +16.8, and +17.5 ppm from HDO) have been assigned to the protons from the β chains in deoxy-Hb A [for details, see Takahashi *et al.* (1980)].

Proximal Histidyl $N_δH$ Resonances. In the ^{1}H NMR spectrum of deoxy-Hb A in H_2O two low-field resonances occur at +58.5 and +71.6 ppm downfield from H_2O (La Mar *et al.*, 1977; Takahashi *et al.*, 1980). These two resonances disappear in the presence of D_2O and have been assigned to the hyperfine-shifted $N_δH$-exchangeable protons of the proximal histidyl residues (F8) of the α and β chains of deoxy-Hb A (Takahashi *et al.*, 1980; La Mar *et al.*, 1980). In Fig. 17, it is clear that the resonance at +71 ppm is missing in Hb M Milwaukee, suggesting that this resonance comes from the β chain and the resonance at +58.5 ppm from the α chain. Hb M Boston has only the resonance at +71 ppm, which confirms the assignment of the +71-ppm resonance to the $N_δH$ proton of the proximal histidine of the β chain and the +58.5-ppm resonance to the corresponding residue in the α chain (Takahashi *et al.*, 1980). The fact that the α- and β-heme resonances are clearly separated indicates that the conformations of the proximal histidyl residues are different in the α and β chains of deoxy-Hb A.

2. Heme Pockets in Ligated Form

The ring-current fields of the porphyrins in Hb cause large shifts in the resonances of nearby protons. Changes in these shifts reflect structural changes in the heme pocket, thus giving information about changes in the tertiary structure of the active center of the Hb molecule. Much effort has, therefore, gone into the attempt to assign these ring-current-shifted resonances in the region −5 to −8 ppm upfield from HDO (or −0.3 to −3.3 ppm upfield from DSS). The closest amino acid residues to the hemes are the methyl groups of E11 valine (distal valine) and the C-2 proton of E7 histidine (distal histidine). Figure 18 shows the heme structure and the positions of the two methyl groups of E11 valine and the C-2 proton of E7 histidine with respect to the porphyrin ring in both carbonmonoxy (full circles) and oxy (broken circles) forms. The following section shows the assignment of these residues to ring-current-shifted resonances. It has been suggested that the conformations of the heme pockets (as manifested by the ring-current-shifted proton resonances) of the isolated α and β chains in the CO form are quite similar to those in the intact HbCO A (Lindstrom *et al.*, 1972b). Thus, for both convenience and ease of spectral assignments, we have investigated the

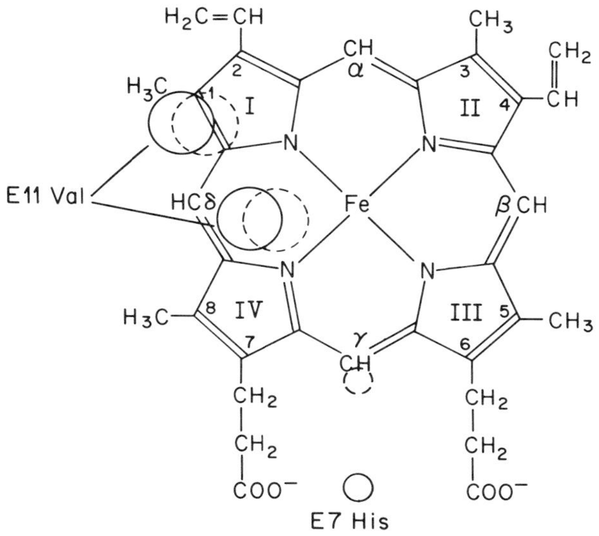

FIG. 18. Structure of the heme group in hemoglobin and the positions of the γ_1- and γ_2-CH_3 groups of the E11 valyl residue and the C-2–H of the E7 histidyl residue in the CO and oxy forms of the α chain of Hb A. The positions for the CO form (full circles) were calculated from the X-ray coordinates for the α chain of HbCO A (Baldwin, 1980), whereas the approximate positions for the oxy form (broken circles) were calculated from our NMR data and theoretical ring-current calculations. [From Dalvit and Ho (1985)].

ring-current-shifted proton resonances of isolated α and β chains and of Hb A in both CO and oxy forms. (For a comparison of CO and O_2 as ligands for Hb A, see Section IV,A.)

a. Carbon Monoxide as Ligand for Hemoglobin.
i. α Chains in CO form. Figure 19A (a) shows the 300-MHz ring-current-shifted proton resonances of isolated α chains in the CO form. The resonance at -1.78 ppm from DSS (or -6.51 ppm from HDO) was previously assigned to the γ_2-CH_3 group of the α62(E11)Val (Lindstrom et al., 1972b). Figure 19A (b) shows the truncated-driven NOE difference spectrum of isolated α chains obtained on preirradiation of the resonance at -1.78 ppm from DSS with a radio frequency (rf) pulse of 100-msec duration (Dalvit and Ho, 1985). Only a few resonances appear in both aliphatic and aromatic resonance regions in the difference spectrum, indicating that the observed NOEs are extremely selective. The resonances that are present in Fig. 19A

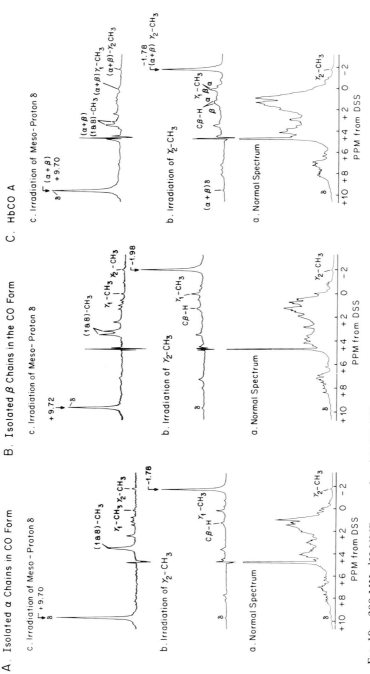

FIG. 19. 300-MHz ^{1}H NMR normal and NOE difference spectra of isolated α and β chains and of HbCO A in the CO form in 0.1 M phosphate in D_2O at pH 7.3 and 29°C. The NOE difference spectra were obtained with a preirradiation pulse of 100 msec at the positions indicated by arrows. $γ_1$-CH_3, $γ_2$-CH_3, and Cβ-H are protons of E11Val. [From Dalvit and Ho (1985)].

(b) originate from the protons located very close to the γ_2-CH$_3$ of α62(E11)Val, and the resonance intensity reflects to a first approximation the proton–proton distance. The protons closest to the γ_2-CH$_3$ group of the E11Val are the protons of the same amino acid residue, namely, γ_1-CH$_3$ and Cβ-H. From the relative intensities of these two resonances, we have assigned the resonances at +0.13 and +1.21 ppm from DSS (or −4.60 and −3.52 ppm from HDO), respectively, to the γ_1-CH$_3$ and CB-H of E11Val. The assignment of the resonance at +9.70 ppm from DSS (or +4.97 ppm from HDO) to the meso proton δ has been confirmed by the truncated-driven NOE difference spectrum of Fig. 19A (c).

Another method used to confirm our spectral assignment of E11Val is the COSY technique (Dalvit and Ho, 1985). A typical COSY experiment can, in principle, provide a complete map of all ^{1}H–^{1}H J connectivities, thus avoiding the lack of selectivity of irradiation in crowded spectral regions. Figure 20 shows the COSY contour plot in the region between +2 and −2.2 ppm from DSS with the connectivities for E11Val. These results are in agreement with the one-dimensional NOE experiments discussed above. Theoretical ring-current calculations are also consistent with our assignment of the γ_1-CH$_3$ of E11Val (Dalvit and Ho, 1985). Thus, the present assignments for E11Val of the α chain in the CO form can be considered definitive.

In the low-field end of the ^{1}H NMR spectrum of isolated α chains of Hb A in the CO form in D$_2$O, there are four singlets (see Dalvit and Ho, 1985). The truncated-driven NOE difference spectra with saturation of these four resonances confirm that they arise from the four meso protons of the porphyrin. In the NOE difference spectrum with preirradiation of the resonance of meso proton γ (see Dalvit and Ho, 1985), a strong NOE is observed at +8.24 ppm from DSS (or +3.51 ppm from HDO). This resonance has been assigned to the C-2–^{1}H of α58(E7)His, the distal histidyl residue. This assignment is consistent with our ring-current calculation for the C-2–^{1}H of α58His. By following the chemical shift of the C-2–^{1}H of α58His as a function of pH in 0.1 M phosphate in D$_2$O at 29°C, we have found that the apparent pK of α58His is 6.0 ± 0.1 (Dalvit and Ho, 1985).

ii. β Chains in CO form. Figure 19B (a) shows the 300-MHz ^{1}H NMR spectrum of isolated β chains of Hb A in the CO form. The resonances in Fig. 19B (a) are not as sharp as those for isolated α chains shown in Fig. 19A (a). This is because isolated β chains exist in solution as tetramers whereas isolated α chains exist as either monomers or dimers (Benesch and Benesch, 1964; Valdes and Ackers, 1977).

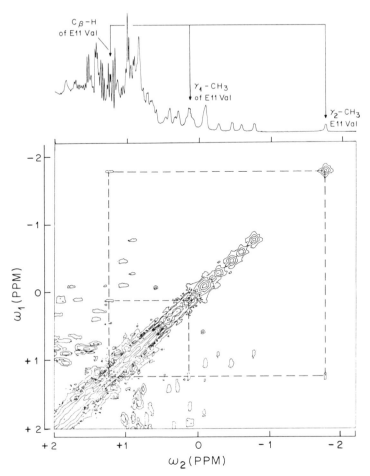

FIG. 20. 300-MHz ^{1}H NMR spectrum and ^{1}H COSY spectrum of isolated α chains of HbCO A in 0.1 M phosphate in D_2O at pH 5.3 and 29°C. The ^{1}H chemical shift scale is referred to DSS. [From Dalvit and Ho (1985)].

The upfield resonance at -1.98 ppm from DSS (or -6.71 ppm from HDO) was assigned to the γ_2-CH_3 of β67(E11)Val (Lindstrom *et al.*, 1972b). The NOE difference spectrum with preirradiation on this resonance for 100 msec is shown in Fig. 19B (b). The observed spectral features are very similar to those shown in Fig. 19A (b). Following the same procedures and considerations used for the α chain, we have assigned the resonances at -0.10 and $+1.36$ ppm from DSS (or -4.83 and -3.37 ppm from HDO) to the γ_1-CH_3 and $C\beta$-^1H of $\beta 67E11Val$, respectively.

The present assignment of the resonance at -0.10 ppm from DSS to the γ_1-CH_3 of $\beta 67E11$ is different from that previously suggested by Lindstrom *et al.* (1972b), who assigned the resonance at -1.1 ppm upfield from DSS (or -5.83 ppm from HDO) to this amino acid residue. We have observed only a very small NOE on the resonance at -1.1 ppm on irradiating the resonance at -1.98 ppm, suggesting that this resonance comes from protons that are relatively far from the γ_2-CH_3 of $\beta 67E11Val$. We believe that the present assignment is more reliable than the previous work, and it is also confirmed by other studies (for details, see Dalvit and Ho, 1985).

iii. HbCO A. Figure 19C (a) shows the 300-MHz ^1H NMR spectrum of HbCO A. The resonance at -1.78 ppm from DSS (or -6.51 ppm from HDO) corresponds to the γ_2-CH_3 groups of E11Val from both the α and β chains (Lindstrom *et al.*, 1972b). Figure 19C (b) shows the truncated-driven NOE difference spectrum with preirradiation at -1.78 ppm upfield from DSS for 100 msec. Comparing the difference spectra of isolated α and β chains with that of HbCO A, we have assigned several resonances to specific groups in the α and β subunits. Figure 19C (c) shows the NOE difference spectrum of HbCO A with preirradiation on the resonance at -9.70 ppm from DSS (or -4.97 ppm from HDO), which is analogous to the meso protons of the hemes of isolated α and β chains. In this difference spectrum, we observe the ring methyl groups 1 and 8 of both α and β subunits of HbCO A and the γ_1- and γ_2-CH_3 groups of E11Val.

b. Oxygen as Ligand for Hemoglobin. We have carried out spectral assignments of the heme substituents and of the distal histidyl and valyl residues of the heme pockets of the isolated α and β chains and of Hb A, all in the oxy form, by using the same techniques that we have used in our spectral assignments for these three samples in the CO form (for details, see Dalvit and Ho, 1985). Figure 21 gives the 300-MHz ^1H NMR normal and NOE difference spectra of isolated α and β chains and of Hb A in the oxy form in 0.1 M phosphate in D_2O at pH 7.3 and 29°C. The most upfield resonance at -2.38 ppm upfield from DSS (or -7.11 ppm from HDO) in Fig. 21 was assigned to the γ_2-CH_3 group of E11Val (Lindstrom and Ho, 1973).

In the NOE difference spectrum of isolated α chains in the oxy form with preirradiation of the resonance of meso proton γ, a strong NOE is observed at $+5.41$ ppm from DSS (or $+0.68$ ppm from HDO) (see Dalvit and Ho, 1985). This resonance is a singlet as is its

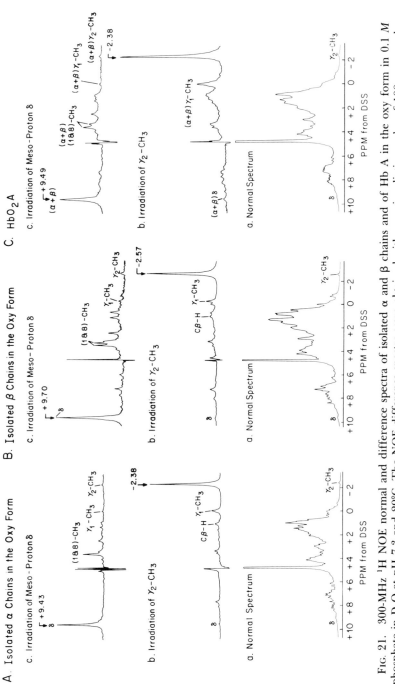

FIG. 21. 300-MHz ^{1}H NOE normal and difference spectra of isolated α and β chains and of Hb A in the oxy form in 0.1 M phosphate in D_2O at pH 7.3 and 29°C. The NOE difference spectra were obtained with a preirradiation pulse of 100 msec at the positions indicated by arrows. γ_1-CH_3, γ_2-CH_3, and Cβ-H are protons of E11Val. [From Dalvit and Ho (1985)].

corresponding one in the CO form and originates from the C-2–^{1}H of α58(E7)His. The assignment of this resonance was assisted by the X-ray coordinates of HbO$_2$ A (Shaanan, 1983) and by a truncated-driven NOE difference spectrum with preirradiation of the resonance of γ$_2$-CH$_3$ of E11Val at −2.38 ppm from DSS (see Dalvit and Ho, 1985). When a similar truncated-driven NOE experiment was carried out for the α chain in the CO form, we did not observe a NOE on the C-2–^{1}H of α58(E7)His (Dalvit and Ho, 1985).

The C-2–^{1}H of α58(E7)His of HbCO A is located 2.7 Å from the heme plane and at a distance of 6.3 Å from the center of the heme plane (Baldwin, 1980), and experiences a ring-current shift of 0.39 ppm downfield. On the other hand, the observed ring-current shift for the C-2–^{1}H of α58(E7)His in the oxy form of isolated α chains is 2.34 ppm upfield. This large upfield shift for the C-2–^{1}H of the distal histidyl residue in the oxy form of the α chain can be explained by a movement of this histidyl residue toward the center of the heme (see Fig. 18). Based on ring-current calculations, we have estimated a distance of ~4 Å of the C-2–^{1}H of α58(E7)His relative to the center of the heme plane for the oxy form, compared with the value of 6.3 Å found for the C-2–^{1}H of α58(E7)His in the CO form (Dalvit and Ho, 1985). The X-ray structure of HbO$_2$ A at 2.1 Å resolution (Shaanan, 1983) also indicates that the C-2–^{1}H of α58(E7)His is closer to the center of the heme plane in HbO$_2$ A than in HbCO A (Baldwin, 1980), i.e., 4.7 versus 6.3 Å. Thus, there is agreement between the NMR results obtained in the solution state and the X-ray diffraction results in the crystalline state. The NMR results [Figs. 19 and 21 as well as the studies reported by Lindstrom and Ho (1973)] show that the γ$_1$- and γ$_2$-CH$_3$ of E11Val of Hb A in the oxy form experience a larger ring-current shift than in the CO form. On the basis of theoretical ring-current calculations and NOE results, we have concluded that in the oxy form, the E11Val is located closer to the normal of the heme than in the CO form. This movement of E11Val toward the normal of the heme in the oxy form has to be less than 0.5 Å (Dalvit and Ho, 1985).

The atomic models of Hb A have clearly shown that the two distal residues, E7His and E11Val, are strategically located in the immediate vicinity of the ligand-binding sites of both α and β chains (Perutz, 1970; Baldwin and Chothia, 1979; Baldwin, 1980; Shaanan, 1983; Fermi *et al.*, 1984). High-resolution X-ray crystallographic structural results on HbO$_2$ A and HbCO A show that there are substantial differences between the distal portions of the α and β heme pockets (Shaanan, 1983; Derewenda *et al.*, 1990). For HbO$_2$ A, the distances between Nε of His (E7) and Cγ$_2$ of Val(E11) and the second bound O$_2$ atom are 2.6 and 3.2 Å, respectively, in the α chain and 3.5 and 3.4 Å in the β chain, respectively (Shaanan, 1983). The distal histidine (E7) of the α chain appears to form a hydrogen bond with the bound O$_2$ and to sterically hinder CO binding (Shaanan, 1983). The corresponding distances in HbCO A are 2.95 and 2.98 Å in the α chain and 3.30 and 3.28 Å in the β chain (Derewenda *et al.*, 1990). In the case of the βO$_2$ chain, the second oxygen atom is too far for a strong hydrogen bond with the distal histidyl residue (E7) (Shaanan, 1983). For the βCO chain, the second bound ligand atom is not well defined, suggesting multiple orientations due to a more open distal pocket in the β chain as compared to the α chain (Derewenda *et al.*, 1990). The conformations of these two amino acid residues [His(E7) and Val(E11)] are thought to play an important role in regulating both the thermodynamics and kinetics of the ligation process of Hb A (for example, see Moffat *et al.*, 1979; Mathews *et al.*, 1989, 1991). The present assignments of the proton resonances of E11Val in isolated α and β chains of Hb A and of the C-2–^{1}H of E7His in isolated α chains of Hb A in CO and oxy forms allow us to use NMR spectroscopy to characterize the conformational differences of E7His and E11Val between the CO and oxy forms in solution and to compare them to those derived from X-ray diffraction data in single crystals. It should be noted that the overall structural features of the heme pockets of HbCO A and HbO$_2$ A as derived from our ring-current-shifted proton resonances and ring-current calculations are consistent with those derived from the X-ray crystallographic structural determinations of these two Hbs.

IV. ^{1}H NMR Investigations of Ligand Binding to α and β Chains of Hemoglobin

The ^{1}H NMR and X-ray crystallography results discussed in Section III have clearly shown that the environments of the heme pockets of the α and β chains of Hb as manifested by the conformations of proximal histidine (F8), distal histidine (E7), and distal valine (E11), as well as the conformation of several porphyrin protons and the electronic structure of the heme group in both deoxy and ligated states, are not equivalent. It has not been clear what this difference in structure in the α and β heme pockets means in terms of differences in function, because there are discrepancies in the published results regarding the relative ligand affinities of the α and β chains of Hb A.

A. Ligand Binding to α and β Chains of Hb A

Most studies on the binding of O_2 indicate that O_2 binds preferentially to the α chains. In a preliminary ^{1}H NMR investigation of the hfs resonances of deoxy-Hb A as a function of oxygenation, we reported that O_2 binds preferentially to the α chains in the presence of excess 2,3-DPG or IHP (Ho and Lindstrom, 1972; Lindstrom and Ho, 1972; Johnson and Ho, 1974). From ^{19}F NMR studies on Hb A trifluoroacetonylated at β93 cysteines, Huestis and Raftery (1972c) found that in the presence of 2,3-DPG, O_2 binds initially to the α chains and that binding to the β chains lags ~10% behind overall ligand binding throughout the binding curve. Gibson (1973) suggested that both the association and dissociation rates of O_2 with the β chains are much faster than with the α chains, but that the equilibrium constants are probably such that the α chains have a higher O_2 affinity than the β chains at low fractional saturations. Indirect agreement with the general conclusion that the α chains have a higher O_2 affinity than the β chains has also been obtained in a study of chain oxidation rates, in which it was found that decreasing the O_2 pressure increases the β chain oxidation rate much more than the α chain rate, indicating that the β chains have a lower O_2 affinity (Mansouri and Winterhalter, 1973). However, using a multidimensional spectroscopic measuring system, Nasuda-Kouyama et al. (1983) reported that the β chains of Hb A have a slightly higher O_2 affinity in the presence of excess IHP compared to the α chains. Di Cera et al. (1987), reporting their O_2 and CO binding curves for Hb A in the presence of excess IHP at pH 6.94 and 25°C, showed that ligation occurs first at the α chains in the T state and then at the β chains after the conformational transition to the R state.

For CO, the situation is somewhat less clear. In preliminary ^{1}H NMR studies using the hfs resonances, we found no evidence for preferential binding of CO to either the α or the β chains of Hb A in 0.1 M phosphate, in the presence of 1–2 M excess of 2,3-DPG or IHP at pH 6.6 (Lindstrom and Ho, 1972; Johnson and Ho, 1974). Using spin-labeled triphosphates, Ogata and McConnell (1972a) found that in the presence of these triphosphates, the α chains exhibit a very slight preference for CO (≤2%) at low fractional saturation. Huestis and Raftery (1972b, 1973) concluded from ^{19}F NMR measurements of their ^{19}F-labeled Hb A and from ^{31}P NMR studies of the 2,3-DPG bound to Hb A that CO follows essentially the same course as O_2 in binding to Hb A, namely, that it is initially bound by the α chains and that the binding by the β chains lags behind the overall ligand binding by 10%. Kinetic studies by Gray and Gibson

(1971a,b) suggested that in the absence of phosphate, the binding rates of the different chains with CO are indistinguishable, but that on addition of inorganic phosphate, 2,3-DPG, or IHP, the β chains exhibit a faster binding rate than the α chains. However, the data on dissociation rates are insufficient to give any information on the relative affinities of the α and β chains under equilibrium conditions.

The evidence for preferential binding by *n*-butyl isocyanide is also confusing. From ^{1}H NMR measurements of the hfs resonances, we found that in the presence of IHP, *n*-butyl isocyanide exhibits preferential binding to the β chains (Lindstrom *et al.*, 1971). This conclusion was extended in a series of kinetic measurements by Gibson and co-workers (Olson and Gibson, 1971, 1972, 1973a,b; Cole and Gibson, 1973), who showed that the association rates of the β chains for *n*-butyl isocyanide are greater than those of the α chains both in the absence and in the presence of 2,3-DPG or IHP. They reported that in the absence of organic phosphates, the equilibrium affinities of the chains are approximately equal. However, in the presence of 2,3-DPG or IHP, and at low ligand saturation, the equilibrium affinity of the β chains is 20–40 times that of the α chains, with the affinities becoming approximately equal at high ligand saturation. Based on their ^{19}F NMR measurements of ^{19}F-labeled Hb A, Huestis and Raftery (1972a) found that in the presence of 2,3-DPG, the α chains are preferentially bound by *n*-butyl isocyanide, but that in the absence of phosphate, the α and β chains exhibit random binding.

In recent work, the functional significance of the structural differences observed between the α and β chains of Hb has been examined by measuring the rate and equilibrium constants for ligand binding to His(E7) and Val(E11) mutants of genetically engineered recombinant Hb (Nagai *et al.*, 1987; Olson *et al.*, 1988; Mathews *et al.*, 1989, 1991; Tame *et al.*, 1991). Nagai *et al.* (1987) and Tame *et al.* (1991) reported the O_2 equilibrium binding curves for mutant Hbs containing Val(E11) → Ala, Leu, or Ile) substitutions in the α and β chains. Their results suggested that the native chains have similar T-state O_2 affinities and that Val(E11) sterically restricts ligand binding in the β chains but not in the α chains. It should be mentioned that the interpretation of these experiments is complicated by the high degree of cooperativity, which prevents the buildup of partially ligated intermediates, as well as by the lack of simple spectral signals that can discriminate between the binding of ligands to the α and β chains. Mathews *et al.* (1989) reported that the ligand-binding site in the R-state β chains appears to be more accessible to CO and O_2 as judged by its larger CO and O_2 association rate constants and by the lack of effects of amino acid substitutions at His(E7) and Val(E11) in the β chains on the rates of CO and O_2 binding. Because the mutations His(E7 → Gly) and Val(E11 → Ala or Ile) also have little effect on the equilibrium binding constants for CO and O_2, they believe that some mechanism other than hydrogen bonding to the distal histidine must account for the high O_2 affinity and low ratio of K_{CO}/K_{O_2} of the R-state β-chains. On the other hand, the distal histidyl residue in the R-state α chain has been found to stabilize the bound O_2 by 1–2 kcal/mol, to discriminate against CO binding, and also to limit the rate of ligand access to the heme iron atom (Olson et al., 1988).

Mathews et al. (1991) investigated the functional differences between the α and β chains in the T state by measuring the rates of CO binding to genetically engineered mutant deoxy-Hbs containing one set of native chains and one set of mutant chains with either His(E7 → Gly), Val(E11 → Ala), or Val(E11 → Ile) substitution. They found that at pH 7 and 20°C, the association rate constant for the binding of the first CO to deoxy-Hb is about the same for native α and β chains ($k'_{T\alpha} = 0.12$ μM^{-1} s^{-1} and $k'_{T\beta} = 0.18$ μM^{-1} s^{-1}). The equilibrium constants for the first step in the CO binding to the native α and β chains have been found to be identical within experimental error [$K_{T\alpha} = K_{T\beta} = (0.8–1.1) \times 10^6$ M^{-1} at pH 7, 20°C] (Mathews et al., 1991; Sharma et al., 1991). Mathews et al. (1991) suggested that the E7 and E11 substitutions affect primarily the intrinsic T-state rate constants for the CO binding rather than the equilibrium rate constant for the quaternary conformational transition.

Because an understanding of the functional properties of the α and β chains of an intact Hb A molecule is important for our eventual understanding of the detailed mechanism for the cooperative oxygenation of Hb A, we have continued ^{1}H NMR investigations of ligand binding to the α and β chains of Hb A.

Due to unique physical (including spectroscopic) and biochemical properties (including stability) of Hb ligated to various nonoxygen ligands such as CO, NO, cyanide, azide, and alkyl isocyanides, these forms of Hb derivatives have been used extensively by a large number of researchers to investigate the molecular basis for the cooperative oxygenation of Hb A. Because of the high affinity of Hb for NO, partially NO-ligated species have been used to gain information about ligation intermediates. However, we shall not describe here the equilibrium and kinetic studies on the binding of NO to Hb [for details, refer to Antonini and Brunori (1971), Cassoly (1978), Hille et al. (1979), and references therein]. The most commonly used nonoxygen ligand is CO. HbCO has several features (including a greater stability of HbCO over HbO_2) that make it an attractive model for gaining insights into the structure–function relationship in Hb. Prior to the early 1970s, little attention was given to the possibility that the Hb–ligand reaction might also be ligand specific (for a review, see Johnson and Ho, 1974). For example, it was assumed that, with the exception of a scaling factor, the reactions of Hb A with O_2 and CO are equivalent (Antonini and Brunori, 1971). [For reviews dealing with various structural and functional properties of HbCO and their similarities or dissimilarities with the properties of HbO_2, refer to Antonini and Brunori (1971), Moffat *et al.* (1979), Ho *et al.* (1982a), Bunn and Forget (1986), and Perutz (1990), and references therein.]

B. Comparison between HbO_2 A and HbCO A

The binding of CO to Hb A is a cooperative process, similar to that for the oxygenation of Hb A (for example, see Antonini and Brunori, 1971; Thomas and Edelstein, 1972; Wyman *et al.*, 1982; Perrella *et al.*, 1986, 1990b). However, there are differences in the binding of CO and of O_2 to Hb A. CO binds about 210 times more strongly to Hb A than does O_2, and this factor is not affected by pH or 2,3-DPG, but is slightly temperature dependent (Wyman *et al.*, 1982; Bunn and Forget, 1986). The Hill coefficient (an empirical measure of the cooperativity of ligand binding to Hb) for the binding of CO to Hb A in 0.1 M KCl at pH 7 and 22°C is 3.4, compared to 2.8 for the binding of O_2 to Hb A (Perrella *et al.*, 1986, 1990b). According to Di Cera *et al.* (1987), at high concentrations of Hb A ($\sim$0.6 mM) in the presence of excess IHP at pH 6.94 and 25°C, the shape of the binding curve for CO is different from that of O_2. There are also differences in the kinetic cooperativity in the binding of CO and of O_2 to Hb A. It has been reported that O_2 binds to the T and R states of Hb A with approximately equal velocity. On the other hand, CO binds much more slowly to the T state of Hb A than to the R state of Hb A. The main difference in their ligation kinetics is that the cooperativity in O_2 binding lies in differences in the "off" rates, whereas that in CO binding lies in the "on" rates (for example, see Sharma *et al.*, 1976; Hille *et al.*, 1977; Szabo, 1978; Moffat *et al.*, 1979).

Structural differences around the heme pockets of both α and β chains have also been observed between single crystals of HbO_2 and HbCO (Heidner *et al.*, 1976; Baldwin, 1980; Shaanan, 1983; Liddington *et al.*, 1988; Derewenda *et al.*, 1990). X-Ray crystallographic results show that O_2 binds to the heme iron of Hb in a bent or off-axis configuration, whereas CO is bonded to the iron atom of heme in a straight line. The steric constraints, however, prevent CO from assuming a perfect perpendicular orientation to the heme plane in HbCO. There is another structural difference between HbO_2 and HbCO. In HbO_2, the distal atom of the O_2 molecule forms a hydrogen bond with the imidazole of the distal histidyl residue (E7) (Shaanan, 1983). This hydrogen bond is not seen in HbCO (Baldwin, 1980; Shaanan, 1983; Derewenda *et al.*, 1990). By using ^{1}H NMR spectroscopy, we have also observed structural differences in the heme pockets between HbO_2 A and HbCO A (Lindstrom *et al.*, 1972b; Lindstrom and Ho, 1973; Dalvit and Ho, 1985). Thus, one can conclude that there are structural differences between the HbO_2 A and HbCO A heme pockets in both the solution state and in single crystals. These structural differences could be the cause for functional differences between these two ligated Hbs.

It should be emphasized that the overall structural features of Hb O_2 A and HbCO A are quite similar, i.e., they both have an R-type quaternary structure. Thus, HbCO A can be used as a good model for investigating the cooperative oxygenation of Hb A. However, one should be aware that there are differences in details between HbO_2 A and HbCO A.

C. *Binding of O_2 and CO to Hemoglobin*

For an evaluation of the extent of α–β nonequivalence that must be taken into account to describe the cooperative ligation of hemoglobin, the probe selected must allow a direct observation of α and β chains of an intact tetrameric Hb molecule at different stages of ligation. The ferrous hfs proton resonances of deoxy-Hb A in the spectral region from +6 to +20 ppm from HDO offer a method to investigate this important problem. By comparing the hfs proton resonances of deoxy-Hb A and appropriate mutant Hbs, we have assigned these hfs resonances to the α and β chains of deoxy-Hb A (see Fig. 16). When deoxy-Hb A combines with O_2 or CO, it becomes diamagnetic, causing these hfs proton resonances to disappear and making this spectral region flat. Thus, monitoring the intensity of the hfs resonances as a function of ligation, we can monitor the binding of a ligand to the α and β chains of Hb A. Our early work was based on measuring the ratio of the area of the resonance at +12 ppm (due to the α chain) to the area of the resonance at +18 ppm (due to the β chain). We found an essentially constant ratio when Hb A combines with CO in the presence and absence of organic phosphates (Johnson and Ho, 1974). We have concluded that there is no preference for the binding of CO to the α and β chains of Hb A. This conclusion is supported by the recent results of Sharma *et al.* (1991). This area ratio also remains constant on oxygenation of Hb A in the absence of organic phosphate, but decreases at partial O_2 saturations in the presence of 2,3-DPG and even more so in the presence of IHP (Johnson and Ho, 1974). Based on these experimental results, we have concluded that (1) there is no preference of the α and β chains of deoxy-Hb A for the binding of CO, (2) in the absence of organic phosphate, the α and β chains have similar affinities for O_2, and (3) in the presence of 2,3-DPG or IHP, the α chains have a higher affinity for O_2 than the β chains, and the difference is enhanced in the presence of IHP.

Our conclusion regarding the preferential binding of O_2 to the α chains of Hb A was challenged by Shulman *et al.* (1975) and Huang and Redfield (1976). Because of the importance of a knowledge of the functional equivalence or nonequivalence of the α and β chains of Hb A and of the controversial nature of the subject, we extended our earlier ^{1}H NMR studies on the binding of O_2 to Hb A by three different approaches.

First, we made major technical improvements. We obtained a satisfactory baseline over the spectral region from +6 to +30 ppm from HDO, used a better method for preparing partially oxygenated Hb samples, and also made improvements in the analysis of our data (Viggiano *et al.*, 1979). The experiments shown in Figs. 22 and 23 were designed to check whether there is a true decrease in the areas of the α-heme resonance at +12 ppm and the β-heme resonance at +18 ppm on oxygenation in the absence and presence of IHP, or if there is a baseline problem (by superimposition of overlapping resonances), or if there is a significant line broadening on oxygenation. ^{1}H NMR spectra were recorded on a sample of Hb A, first on a fully deoxygenated sample (Fig. 22A), then on the sample oxygenated to 23.5% (Fig. 22B) and 31.6% (Figs. 22C and 23A) in the absence of organic phosphate, then on the sample with IHP added to the same Hb solution (Fig. 23B), and finally on the sample with the O_2 saturation increased to 48% in the presence of 11 mM IHP (Fig. 23C). The spectra are prescaled to give the same maximum height of the β peak at +18 ppm. Without organic phosphates, the 18-ppm and 12-ppm peaks for the fully deoxy-Hb A and Hb at 23.5 and 31.6% O_2 saturations (Figs. 22B and 22C) are essentially identical. These results clearly show that there is no evidence for any significant line broadening (<3.3 Hz/1% O_2 saturation for oxygenation-dependent line broadening) in these spectra. Addition of 11 mM IHP at 31.6%

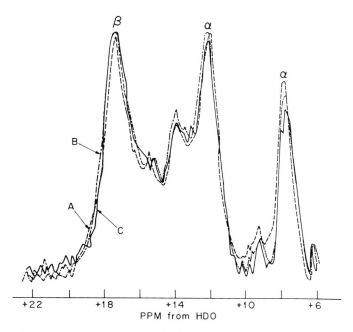

FIG. 22. Effects of oxygenation on the linewidth of hyperfine-shifted proton resonances of ~12% Hb A in 0.1 M Bis–Tris in D_2O at pH 6.6 and 27°C. (A) Fully deoxy sample (----); (B) 23.5% oxygenation (–·–); (C) 31.6% oxygenation (—). The intensities of spectra B and C have been prescaled to give the same intensity as spectrum A. [From Viggiano et al. (1979)].

O_2 saturation (Fig. 23B) causes a shift of the β peak by about 0.6 ppm downfield (as first reported by Ho et al., 1973) and an unquestionable decrease in the relative intensity of the +12-ppm α peak. Increasing the O_2 saturation to 48% in the presence of IHP further decreases the intensity of the +12-ppm peak (Fig. 23C).

Figure 24 shows a representative series of the hfs proton resonances of Hb A as a function of oxygenation in D_2O at neutral pH in the absence and presence of organic phosphates. These spectra are in good agreement with the description of Johnson and Ho (1974) for the spectral region from +10 to +20 ppm from HDO. The much improved baseline in the spectra shown in Fig. 24 clears up some of the problems associated with the spectral region from +6 to +10 ppm. The phenomenon of a higher O_2 affinity of the α chains as compared to the β chains in the presence of IHP first reported by

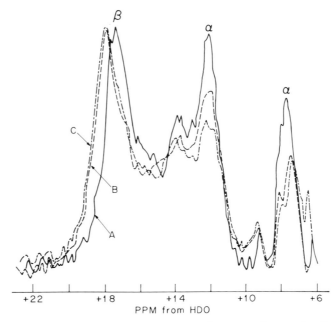

FIG. 23. Effects of inositol hexaphosphate and oxygenation on the linewidth of the hyperfine-shifted proton resonances of ~12% Hb A in 0.1 M Bis–Tris in D_2O at pH 6.6 and 27°C. (A) 31.6% oxygenation (—) (same as spectrum C of Fig. 22); (B) 31.6% oxygenation in the presence of 11mM IHP (----); (C) 48% oxygenation in the presence of 11 mM IHP (–·–). The intensities of the 18-ppm resonance in spectra B and C have been prescaled to give the same intensity as that of the 18-ppm peak in spectrum A. [From Viggiano et al. (1979)].

us (Ho and Lindstrom, 1972; Lindstrom and Ho, 1972) can readily be seen by looking at the intensities of the peaks at +18 ppm (β chain) and +12 ppm (α chain), as shown in Fig. 24E. The flatness of the baseline strongly suggests that there is a real change in the area ratio of these resonances on oxygenation, not a change in the linewidth associated with a kinetic effect. A deeper insight into the behavior of the hfs proton resonances of Hb A on oxygenation can be obtained from the difference spectra shown in Fig. 24B (without organic phosphate), Fig. 24D (in the presence of 2,3-DPG), and Fig. 24F (in the presence of IHP). These difference spectra are representative of the resonances lost on oxygenation so they are flat at 0% oxygenation and give the deoxy spectrum at 100% oxygenation. Thus, the intensities of the resonances in the difference spectra give the corresponding oxygenated species. The three prominent reso-

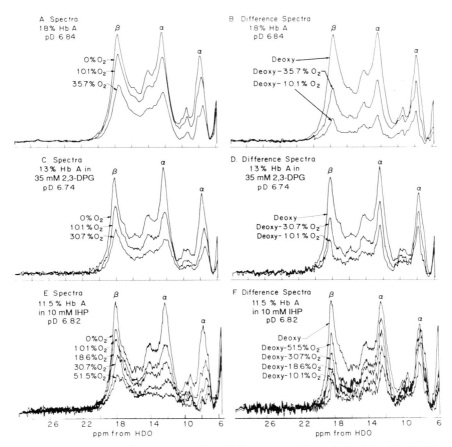

FIG. 24. 250-MHz ^{1}H NMR studies of the oxygenation of Hb A in 0.1 M Bis–Tris in D_2O at 27°C. (A) 18% Hb A at pH 6.44; (B) difference spectra of 18% Hb A at pH 6.44; (C) 13% Hb A in 35 mM 2,3-DPG at pH 6.34; (D) the difference spectrum of 13% Hb A in 35 mM 2,3-DPG at pH 6.34; (E) 11.5% Hb A in 10 mM IHP at pH 6.42; (F) difference spectra of 11.5% Hb A in 10 mM IHP at pH 6.42. [Adapted from Viggiano et al. (1979)].

nances ($\sim +18$, $\sim +12$, and $\sim +8$ ppm) are better resolved and sharper in the difference spectra. There is no observable line broadening at low O_2 saturations.

Second, in order to calibrate the intensity of each resonance over the spectral region from $+10$ to $+20$ ppm from HDO to the number of protons per heme for each resonance, a reference standard is required. We have found that an NMR shift reagent, tris(6,6,7,7,8,8,8-

heptafluoro-2,2-dimethyl-3,5-octanedionato) europium, complexed with the OH of tert-butanol, gives a single proton resonance at ~+27 ppm downfield from HDO (Ho et al., 1982b). There is no other observable proton signal from +10 to +30 ppm from HDO. Figure 25 shows the 250-MHz ^{1}H NMR spectra of deoxy- and oxy-Hb A in 0.1 M Bis–Tris plus 10 mM IHP at pH ~6.6 and 27°C with the external intensity standard over the spectral region from +10 to +30 ppm from HDO. By using this intensity standard, we have calculated the number of protons in each of the Hb resonances between +10 and +27 ppm from HDO. The total number of protons per heme over this spectral region is 22, with an accuracy of approximately ±10% (Ho et al., 1982b). A computer program based on the algorithm formulated by Marquandt (1963) for the least-squares estimation of nonlinear parameters was used to simulate the ferrous hfs proton resonances of Hb A (for details, see Ho et al.,

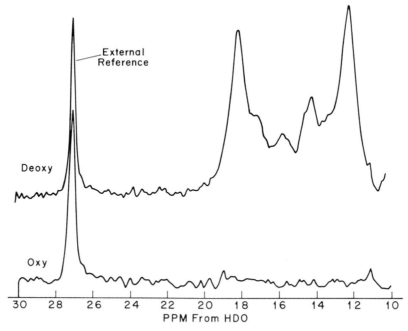

FIG. 25. 250-MHz ^{1}H NMR spectra of Hb A in both deoxy and oxy forms in 0.1 M Bis-Tris plus 10 mM IHP in D_2O at pH ~6.6 at 27°C. The resonance at ~27 ppm was that of the NMR shift reagent. This reference signal was used for the intensity calibration of the ferrous hyperfine-shifted proton resonances. [From Ho et al. (1982b)].

1982b). The parameters obtained from the least-squares program were then used to generate a computer-simulated spectrum of deoxy-Hb A over the spectral region from +10 to +30 ppm from HDO as shown in Fig. 26B. Figure 26C is the difference spectrum obtained by subtracting Fig. 26A (experimental spectrum) from Fig. 26B (the simulated spectrum). This difference spectrum is only slightly above the noise level of the experimental spectrum, suggesting that the spectral parameters obtained by the fitting routine explain the experimentally observed spectrum reasonably well. The overlapping peaks in the experimental ^{1}H NMR spectra are resolved using the computer-generated spectra. Figure 27 gives a series of ^{1}H NMR as well as computer-simulated spectra of Hb A in 0.1 M Bis–Tris plus 10 mM IHP as a function of O_2 saturations at pH ~6.6 and 27°C. Figure 27A shows the experimental ^{1}H NMR data and Fig. 27B gives the corresponding least-squares simulated spectra. [Figure 27C is taken from Fig. 4E of Viggiano *et al.* (1979) for comparison.] The shoulder resonance at +16.9 ppm has been shown to be a measure

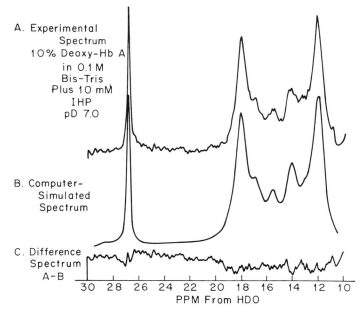

FIG. 26. 250-MHz ^{1}H NMR spectra of deoxy-Hb A at 27°C. (A) Experimental spectrum; (B) computer-simulated spectrum; (C) the difference spectrum between A and B. [From Ho *et al.* (1982b)].

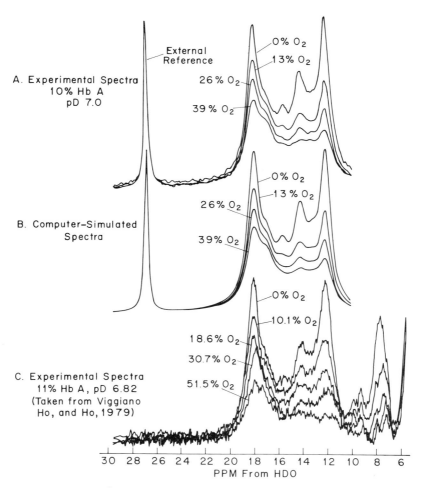

FIG. 27. 250-MHz ^{1}H NMR spectra of Hb A in 0.1 M Bis–Tris plus 10 mM IHP in D$_2$O as a function of oxygenation at 27°C. (A) Experimental spectra of 10% Hb A at pH 6.6; (B) computer-simulated spectra; (C) experimental spectra of 11% Hb A at pH 6.42. [From Ho et al. (1982b)].

of the amount of oxygenation in the β chain. This resonance is not affected by structural changes of the Hb molecule on oxygenation, unlike the +12-ppm α-chain and +18-ppm β-chain resonances (Viggiano et al., 1979; Ho et al., 1982b). There is also excellent agreement among several independent sets of experimental results on the variation of the ferrous hfs proton resonances as a function of oxygen-

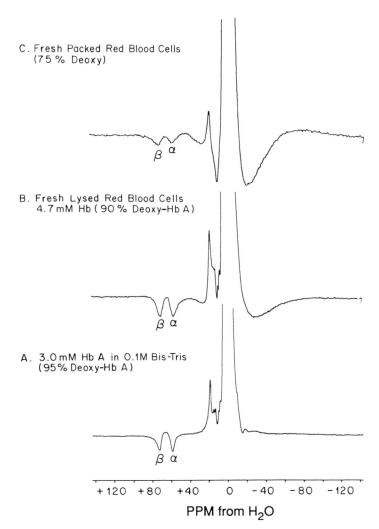

FIG. 28. 300-MHz hyperfine-shifted exchangeable proton resonances of proximal histidyl residues of the α and β chains of deoxy-Hb A. (A) 3.0 mM Hb A in 0.1 M Bis–Tris (95% deoxy-Hb A); (B) freshly lysed red blood cells, 4.7 mM Hb A (90% deoxy-Hb A); (C) freshly packed red blood cells (75% deoxy-Hb A). The broad dips in spectra B and C are due to membranes and other cell components. [From Yao *et al.* (1986)].

ation of Hb A (Ho et al., 1982b). The computer-simulated spectra show no significant line broadening for all observable hfs proton resonances on oxygenation (no more than 4 Hz/1% O_2 saturation). Thus, the experimental results have confirmed one of the major conclusions reached by our laboratory from ^{1}H NMR studies of Hb A (Ho and Lindstrom, 1972; Lindstrom and Ho, 1972; Johnson and Ho, 1974; Viggiano and Ho, 1979; Viggiano et al., 1979).

Third, by using the soft $1\overline{2}1$ pulse sequence to suppress the H_2O signal, we have investigated the binding of O_2 to Hb A in H_2O and in red blood cells by monitoring the intensities of the two hyperfine-shifted exchangeable proton resonances at about $+60$ and $+70$ ppm from H_2O (Yao et al., 1986). As shown in Fig. 17, the resonance at $\sim +60$ ppm has been assigned to the $N_\delta H$ of the proximal histidyl residue of the α chain and the one at $\sim +70$ ppm to that of the proximal histidyl residue of the β chain of deoxy-Hb A. Figure 28 shows the 300-MHz ^{1}H NMR spectra of deoxy-Hb A in Bis–Tris solution, in freshly lysed cells, and in packed fresh red blood cells. There is a distinct advantage in using the spectral region between

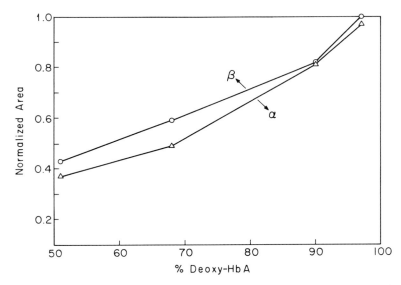

FIG. 29. Binding of oxygen to the α and β chains of Hb A in fresh hemolysate (4.5 mM Hb A, pH 6.6) as determined by the hyperfine-shifted exchangeable proton resonances, $N_\delta H$, of proximal histidyl residues of the α and β chains of Hb A. [From Yao et al. (1986)].

+60 and +70 ppm from H_2O, i.e., these two resonances are far removed from the main protein signals as well as being relatively well resolved with respect to each other, making possible precise measurements of the O_2 binding properties of the Hb A α and β chains. As shown in Fig. 29, there is a clear indication that the Hb A α chain binds O_2 more strongly than the β chain, consistent with the results obtained by monitoring the ferrous hfs proton resonances of deoxy-Hb A in the spectral range of +10 to +20 ppm from HDO as a function of oxygenation.

The results discussed here have confirmed our earlier conclusions that (1) in the absence of organic phosphate (i.e., in 0.1 M Bis–Tris buffer), there is no preferential O_2 binding to the α or β chains of Hb A, and (2) in the presence of organic phosphate, the α chains have a higher affinity for O_2 than do the β chains.

D. Binding of Alkyl Isocyanides to Hemoglobin

Alkyl isocyanides have been used quite extensively to probe the stereochemistry of the heme pockets of the α and β chains of Hb A, because the size of the alkyl substituents can be varied in a systematic manner [for details, refer to Antonini and Brunori (1971) and Olson and co-workers (Mims *et al.*, 1983b)].

As mentioned in Section II, the magnitude of the ring-current shift is extremely sensitive to the geometry of the proton with respect to the heme ring. Thus, changes in the position of the methyl resonances of the E11Val should reflect amino acid movement at the ligand-binding site (see Fig. 18). In collaboration with Olson and co-workers, we have carried out a systematic ^1H NMR investigation of isonitrile binding to Hb A, the isolated α and β chains of Hb A, and sperm whale myoglobin (Mb), which is homologous to a single Hb chain. We have also attempted to correlate the shifts of the E11Val resonances with the observed free energies of ligand binding. In addition, the magnitudes of the ring-current shifts of the isonitrile protons have allowed us to estimate the positions of the atoms in the bound-ligand molecule. From these data, three-dimensional structural models of the packing at the sixth coordination positions in the α and β subunits of Hb A and in Mb were constructed. By varying the size of the alkyl groups of isocyanide, we can also gain some insights into the packing and flexibility of the distal heme pockets of the α and β chains of Hb A (for details, see Mims *et al.*, 1983a).

Figure 30 gives the 600-MHz ring-current-shifted ^1H NMR spectra of isolated α and β chains, Hb A, and Mb complexed with CO and

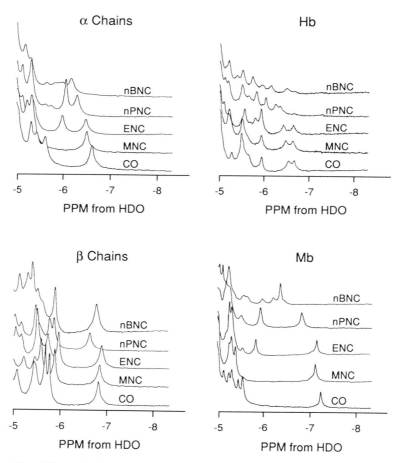

FIG. 30. Effects of alkyl isocyanides (*n*-series) on the 600-MHz ring-current-shifted proton resonances of isolated α and β chains of Hb A, Hb A, and sperm whale myoglobin (Mb) in 0.2 M phosphate in D_2O at pH 6.6 and 21°C: CO, carbon monoxide; MNC, methyl isocyanide; ENC, ethyl isocyanide; nPNC, *n*-propyl isocyanide; nBNC, *n*-butyl isocyanide. [Adapted from Mims *et al.* (1983a)].

with the n-series of alkyl isocyanides, consisting of methyl, ethyl, *n*-propyl, and *n*-butyl isocyanides. The positions and ring-current shifts of the furthest upfield peak, assigned to the γ_2-CH_3 protons of E11Val (Lindstrom *et al.*, 1972b; Dalvit and Ho, 1985), are given in Fig. 31A. The valine methyl (E11) resonance for α chains is systematically shifted to lower field as the ligand length is increased, as shown in Figs. 30 and 31A. The shift in going from bound CO to ethyl isocyanide is

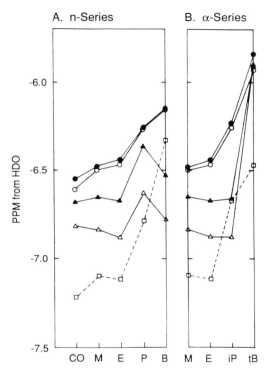

FIG. 31. A comparison of the ^{1}H chemical shifts for the E11 valine γ_2-CH$_3$ protons of the various isonitrile complexes of isolated α chains (○) and β chains (△) of Hb A, the α (●) and β (▲) chains as identified within an intact Hb A molecule, and sperm whale myoglobin (□). Ligands: CO, carbon monoxide; M, methyl isocyanide; E, ethyl isocyanide; P, n-propyl isocyanide; B, n-butyl isocyanide; iP, isopropyl isocyanide; and tB, tert-butyl isocyanide. [From Mims et al. (1983a)].

small, −6.61 to −6.47 ppm from HDO; much larger changes have been observed in going from ethyl to n-butyl isocyanide, −6.47 to −6.16 ppm from HDO. This downfield shift indicates a movement of the E11Val away from the heme center as the isonitrile becomes longer. In the case of β chains, there is no consistent shift of the valine peak to lower field (Fig. 31A). The two upfield peaks of the Hb A spectra exhibit shifts with increasing ligand size that are analogous to those for the isolated chains (Fig. 31A). The highest field shift for Mb has also been assumed to be due to the E11Val residue, which is located over the heme ring in the deoxy form (Takano, 1977). The changes in the Mb spectra are similar to those of the α chains. Small downfield shifts of the valine resonance are observed in going from CO to ethyl isocyanide and much larger shifts are seen in going from ethyl to *n*-butyl isocyanide (Fig. 31A).

Ring-current-shifted ^{1}H NMR spectra for these heme proteins containing a series of C-1-substituted (or α-series) ligands, namely, methyl, ethyl, isopropyl, and *tert*-butyl isocyanide, are given in Fig. 32. The α-chain spectra show a marked downfield shift of the valine peak as methyl groups are added to ethyl isocyanide (Fig. 31B). For the *tert*-butyl isocyanide complex, the valine resonance is indistinguishable

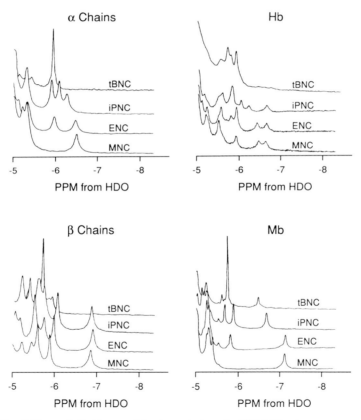

FIG. 32. Effects of alkyl isocyanides (α-series) on the 600-MHz ring-current-shifted proton resonances of isolated α chains and β chains of Hb A, Hb A, and sperm whale myoglobin (Mb) in 0.2 M phosphate in D_2O at pH 6.6 and 21°C: MNC, methyl isocyanide; ENC, ethyl isocyanide; iPNC, isopropyl isocyanide; and tBNC, *tert*-butyl isocyanide. [Adapted from Mims *et al.* (1983a)].

from the large unidentified peak at -5.9 ppm from HDO. In the α chains then, both the substituted and the longer straight-chain ligands seem to push the E11Val away from the heme ring. The situation in the β chains is somewhat different. Increasing substitution of the isonitrile has little effect on the position of the βE11Val resonance until *tert*-butyl isocyanide is bound. The valine peak in the *tert*-butyl spectrum is also probably part of the resonance at -5.9 ppm, which is significantly downfield from the valine resonance for the β-chain isopropyl isocyanide complex (Figs. 31B and 32). Thus, the β-heme pocket seems to be able to accommodate all but the most substituted isonitriles without movement of the E11 valine methyl groups. Similar shifts in the valine peaks are seen, although less clearly, in the Hb A spectra. Myoglobin again shows a pattern akin to that of the α chains. The valine resonances in the methyl and ethyl isocyanide spectra are in roughly the same position, whereas those seen in the isopropyl and *tert*-butyl spectra are shifted markedly to lower fields (Fig. 31B).

The ring-current-shifted ^{1}H NMR spectra for the isolated α and β chains do not sum perfectly to give a Hb A spectrum. As shown in Fig. 31, the E11Val γ_2-CH$_3$ resonance for α chains is the same regardless of whether the α chain is isolated or incorporated into Hb A. In contrast, the isolated β-chain valine resonance is shifted about 0.14 ppm upfield from the position observed for this subunit in intact Hb A. The isolated β chains at the concentration used for ^{1}H NMR studies exist as a tetramer and also exhibit ligand-binding properties that are measurably different from either the monomeric form or the subunit within the liganded R state of Hb (Valdes and Ackers, 1987a,b). Evidently, the formation of the β_4 tetramer can cause a change in the conformation of the E11Val groups. However, the dependencies of the valine shifts on ligand length and substitution are the same for β tetramers or β chains within Hb A. Thus, both isolated subunits and those within Hb A appear to react similarly in accommodating ligand molecules.

As shown in Figs. 30 and 32, a number of new resonances appear in the ring-current-shifted region of the ^{1}H NMR spectra when ligands containing two or more alkyl carbon atoms are bound to the heme proteins. In order to ascertain whether these extra resonances arise from the ligand molecules or the amino acid residues in the heme pocket, we have synthesized three perdeuterated ligands: ethyl, isopropyl, and *n*-butyl isonitrile. Because the deuterium atom is not visible in ^{1}H NMR spectra, it should be obvious from the spectra of protein containing bound deuterated isonitrile if the extra peaks seen in the protonated complexes are due to ligand hydrogen atoms. These results show unambiguously that the additional resonances are indeed due to the isonitrile protons and not to the amino acid residues (see Mims *et al.*, 1983a). The magnitudes of the ring-current shifts of the alkyl protons on isonitriles bound to isolated α and β chains and myoglobin were calculated. The results suggest a linear geometry for the Fe=C=N—C bonds in β chains and a bent geometry for α chains.

Correlation between NMR Spectral Results and Equilibrium Binding Parameters

The ^{1}H NMR studies of isonitrile–heme protein complexes described above and the equilibrium constants for the reactions of isonitriles with Hb A and isolated chains reported by Reisberg and Olson (1980a,b,c) as well as a similar series of experiments with Mb reported by Stetzkowski *et al.* (1979) offer a unique opportunity to correlate the structural results obtained by NMR and the thermodynamic studies. Comparison of these free energy parameters with the ring-current shift data has led to three important conclusions (Mims *et al.*, 1983a). First, the free energies of the bound state, G_{PX}, for the α and β chains are found to be nearly equal to each other for the methyl, ethyl, isopropyl, and *tert*-butyl isocyanide complexes. Thus, the bent and linear geometries are energetically quite similar. This points out the difficulty in predicting affinity changes on the basis of slight alterations in the structural parameters of the bound-ligand molecule. One might predict *a priori* that the α-chain ethyl isocyanide complex would be less stable due to bending of the Fe=C=N—C-1 bonds; however, the experimental data suggest that the α-chain complex is as stable as or slightly more stable than the β-subunit complex. Second, the bond potentials for methyl and ethyl isocyanides are found to be approximately equal to each other, regardless of which protein is investigated. This shows that there is little or no steric resistance to placing the C-2 atom into the ligand-binding sites of Mb and Hb A. The fact that the ring-current shift of the E11Val methyl protons is approximately the same for the CO, methyl, and ethyl isocyanide complexes suggests that this residue also offers little resistance to the addition of the C-1 atom, in agreement with the geometries predicted by the ring-current calculations (Mims *et al.*, 1983a). Third, G_{PX} is found to increase roughly 1.5 kcal/mol/methyl group in going from ethyl to *n*-butyl isocyanide. Thus, the heme pocket structure where the C-3 atom is placed would be rigid and not easily displaced. This steric hindrance accounts for the decrease in affinity observed for all three proteins in going from ethyl to *n*-butyl isocyanide (Reisberg and Olson, 1980a,b; Olson et al., 1982). In the isolated α chain and Mb, this increase in G_{PX} in going from ethyl to n-butyl isocyanide appears to be directly correlated with movements of the valine E11 methyl group away from the heme center (Fig. 31A). We have interpreted this in terms of a conformation in which the C-3 and C-4 atoms of the longer isonitriles can curve around the ligand-binding site, forcing the displacement of the E11 valine residue. Examination of the published X-ray structures of these proteins suggests that the C-3 and C-4 atoms could not project up and away from the heme center without interacting directly with the distal His(E7) residue (Mims et al., 1983a). Thus, this residue appears to restrict the binding of the longer isonitriles, causing them to fold back and displace the valine methyl group. The situation for the isolated β chain is less clear. The free energy dependence on ligand size and shape appears to be the same as that observed for the isolated α chain and Mb, but the NMR results are not straightforward enough to allow a specific structural interpretation (Mims et al., 1983a).

An important conclusion from these studies is that bulky side chains of alkyl isocyanides can be incorporated into the distal sides of the heme pockets of both the α and β chains of Hb A in spite of the findings that the heme pockets are tightly packed according to the X-ray crystallographic structural results of deoxy-Hb A, HbO_2 A, and HbCO A (Perutz, 1970; Shaanan, 1983; Fermi et al., 1984; Liddington et al., 1988; Waller and Liddington, 1990). This implies that the heme pockets are quite flexible and can accommodate larger ligands.

An early example of using ^1H NMR to investigate preferential ligand binding to the α and β chains of Hb A is given in Fig. 33 (Lindstrom et al., 1971). In the presence of 2 mM IHP plus 0.05 mM Bis–Tris in D_2O at pH 6.7 and 26°C, deoxy-Hb A shows three prominent ferrous hfs proton resonances positioned at +17.8, +12.1, and +7.4 ppm from HDO (Fig. 33). The resonance at +17.8 ppm has been assigned to the β chain and the resonances at +12.1 and +7.4 ppm, to the α chain (see Fig. 16). When n-butyl isocyanide (BIC) is added to deoxy-Hb A in the presence of IHP, the decrease in intensity is strikingly unequal for the peaks at +17.8 and +12.1 ppm (Fig. 33). At a BIC/heme ratio of 1.2, the peak at +17.8 ppm is clearly less intense than the peak at +12.1 ppm, and when the ratio reaches 1.9, the peak at +17.8 ppm is virtually nonexistent whereas the resonances at +12.1 and +7.4 ppm are still clearly visible. The results indicate that BIC binds to the β heme in preference to the α heme of Hb A in the presence of IHP (Lindstrom et al., 1971).

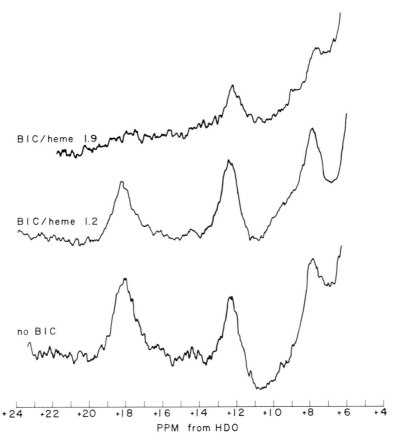

FIG. 33. 90-MHz ^{1}H NMR spectra of 6.25 mM deoxy-Hb A in 0.1 M Bis–Tris containing 2 mM inositol hexaphosphate in D_2O at pH 6.7 and 26°C as a function of n-butyl isocyanide (BIC) concentration. [Adapted from Lindstrom *et al.* (1971)].

E. Binding of Azide and Cyanide to Methemoglobin

Both cyanide and azide react with met-Hb to form cyanomet-Hb (met-HbCN) and azidomet-Hb (met-HbN$_3$), respectively (see Antonini and Brunori, 1971). These reactions have been studied extensively in both equilibrium and kinetic studies. There are conflicting reports concerning the value of the Hill coefficient for these two reactions (n = 1 to 2) (Antonini and Brunori, 1971).

For a historical perspective, the first demonstration that the environment of the heme pocket in the α and β chains of Hb A is not equivalent came from a comparison of the four hfs proton resonances of Hb A, Hb F ($\alpha_2\gamma_2$), and horse Hb in the cyanomet form (Davis et al., 1969a). Figure 34 shows the 100-MHz ^{1}H NMR spectra of the low-spin cyanomet form of Hb A, Hb F, and horse Hb in 0.1 M phosphate in D_2O at pH 6.6 at 27°C over the spectral range from +7.0 to +20.0 ppm from HDO. The resonances at +16 to +18 ppm downfield from HDO have been assigned to some of the methyl groups on the porphyrin in Hb (Wüthrich et al., 1968; Davis et al., 1969a). The fact that the methyl resonance of Hb F and horse Hb

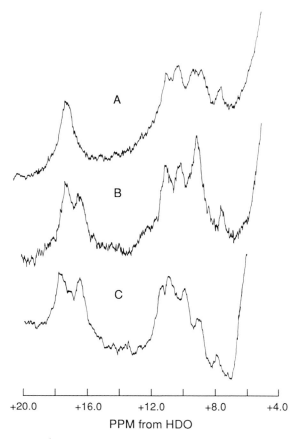

FIG. 34. 100-MHz ^{1}H NMR spectra of cyanomethemoglobins in 0.1 M phosphate in D_2O at pH 6.6 and 27°C. (A) Hb A; (B) Hb F ($\alpha_2\gamma_2$); (C) horse Hb. The frequency scale is referenced to HDO. [Adapted from Davis et al. (1969a)].

is split whereas that of Hb A is not split suggests that there are differences in the interaction between the heme groups and the surrounding amino acid residues among these three Hbs. This finding is consistent with the X-ray structural analysis of horse met-Hb at 2.8 Å reported by Perutz *et al.* (1968). They have found that amino acid residues 70 (serine) and 71 (phenylalanine) in the β chain of horse Hb are within 4 Å of the two methyl groups in pyrrole rings IV and I, respectively. According to the amino acid sequences of human β and γ chains (Braunitzer *et al.*, 1964) and the atomic model of Hb proposed by Perutz *et al.* (1968), the amino acid residues near the heme, which are replaced in human β chain and γ chain compared to those of the horse β chain, are alanine (human β chain, amino acid residue 70) and leucine (human γ chain, amino acid residue 71). According to the amino acid sequences of human α chain and horse α chain and the atomic model of Hb, there is no amino acid replacement in the α chain which is within 4 Å of the heme group. Thus, the splitting of the methyl resonance at +17.0 ppm for Hb F and horse Hb suggests that the serine oxygen in amino acid residue 70 of the β chain could be in contact with the methyl group of the pyrrole ring IV in these two proteins, thus shifting the resonance of the methyl group of the β chain.

There are also differences in the ^{1}H NMR spectra among these three proteins over the spectral range from +8.00 to +12.00 ppm from HDO as shown in Fig. 34. In particular, the resonances in the region of about +9.0 ppm from HDO in Hb F are quite different from those in horse Hb and Hb A. Our NMR results suggest that the interactions between the leucine at residue 71 and the methyl group of the pyrrole ring I in the γ chain are different from those of phenylalanine at amino acid residue 71 with corresponding methyl groups in both horse and human β chains. The results shown in Fig. 34 suggest that the amino acid residues in the vicinity of the heme group could alter the electronic distribution of this group and, thus, could give rise to differences in the NMR spectra of these proteins. Hence, the hfs resonances of Hbs are sensitive indicators of the detailed conformation of the heme pockets in Hb.

In a related study, we have investigated the hfs proton resonances of Hb A, Hb F, Hb Zürich (β63His → Arg), and Mb, all in the azidomet form in 0.1 M phosphate in D_2O at pH 6.9 and 31°C, as shown in Fig. 35 (Davis *et al.*, 1969b). The prominent lines from +15 to +23 ppm downfield from HDO in these spectra have been assigned to some of the methyls of the porphyrin on the basis of relative line intensities and comparison with the NMR spectra of met-MbN_3

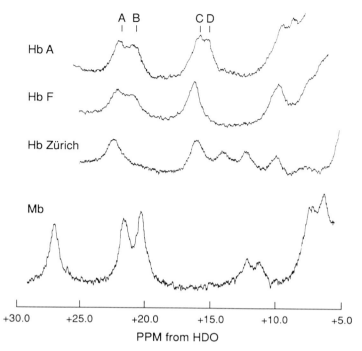

FIG. 35. 100-MHz ^{1}H NMR spectra of azidomethemoglobins Hb A, Hb F ($\alpha_2\gamma_2$), and Hb Zürich (β63His → Arg), and azidometmyoglobin in 0.1 M phosphate in D_2O at pH 6.6 and 31°C. [Adapted from Davis et al. (1969b)].

and met-HbCN (Wüthrich et al., 1968; Davis et al., 1969b). In the spectrum of met-HbN$_3$ A, the relative intensities of the lines at +22.4, +21.3, +16.5, and +15.7 (labeled A, B, C, and D, respectively) are approximately 1:1:1:1. For met-HbN$_3$ F, the relative intensities of the lines at A, B, and C are approximately 1:1:2. The ^{1}H NMR spectrum of horse met-HbN$_3$ is identical to that of met-HbN$_3$ F in this spectral region. In the case of met-HbN$_3$ Zürich [β63(E7)His → Arg], the relative intensities of the peaks at A and C are about 1:1 and the B and D peaks are missing. The environment of the azide–heme complex in the abnormal β chain is altered by the substitution of arginine for histidine in the β63, but the α-heme environment remains unaffected. Thus, the resonances at +22.4 and +16.5 ppm from HDO have been assigned to the α-heme protons and those at +21.3 and +15.8 ppm to the β-heme protons of met-HbN$_3$ A.

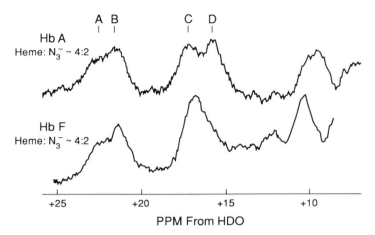

FIG. 36. 100-MHz ^{1}H NMR spectra of human normal adult and fetal methemoglobins in which the molar ratio of heme to azide is approximately 4:2 in 0.1 M phosphate in D_2O at pH 6.6 and 31°C. [Adapted from Davis et al. (1969b)].

The first evidence for the functional nonequivalence of α and β chains was derived from our ^{1}H NMR studies on the binding of N_3^- to met-Hb A (Fig. 36). This azide-ion titration experiment, which shows that peaks A and C decrease in intensity relative to peaks B and D, indicates that A and C arise from a heme group whose binding with N_3^- is less than the heme group associated with B and D. Thus, this conclusion suggests that the affinity of the β chain for N_3^- is higher than that of the α chain of met-Hb A (Davis et al., 1969b).

V. ^{1}H NMR Investigations of Partially Oxygenated Species of Hemoglobin: Evidence for Nonconcerted Structural Changes during the Oxygenation Process

To monitor the structural changes associated with the cooperative oxygenation process for Hb A, we depend on measuring the intensities of two types of proton resonances, i.e., the three ferrous hfs proton resonances at about $+12$, $+16.9$, and $+18$ ppm and the two exchangeable proton resonances at about $+6.4$ and $+9.4$ ppm from H_2O. By monitoring the intensities of these resonances as a function of oxygenation, the amount of O_2 bound to the α and β chains of the Hb molecule can be determined and the relationship between tertiary and quaternary structural changes under a given set of experimental conditions can be investigated. In this section, we first discuss the NMR results on mutant Hbs that indicate that these resonances are good monitors of the cooperative oxygenation process, and then the results on the variation of the intensities of these resonances as a function of oxygenation as well as their implication for the molecular mechanism of the cooperative oxygenation of Hb A.

Figure 37 shows that the hfs proton resonance pattern of deoxy-Hb A is very different from those of isolated deoxy α and β chains, and the resonance patterns of isolated α and β chains are also different from each other (see Perutz et al., 1974). The spectra of the isolated deoxy α and β chains of Hb A do not show resonances at +18 or +12 ppm from H_2O, but on mixing of stoichiometric amounts of α and β chains, the ^1H NMR spectrum of the mixture is identical to that of Hb A (see Perutz et al., 1974). These results suggest that these resonances are very sensitive to the detailed electronic structure of the heme group in each protein and/or the detailed conformation or environment of the heme group.

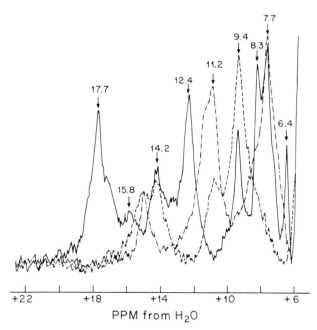

FIG. 37. 250-MHz ^1H NMR spectra of Hb A and isolated α and β chains of Hb A in the deoxy form in 0.1 M Bis–Tris at pH 7.0 and 27°C: —, Hb A; –·–, isolated α; and ----, isolated β chains. [C. Ho, unpublished results, similar results are shown in Perutz et al. (1974)].

We have found that the β peak at ~+18 ppm is missing (or is broadened) in those high-affinity Hb mutants [such as Hb Kempsey (β99Asp → Asn), Hb Philly (β35Tyr → Phe), Hb Radcliff (β99Asp → Ala), Hb Osler (β145Tyr → Asp), and Hb McKees Rocks (β145Tyr → Term)] that do not show cooperativity in their binding of O_2. Cooperativity can be restored on adding IHP (Perutz et al., 1974; Ho et al., 1975; Asakura et al., 1976; Weatherall et al., 1977; Viggiano et al., 1978), and under such conditions, the hfs proton resonance pattern is returned to that of deoxy-Hb A. Figure 38 gives an illustration showing the effects of IHP on the ferrous hfs proton resonances of a noncooperative Hb, deoxy-Hb Kempsey (β99Asp → Asn). The spectral pattern for the hfs proton resonances of deoxy-Hb Kempsey in the absence of IHP resembles that of the isolated deoxy α or β chain more than that of deoxy-Hb A (Fig. 37). In fact, crystals of deoxy-Hb Kempsey have been found to have a quaternary structure similar to that of HbO_2 A (Perutz et al., 1974). On the addition of IHP, the ferrous hfs proton resonances of deoxy-Hb Kempsey are

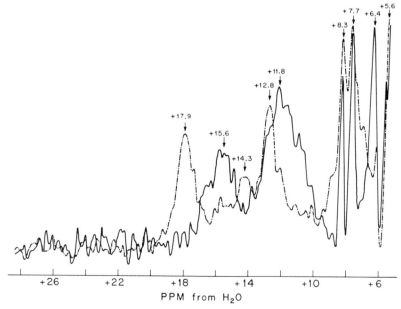

FIG. 38. 250-MHz ^{1}H NMR spectra of deoxy-Hb Kempsey in 0.1 M Bis–Tris at pH 7.09 and 27°C: —, Hb Kempsey; and –·–, Hb Kempsey + 15 mM IHP. [C. Ho, unpublished results; similar results are shown in Ho et al. (1975)].

converted to a pattern similar to that of deoxy-Hb A (Fig. 38). The Hill coefficient (n) for Hb Kempsey is ~1.1 and is increased to ~1.6 in the presence of IHP (Lindstrom et al., 1973; Bunn and Forget, 1986). Thus, the ferrous hfs proton resonances appear to be sensitive to the cooperative transition in Hb. For another example, the spectrum of deoxy-Hb A is quite insensitive to pH changes, but the spectrum of des-Arg(α141)-Hb A loses both the +18-ppm β peak and +12-ppm α peak in going from pH 7 to 9 as the mixture of T and R states is shifted toward the R state (Perutz et al., 1974). These results strongly suggest that the hfs proton resonances of Hb A are good monitors of the cooperative oxygenation process and the observed spectral changes are energetically significant.

Finally, there is another piece of experimental evidence supporting the suggestion that the ferrous hfs proton resonances are sensitive to the quaternary structural state of hemoglobin. In 1971, we investigated the ferrous hfs proton resonances of four mutant deoxyhemoglobins, Hb Chesapeake [α92(FG4)Arg $\rightarrow$ Leu], Hb J Capetown [α92(FG4)Arg $\rightarrow$ Gln], Hb Yakima [β99(G1)Asp $\rightarrow$ His], and Hb Kempsey [β99(G1)Asp $\rightarrow$ Asn], which have amino acid substitutions in the $\alpha_1\beta_2$ subunit interface (Davis et al., 1971) The locations of these mutant Hbs are shown in Fig. 11. Figure 39 shows the effects of these

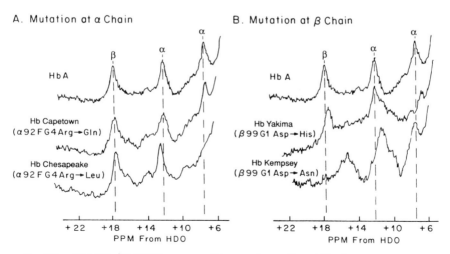

FIG. 39. 90-MHz ^{1}H NMR spectra of ferrous hyperfine-shifted proton resonances of deoxyhemoglobins containing mutations in the $\alpha_1\beta_2$ subunit interface in 0.1 M phosphate in D_2O at pH 6.6 and 25°C. [Adapted from Davis et al. (1971) and Ho and Russu (1985)].

amino acid substitutions in the $\alpha_1\beta_2$ interface in either the α or β chain on the environments of these heme pockets of both α and β chains of intact Hb. The Hill coefficients for these four mutants are ~1.4 for Hb Chesapeake, ~2.2 for Hb J Capetown, ~1.1 for Hb Yakima, and ~1.1 for Hb Kempsey (see Davis et al., 1971; Imai, 1982). In an α-chain mutant, Hb Chesapeake, there is no resonance at ~+7.4 ppm (an α-heme resonance); the ~+12-ppm resonance (an α-heme resonance) is shifted downfield by ~0.3 ppm and the β-heme resonance at ~+17.9 ppm is shifted upfield by ~0.2 ppm. In the other α-chain mutant, Hb J Capetown, the ~+12-ppm resonance shows no change in its position, the highest field α-heme resonance is shifted upfield by ~0.5 ppm and the β-heme resonance at ~+17.9 ppm is shifted upfield by ~0.1 ppm. These smaller variations in Hb J Capetown are expected in light of its nearly normal functional properties. In a β-chain mutant, Hb Yakima, the α-heme resonance at ~+7.4 ppm is missing (or shifted upfield with greatly diminished intensity) and the β-heme resonance at ~+17.9 ppm is shifted upfield by ~0.3 ppm. In the other β-chain mutant, Hb Kempsey, we have observed very large changes in the hfs proton resonances of both the α and β chains.

Based on the structural information about the heme pockets and the so-called "switch" region (Perutz, 1970; Baldwin and Chothia, 1979), the observed changes in the hfs proton resonances of these four mutant Hbs are not unexpected. Figures 3 and 4 provide some information about the structural features in the $\alpha_1\beta_2$ subunit of deoxy-Hb A. In the case of Hb Chesapeake and Hb J Capetown, the position of the mutation at $\alpha92(FG4)$ is spanned by $\alpha91(FG3)$ and $\alpha93(FG5)$ and is close to $\alpha42(C7)$ (Figs. 3, 4, and 11). These residues make contact with the heme methyl groups of the α chain (Perutz, 1970; Baldwin and Chothia, 1979). In the case of Hb Yakima and Hb Kempsey, the amino acid substitution at $\beta99(G1)$ is next to $\beta98(FG5)$, a residue that makes contact with the heme group of the β chain (Figs. 3, 4, and 11; Perutz, 1970; Baldwin and Chothia, 1979). Thus, the α-heme resonances would be perturbed if a substituion at $\alpha92(FG4)$ alters the conformations and the heme contacts of the amino acid residues at FG3, FG5, or C7 of the α chains (see Figs. 4 and 62). Similarly, the β-heme resonances would be perturbed if a mutation changes the conformation of FG5 of the β chains. These conformational effects are not limited to the heme pockets of the chains containing the structural modifications, but they are, in fact, extended to the adjacent normal chains (see Figs. 4 and 62). One can compare the effects of the conformation of the α-chain heme pocket induced by the α92Arg → Leu substitution in Hb Chesapeake to those induced by the β99Asp → His substitution in Hb Yakima. This type of conformational effect depends on the nature of the replacing amino acid residue, as clearly shown in these four mutants. The ^{1}H NMR results shown in Fig. 39 clearly indicate that the structural changes in one chain can be transmitted across the $\alpha_1\beta_2$ subunit interface to affect the properties of the adjacent heme. They give support to the suggestion that in Hb A, the heme environments of the unligated chains can be influenced by the state of ligation of the hemes in the neighboring chains, as first pointed out by us in 1971 (Davis *et al.*, 1971).

Now, let us consider the detailed variation of the intensities of the hfs proton resonances at about $+12$, $+16.9$, and $+18$ ppm from HDO as a function of oxygenation of Hb A. Figure 40 gives the relative intensities of the α-heme resonance at $+12$ ppm and the β-heme resonance at $+18$ ppm compared with the percentage of fully deoxy-Hb tetramers as a function of oxygenation in the absence and presence of organic phosphates. A remarkable feature is that the α- and β-heme resonance intensities lie below the diagonal, indicating that the sum of deoxy α and β chains giving rise to the $+12$- and $+18$-ppm resonances is smaller than that predicted from the total O_2 saturation under all conditions. In the presence of organic phosphates, the area of the α peak at $+12$ ppm follows the behavior of the fully deoxy-Hb A populations (Figs. 40B and 40C). This means that the presence of even one O_2 molecule in the Hb tetramer is enough to cause a decrease in the area of the $+12$-ppm α resonance

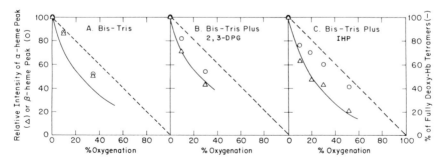

FIG. 40. A comparison of the relative intensities of the α-heme resonance at 12 ppm (△) and the β-heme resonance at 18 ppm (○) with the percentage of fully deoxy-Hb tetramers (—) as a function of oxygenation. (A) In 0.1 *M* Bis–Tris; (B) in 0.1 *M* Bis–Tris plus 35 m*M* 2,3-DPG; (C) in 0.1 *M* Bis–Tris plus 10 m*M* IHP. Fraction of fully deoxy-Hb tetramers was calculated from the data of Tyuma *et al.* (1973). [Adapted from Viggiano *et al.* (1979)].

but not the +18-ppm β resonance. The area of the +18-ppm β peak has been found to be significantly larger than that of the +12-ppm α peak. Thus, the structural changes at partial O_2 saturations in the presence of organic phosphates as seen by ^{1}H NMR spectroscopy cannot be concerted as required by two-state allosteric models, even if extended to take into account the difference in the affinity between the α and β chains as proposed by Ogata and McConnell (1971). However, there are quantitative differences in the areas of these peaks in the presence of Bis–Tris buffer, Bis–Tris buffer plus 2,3-DPG, and Bis–Tris buffer plus IHP. This means that the structural changes that occur in the partially ligated-Hb tetramers [i.e., $Hb(O_2)_1$, $Hb(O_2)_2$ and $Hb(O_2)_3$] can affect the deoxy chains so as to cause a disappearance of the measured resonances at +18 and +12 ppm. This supports the conclusion that the +12-ppm α resonance and the +18-ppm β resonance are characteristic of the deoxy state or deoxy-like quaternary structure (Perutz et al., 1974; Ho et al., 1975).

The two peaks from the β chains, one at +18 ppm and the other at +16.9 ppm, do not change their area ratio on oxygenation. We have found that, if an O_2 molecule is bound to a β chain, these two resonances disappear. However, the experimental results show that if structural changes occur that affect a β-deoxy chain in partially ligated Hb tetramers, the +18-ppm resonance disappears (or shifts) but not the +16.9-ppm peak (Viggiano et al., 1979; Ho et al., 1982b). Figure 41 summarizes the variation of the intensities of the three ferrous hfs proton resonances of Hb A in 0.1 M Bis–Tris plus 10 mM IHP as a function of oxygenation at pH ~6.6 and 27°C. There is good agreement among several independent sets of experimental results (including the computer-simulated spectra). The data show that the +16.9-ppm β-chain resonance is a measure of the amount of oxygenation in the β chain and that this peak is not affected by structural changes of the Hb molecule on oxygenation, unlike the +12-ppm α-chain and +18-ppm β-chain resonances (see Fig. 40). The experimental results allow us not only to calculate various populations of the partially oxygenated species, namely Hb, $Hb(O_2)_1$, $Hb(O_2)_2$, $Hb(O_2)_3$, and $Hb(O_2)_4$, but also the fractions of the α and β chains that are oxygenated in the Hb tetramer as a function of oxygenation. In the presence of IHP, it is clear that the α chains have a much higher O_2 affinity than the β chains. For example, the ratios of oxygenated α chains over β chains of Hb A are 31.5, 10.8, and 8.3 at 13, 26, and 39% of total oxygenation, respectively. As a first approximation, one can assume that at low O_2 saturations, both the

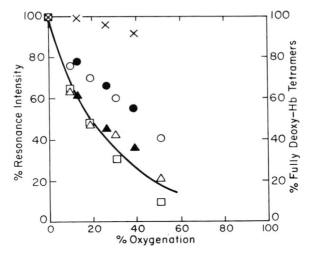

FIG. 41. Variation of the ferrous hyperfine-shifted proton resonances of Hb A in 0.1 M Bis–Tris plus 10 mM IHP as a function of oxygenation at pH ~6.6 and 27°C: △, α-heme resonance at +12 ppm from HDO (data from Viggiano *et al.*, 1979); ▲, α-heme resonance at +12 ppm from HDO (data from Ho *et al.*, 1982b); □, α-heme resonance at +12 ppm from H_2O (data from Viggiano and Ho, 1979); ○, β-heme resonance at +18 ppm from HDO (data from Viggiano *et al.*, 1979); ●, β-heme resonance at +18 ppm from HDO (data from Ho *et al.*, 1982b); ×, β-heme resonance at +16.9 ppm from HDO (data from Ho *et al.*, 1982b); the curve is the fraction of fully deoxy-Hb tetramers calculated from the data of Tyuma *et al.* (1973). [From Ho *et al.* (1982b)].

singly and doubly oxygenated species all have O_2 molecules bound on the α chains. The behavior of the +18-ppm β-chain resonance, which is sensitive to both ligation and structural changes, shows that the number of modified β chains is less than the number of modified α chains (Viggiano *et al.*, 1979; Ho *et al.*, 1982b). The relative intensity of the +12-ppm α-chain resonance follows the population of the fully deoxy-Hb tetramers, suggesting that all of the oxygenated species either have O_2 bound to the α chains or have some structural modifications in the unligated α chains, as shown in Figs. 40 and 41. Thus, our ^{1}H NMR results suggest that a strong cooperativity must exist in the oxygenation of the α chains and that the oxygenation of one α chain affects the ligand affinity of the other unligated α chain within a tetrameric Hb molecule (Viggiano and Ho, 1979; Viggiano *et al.*, 1979; Ho *et al.*, 1982b). This suggests that there are more than two affinity states in the Hb molecule on ligand binding.

From an analysis of the population of various Hb species and the intensities of the three hfs proton resonances as a function of oxygenation, the four Adair constants for the oxygenation of Hb A in the presence of IHP can be calculated directly (Ho et al., 1982b). Table III gives a comparison of the Adair constants obtained by Tyuma et al. (1973) with our ^{1}H NMR data. It is clear that the K_1 and K_4 values are essentially identical in both sets of data. However, the K_2 and K_3 values obtained by the NMR method are about twice those obtained by Tyuma et al. (1973). It is possible that the K_2 and K_3 values obtained by NMR may be more reliable than those reported by Tyuma et al. (1973), because the NMR method may give a more direct estimate of the partially oxygenated species (Ho et al., 1982b).

Representative 360-MHz ^{1}H NMR spectra of Hb A over the spectral region from +5 to +10 ppm from H$_2$O as a function of O$_2$ saturation in H$_2$O at pH 7 are shown in Fig. 42. From the variation of the intensities of the resonances at about +9.4 ppm (quaternary structural probe) and at about +6.4 ppm as a function of oxygenation, we can gain some information about the tertiary and quaternary structural transitions of the Hb molecule. In the absence of organic phosphate, the intensity of the +9.4-ppm quaternary structure probe maintains an almost constant ratio to the intensities of both the +18-ppm β-heme resonance and the +12-ppm α-heme resonance on oxygenation (Figs. 40A and 43), but the +6.4-ppm resonance decreases with respect to the +9.4-ppm resonance (Fig. 44). This finding suggests that there is no concerted structural change

TABLE III
Adair Constants for Oxygenation of Human Normal Adult Hemoglobin in the Presence of Inositol Hexaphosphate[a]

Adair constant	Results of Tyuma et al. (1973)[b] (mm^{-1})	NMR results[c] (mm^{-1})
K_1	4.5×10^{-3}	4.2×10^{-3}
K_2	8.0×10^{-3}	1.7×10^{-2}
K_3	3.7×10^{-3}	5.6×10^{-3}
K_4	4.2×10^{-1}	4.1×10^{-1}

[a] From Ho et al. (1982b).
[b] The measurements were performed at 25°C with 1.5×10^{-5} M Hb A in 0.05 M Bis–Tris plus 1.7×10^{-3} M IHP at pH 7.4.
[c] The measurements were performed at 27°C with 1.5×10^{-3} M Hb A in 0.1 M Bis–Tris plus 1×10^{-2} M IHP in D$_2$O at pD 7.0.

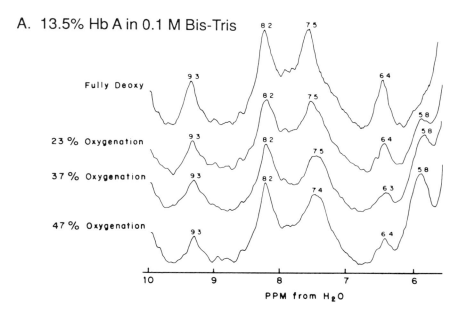

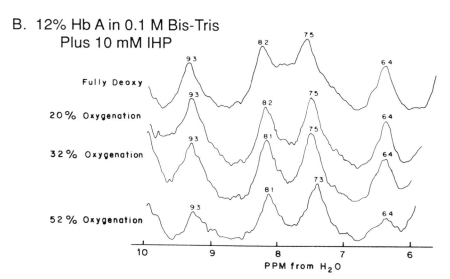

FIG. 42. 360-MHz ^{1}H NMR spectra of Hb A as a function of oxygenation in 0.1 M Bis–Tris in H_2O at pH 7.0 and 25°C. [Adapted from Viggiano and Ho (1979)].

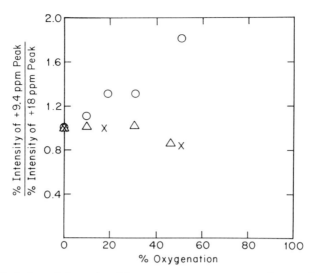

FIG. 43. Variation of the ratio of the quaternary structure probe at +9.4 ppm to the β-heme resonance at +18 ppm as a function of oxygenation of ~12% Hb A in 0.1 M Bis–Tris in H_2O at pH 7.0 and 27°C: ×, in 0.1 M Bis–Tris; △, in 0.1 M Bis–Tris + 35 mM 2,3-DPG; ○, in 0.1 M Bis–Tris + 10 mM IHP. [Adapted from Viggiano and Ho (1979)].

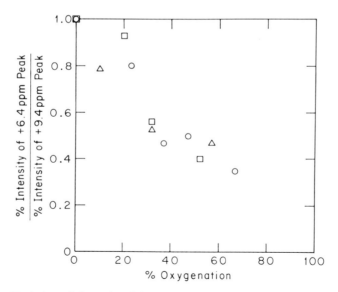

FIG. 44. Variation of the ratio of the tertiary structure probe at +6.4 ppm to the quaternary structure probe at +9.4 ppm as a function of oxygenation of Hb A in 0.1 M Bis–Tris in H_2O at pH 7.0: ○, 13.5% Hb A in 0.1 M Bis–Tris at 25°C; △, 12.5% Hb A in 0.1 M Bis–Tris + 35 mM 2,3-DPG at 27°C; □, 12% Hb A in 0.1 M Bis–Tris + 10 mM IHP at 25°C. [Adapted from Viggiano and Ho (1979)].

if O_2 is distributed randomly between the α and β chains, as shown and discussed earlier. This conclusion is independent of the specific spectral assignment of the +6.4-ppm resonance.

In the presence of organic phosphates, concerted structural changes alone cannot account for the observed spectral features of the four resonances (i.e., about +6.4, +9.4, +12, and +18 ppm from H_2O) on oxygenation. The ratio of the intensities of the +12- and +18-ppm resonances is not constant on oxygenation (Figs. 40B and 40C), in contrast to the case in the absence of organic phosphate (Fig. 40A). This shows a specific effect of these allosteric effectors on the α and β chains of Hb A. In the presence of IHP, the ratio of the intensities of resonances at +9.4 ppm (the quaternary structure probe) and at +18 ppm (β heme) increases with oxygenation, almost doubling from 0 to 50% saturation (Fig. 43). At the same time, the deoxy structural probe of the β chain at +6.4 ppm disappears almost completely. In order to retain the model of a concerted allosteric mechanism, there must be a strong preferential ligation of the β chains. However, this is not in agreement with the relative change in intensities between the +18-ppm β-heme resonance and the +12-ppm α-heme resonance on oxygenation, as shown in Figs. 24 and 40. In the presence of 2,3-DPG (Fig. 43), the ratio of the areas of the +9.4-ppm peak and the +18-ppm peak is constant (up to approximately 40% O_2 saturation), as in the case in the absence of organic phosphate (Fig. 40), whereas the +12-ppm α-heme resonance decreases in intensity with respect to the +18-ppm β-heme resonance, as it does in the presence of IHP (Fig. 40). Within the accuracy of these results, it is difficult to give an exact quantitative relationship to the variation of the +9.4- and +6.4-ppm resonances as a function of oxygenation. It is concluded that the intensity of the +6.4-ppm tertiary structure probe decreases with respect to the +9.4-ppm quaternary structure probe at any partial O_2 saturation. In other words, the intrasubunit hydrogen bond between valine at β98(FG5) and the penultimate tyrosine at β145(HC2) can be broken (as manifested by the disappearance of the +6.4-ppm peak) before the ligation of the β chains and also while the intersubunit hydrogen bond between tyrosine at α42(C7) and aspartic acid at β99(G1) (as manifested by the presence of the +9.4-ppm resonance) is still intact. The relative intensities of the four resonances (at about +6.4, +9.4, +12, and +18 ppm) exhibit a complex variation as a function of oxygenation, especially in the presence of organic phosphates. It is quite clear that there are no concerted structural changes on oxygenation of Hb, as shown by the spectral changes of these four proton resonances.

Because the two exchangeable proton resonances (at about +6.4 and + 9.4 ppm from H_2O) only exist in H_2O and most of the detailed investigations on the three ferrous hfs proton resonances (at about +12, +16.9, and +18 ppm from HDO) have been carried out in D_2O, one needs to investigate the effect of D_2O on the cooperative oxygenation of Hb A. We have found that (1) the pH dependence of the O_2 equilibrium of Hb A (the Bohr effect) in D_2O is shifted by approximately 0.4 pH unit toward higher pH when compared to H_2O, (2) the Hill coefficients are essentially the same in D_2O and H_2O over the pH range 6.0 to 8.2, and (3) the variation of the ferrous hfs proton resonances of Hb A as a function of oxygenation is the same whether it is in D_2O or in H_2O (Viggiano *et al.*, 1979; Ho *et al.*, 1982b; Russu *et al.*, 1982).

VI. ^{1}H NMR Investigations of Structures and Properties of Symmetric Valency Hybrid Hemoglobins: Models for Doubly Ligated Species

Because the oxygenation process is cooperative, Hb A greatly prefers to exist in either the deoxy or the fully oxy form. Thus, the amounts of partially ligated species, i.e., Hb with one O_2, two O_2, and three O_2 bound, are quite limited under the usual experimental conditions. Two approaches have been used to investigate the structures and properties of partially ligated species. The first approach is to monitor the hfs proton resonances and the exchangeable proton resonances as a function of ligation and then to correlate these spectral changes to the structures and properties of partially ligated species. Some very valuable information has been obtained by this approach, as described in Section V. However, the structural information about the partially ligated species derived from this approach is indirect. The second approach is to investigate the structures and functional properties of partially ligated species of Hb using valency hybrid Hbs. In this section, we discuss two types of valency hybrid Hbs, namely, naturally occurring ones (i.e., the M-type Hbs) and synthetic ones [i.e., $(\alpha^{+CN}\beta)_2$ and $(\alpha\beta^{+CN})_2$]. Both types have been used as models for understanding the nature and roles of intermediate species formed during the cooperative ligation process (for example, see Banerjee *et al.*, 1973; Shulman *et al.*, 1975; Fung *et al.*, 1976, 1977; Nagai, 1977, and references therein). In the M-type Hbs, the heme groups in either the α or β chains are selectively oxidized, whereas the other hemes remain in the ferrous state, capable of binding O_2 or CO (such as $\alpha_2^+\beta_2$ or $\alpha_2\beta_2^+$). In synthetic valency hybrid Hbs such as $(\alpha^{+CN}\beta)_2$ and $(\alpha\beta^{+CN})_2$, the heme groups in either the α or β chains are selectively oxidized and bound to CN^- while the other hemes remain in the ferrous state, capable of binding O_2 or CO as in the M-type Hbs. These two types of valency hybrid Hbs have been assumed to be similar in structure to partially oxygen-ligated molecules of Hb A, such as $\alpha_2^{O_2}\beta_2$ or $\alpha_2\beta_2^{O_2}$ (Enoki and Tomita, 1968; McConnell et al., 1968; Banerjee and Cassoly, 1969; Shulman et al., 1969, 1975; Brunori et al., 1970; Cassoly et al., 1971; Haber and Koshland, 1971; Ogawa and Shulman, 1971, 1972; Ogata and McConnell, 1972b; Minton, 1974).

A. M-Type Hemoglobins

Of special interest to this review is Hb M Milwaukee (β67E11Val → Glu). According to X-ray diffraction studies, the E11 valine residue is positioned directly above the heme group adjacent to the distal histidine (E7), and the γ-carboxylate oxygen of the glutamyl residue binds to the ferric β iron as the sixth ligand (Perutz et al., 1972). This substitution allows the β hemes to become oxidized in erythrocytes and the patient suffers from methemoglobinemia (Pisciotta et al., 1959). No gross distortion in either the tertiary structure of the β subunits or the quaternary structure of the molecule as a whole is apparent at a resolution of 3.5 Å (Perutz et al., 1972). The O_2 affinity of Hb M Milwaukee was found to be less than that of Hb A (at 10°C in 0.1 M phosphate at pH 7.2, the O_2 pressure in equilibrium with half-saturated Hb, P_{50} = 21 torr, compared to 2.9 torr for Hb A) and the Hill coefficient is about 1.3 (Udem et al., 1970).

By means of ^{1}H NMR spectroscopy, we have studied several features of the structure of Hb M Milwaukee and of structural changes accompanying oxygenation (for details, see Fung et al., 1976, 1977). In deoxy-Hb M Milwaukee, the α- and β-heme irons have different paramagnetic properties; the hfs proton resonances of the high-spin ferrous α hemes (with spin 2 for ferrous iron) appear in a spectral region different from that of the high-spin ferric β-hemes [with spin 5/2 for ferric iron (Lindstrom et al., 1972a)]. Figures 8D and 8F (b) give a comparison of the hfs proton resonances due to high-spin ferric met-Hb A in the spectral range from +10 to +90 ppm from HDO and to high-spin ferrous deoxy-Hb A in the spectral range from +6 to +20 ppm from HDO. Using these ferric hfs proton resonances, it is possible to observe the structural alterations at the O_2-binding site of one type of subunit (β^+ hemes), which are associated with the binding of O_2 to the neighboring subunits (α hemes).

Thus, the ^{1}H NMR studies of Hb M Milwaukee have allowed us to explore not only the structures of fully deoxygenated and fully oxygenated Hb M Milwaukee, but also the changes at the β^+ hemes at intermediate stages of O_2 binding to the normal α hemes. These measurements have provided a direct probe of the structural manifestation of the "heme–heme" interaction in Hb M Milwaukee (for details, see Fung et al., 1976, 1977).

Figure 45 gives the 250-MHz ^{1}H NMR spectra of exchangeable proton resonances and ferrous hfs proton resonances of Hb M Milwaukee as a function of IHP in H_2O at pH 6.6 and 30°C. The two exchangeable proton resonances at +9.4 and +6.4 ppm from H_2O are present in the deoxy quaternary structure (Fig. 45F), but not in the oxy quaternary structure (Fig. 45A). When the fully ligated sample is titrated with IHP, the deoxy spectroscopic probes at +9.4 and +6.4 ppm reappear in the spectra when a sufficient amount of IHP is added (Figs 45C–45E). The flat region beyond 10 ppm downfield indicates that the α chains are still fully ligated. This suggests that ligated Hb M Milwaukee in the presence of IHP has a deoxylike quaternary structure. This is in agreement with the optical spectroscopic data of Perutz et al. (1972), which suggest that IHP is capable of locking Hb M Milwaukee in the deoxylike quaternary structure regardless of the ligation state of the α hemes.

Both the deoxygenated and oxygenated samples of Hb M Milwaukee give proton resonances in the spectral region between +30 and +60 ppm downfield from HDO. These are ferric hfs proton resonances of the β^+ hemes. As shown in Fig. 46, distinct resonances are observed at about +48, +53, and +55 to +57 ppm when the α chains are deoxygenated in 0.1 M Bis–Tris buffer in D_2O at pD 7.0. When the α chains are fully saturated with O_2 or CO, three major resonances at about +45, +53, and +57 ppm are present. We refer to spectra resembling that shown in Fig. 46A as type A and spectra resembling that shown in Fig. 46F as type F. On stepwise addition of O_2 to the α chains, the β^+ heme spectrum changes gradually from type A to type F, as shown in spectra B–E of Fig. 46, with the $\sim +48$-ppm resonance being replaced by the $\sim +45$-ppm resonance, and the ratio of the intensity of the resonance at +55 to +57 ppm to that of the resonance at +53 ppm decreases from greater than one to less than one. In this series of spectra, it is important to note that, while the intensities of various peaks change significantly with changing O_2 saturation, their positions do not. This is a good indication that the chemical species that give rise to the various resonances interconvert slowly on the NMR time scale, i.e., the mean lifetime of a species is greater than about 0.4 msec (Pople et al., 1959).

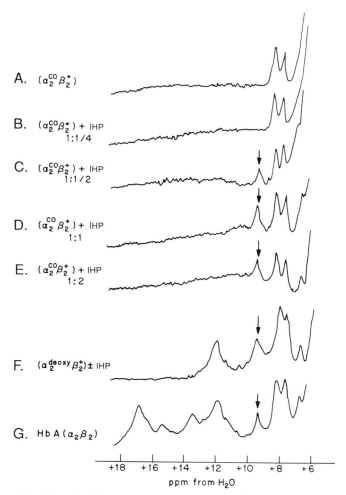

FIG. 45. 250-MHz ^{1}H NMR spectra of exchangeable proton resonances and deoxy hyperfine-shifted proton resonances of 15% Hb M Milwaukee (β67Val → Glu, $\alpha_2\beta_2^+$) as a function of the concentration of inositol hexaphosphate in 0.1 M Bis–Tris in H$_2$O at pH 6.6 and 30°C. The spectrum of Hb A is included for comparison. [Adapted from Fung *et al.* (1977)].

Figure 47 shows the effects of organic phosphates on the ferric hfs proton resonances of oxy-Hb and carbonmonoxy-Hb M Milwaukee. Spectrum A shows that the addition of IHP to deoxy-Hb M Milwaukee has litte effect. However, spectra B and C show that IHP converts the type F spectrum of ligated Hb M Milwaukee to a type A spectrum. In order to be assured that the α chains were still fully ligated when

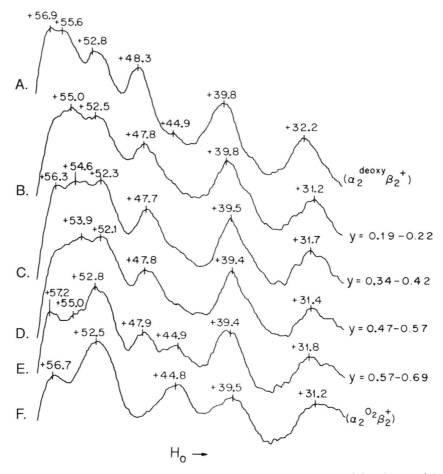

FIG. 46. 250-MHz ferric hyperfine-shifted proton resonances of the abnormal β chains of 15% Hb M Milwaukee (β67Val → Glu, $\alpha_2\beta_2^+$) as a function of oxygenation in 0.1 M Bis–Tris in D_2O at pH 6.6 and 30°C. The proton chemical shifts are referenced to HDO. The range of the fractional saturation (y) was based on P_{50} = 5 torr or P_{50} = 10 torr. [Adapted from Fung et al. (1977)].

excess IHP was added, the deoxy ferrous hfs proton resonance region was monitored simultaneously, with no resonances being found in that region. The addition of excess 2,3-DPG to HbCO M Milwaukee results in a hybrid spectrum (D); between +40 and +50 ppm, it resembles type A and between +50 and +60 ppm it resembles type F. A possible interpretation of this observation is as follows: different

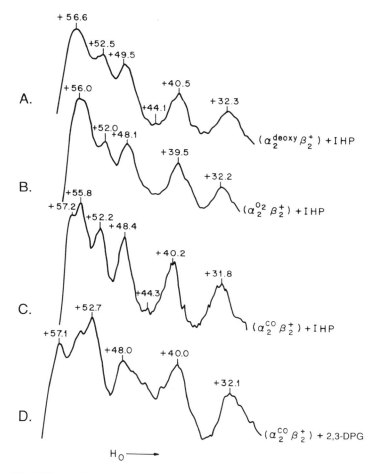

FIG. 47. Effects of organic phosphates on the ferric hyperfine-shifted proton resonances of the abnormal β chains of 15% Hb M Milwaukee (β67Val → Glu, $\alpha_2\beta_2^+$) with and without ligation of O_2 or CO in 0.1 M Bis–Tris in D_2O at pH 6.6 and 30°C. The proton chemical shifts are referenced to HDO. [Adapted from Fung et al. (1977)].

parts of the ferric hfs proton resonance spectrum reflect different sectors of the heme environment, which in turn reflect interactions between a particular β^+ chain and different neighboring subunit chains. Thus, the binding of 2,3-DPG could constrain the interface between β^+ chains to remain in a deoxylike conformation even when the α chains are ligated with O_2 or CO, whereas the interface between α and β⁺ subunits may be altered on ligand binding of the α chains. In contrast, when the larger and more highly charged IHP molecule is bound to oxy-Hb or carbonmonoxy-Hb M Milwaukee, it may constrain both the β⁺–β⁺ and the α–β⁺ interfaces to remain in a deoxylike conformation, thus accounting for both the ferric hfs and exchangeable proton resonance spectra of HbCO M Milwaukee plus IHP. According to this interpretation, the structure of HbCO M Milwaukee plus 2,3-DPG may be classified neither as deoxylike nor oxylike.

Partially oxygenated solutions of Hb M Milwaukee contain an equilibrium mixture of deoxy, singly, and doubly oxygenated species. Because the hfs proton resonance spectra of the β⁺ chains in these solutions indicate that these species interconvert slowly, as discussed earlier, one may assume that, to a good approximation, each spectrum could be approximately described as the time average of the spectra of the three-ligand species,

$$a(x,y) = f_0(y)a_0(x) + f_1(y)a_1(x) + f_2(y)a_2(x) \tag{11}$$

where $a(x,y)$ is the observed amplitude of the spectrum at chemical shift x and fractional saturation y, $f_i(y)$ is the fraction of Hb M Milwaukee to which i molecules of O_2 are bound at fractional saturation y, and $a_i(x)$ is the amplitude of the spectrum of the pure species binding molecules of O_2 at chemical shift x. Using the least-squares procedure described in Fung et al. (1977), two possible ferric hfs proton resonance spectra of the singly oxygenated intermediate corresponding to limiting low and high estimates of the O_2 pressure in equilibrium with half-saturated Hb M Milwaukee, P_{50}, can be calculated. These are shown in spectra B and C in Fig. 48 together with the corresponding spectra of deoxy and doubly oxygenated species (A and D, respectively) for comparison. Spectra B and C are quite similar to each other and exhibit the same qualitative spectral features. Features common to both calculated curves are therefore taken to be characteristic of the spectrum of singly oxygenated Hb M Milwaukee, independent of the precise value of P_{50}. The ferric hfs proton resonance spectrum of Hb M Milwaukee is a sensitive probe of the conformation of the β⁺ hemes and their immediate environments. If the two-heme conformation postulate is correct, then it follows that the ferric hfs proton resonance spectrum of the β⁺ hemes in singly oxygenated Hb M Milwaukee should be the same as the equilibrium average of the corresponding spectra in fully deoxygenated (zero oxygenated) and fully oxygenated (doubly oxygenated)

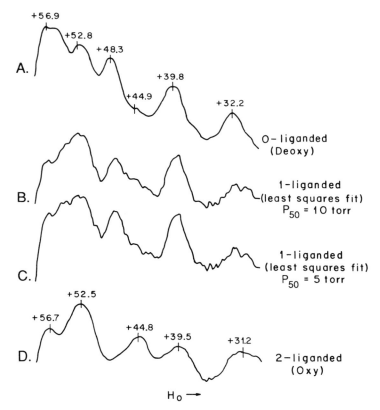

FIG. 48. 250-MHz ferric hyperfine-shifted proton resonances of the abnormal β chains of deoxy- and oxy-Hb M Milwaukee (β67Val → Glu, $\alpha_2\beta_2^+$; A and D, respectively), and calculated spectra of a singly oxygenated intermediate of Hb M Milwaukee (B and C). Spectra B and C were calculated assuming a Hill coefficient $n = 1$. Test calculation values of n up to 1.4 showed no significant differences in the shape of the calculated spectra. The proton chemical shifts were referenced to HDO. [Adapted from Fung et al. (1977)].

Hb M Milwaukee. It is clear that the ferric hfs proton resonance spectrum of singly oxygenated Hb M Milwaukee (Fig. 48, spectrum B or C) cannot be obtained as an average of the corresponding spectra of fully deoxygenated (spectrum A) and fully oxygenated (spectrum D) Hb M Milwaukee. Because the singly oxygenated species does not have the twofold symmetry of the deoxygenated and doubly oxygenated species, the data do not exclude the possibility that one of the two β^+ hemes exists in a deoxylike conformation. However, it seems clear that the conformation of at least one of the two β^+ hemes in singly oxygenated Hb M Milwaukee cannot be described as an equilibrium average of the deoxylike and oxylike conformations. Thus, the data indicate that during the course of oxygenation, unligated binding sites may adopt intermediate as well as deoxylike and oxylike conformations. It follows that the postulate of two-heme affinity states lacks structural justification, and that "two-heme-state" models provide an oversimplified representation of the structure–function relationship in Hb M Milwaukee.

B. Synthetic Symmetric Valency Hybrid Hemoglobins

Shulman and co-workers have carried out a detailed ^{1}H NMR study on synthetic valency hybrids $(\alpha^{+CN}\beta)_2$ and $(\alpha\beta^{+CN})_2$ and Hb M Iwate ($\alpha87\text{His} \rightarrow \text{Tyr}, \alpha_2^+\beta_2$) (Shulman et al., 1969, 1975; Ogawa and Shulman, 1971, 1972; Mayer et al., 1973). They made use of hfs resonances of protons on and near the hemes in the low-spin ferric and high-spin ferrous states as well as the ring-current-shifted proton resonance at about -1.8 ppm upfield from 2,2-dimethyl-2-silapentane 5-sulfonate (DSS) (or about -6.4 ppm upfield from HDO at room temperature) to monitor structural changes around the hemes. In addition, an exchangeable proton resonance at about $+14$ ppm downfield from DSS (or about $+9.4$ ppm downfield from H_2O) has been used as a probe for the deoxy quaternary structure. They have also investigated the effects of 2,3-DPG and IHP on ^{1}H resonances of these three Hbs. From their ^{1}H NMR studies on $(\alpha^{+CN}\beta)_2$ and $(\alpha\beta^{+CN})_2$, they concluded that (1) the α and β chains in Hb are not equivalent in their conformational properties and (2) these symmetric valency hybrid Hbs can take two different quaternary structures (T and R) without changing the degree of ligation (Ogawa and Shulman, 1971, 1972). Hb M Iwate ($\alpha87\text{His} \rightarrow \text{Tyr}, \alpha_2^+\beta_2$) is a mutant Hb in which only the two normal β chains can combine O_2 or CO. This mutant Hb has a very low O_2 affinity and its O_2 binding curve in 0.1 M phosphate at pH 7.0 and 20°C exhibits a Hill coefficient of 1.0 (Hayashi et al., 1966). Mayer et al. (1973) reported that Hb Iwate remains in the deoxy quaternary structure when the β chains are ligated to O_2, CO, or CN$^-$, as well as when they are not ligated. Because of these features, Mayer et al. (1973) used this mutant Hb to confirm the proton resonance markers for the quaternary structures derived from the symmetric valency Hbs reported by Ogawa and Shulman (1971, 1972). Based on the results on $(\alpha^{+CN}\beta)_2$, $(\alpha\beta^{+CN})_2$, and Hb Iwate, Shulman and co-workers concluded that a two-state allosteric model is a suitable "first-order" description of hemoglobin function (Mayer et al., 1973; Shulman et al., 1975).

It should be mentioned that there are differences in approaches between our studies on Hb M Milwaukee (Fung et al., 1976, 1977) and those on $(\alpha^{+CN}\beta)_2$, $(\alpha\beta^{+CN})_2$, and Hb M Iwate reported by Shulman and co-workers (Ogawa and Shulman, 1971, 1972; Mayer et al., 1973). First, Shulman and co-workers did not investigate the spectral changes as a function of O_2 or CO saturation in their studies on $(\alpha^{+CN}\beta)_2$, $(\alpha\beta^{+CN})_2$, and Hb M Iwate. Second, the O_2 binding of Hb M Milwaukee is a cooperative process with a Hill coefficient of approximately 1.3 (Udem et al., 1970), whereas the oxygenation of Hb M Iwate is a noncooperative process with a Hill coefficient of 1.0 (Hayashi et al., 1966). Hence, one cannot obtain information regarding ligation intermediates from a noncooperative system. Third, the hfs proton resonances associated with deoxy hemes and cyanomethemes used by Shulman and co-workers overlap each other in the spectral region examined (+10 to +25 ppm from DSS or +5 to +20 ppm from HDO) [for a comparison of the proton resonances of deoxy-Hb A and met-HbCN A, see Figs. 8E (a) and 8F (b)]. On the other hand, the spectral properties of the ligation intermediates of Hb M Milwaukee are derived from the variations of the hfs proton resonances of the ferric β chains on the oxygenation of the α chains. As shown in Figs. 8D, 17, 46, 47, and 48, the spectral region for the high-spin ferric β chains is very different from that for the high-spin ferrous α chains [see Figs. 8F (b) and 16]. The results of Shulman and co-workers on symmetric valency hybrid Hbs and Hb M Iwate have provided some valuable insights into ^{1}H NMR studies of the Hb problem. However, due to their experimental approach and the choice of Hb M Iwate, their results do not provide sufficient information regarding various structural changes associated with the cooperative ligation process of Hb.

VII. ^{1}H NMR INVESTIGATIONS OF STRUCTURES AND PROPERTIES OF ASYMMETRIC VALENCY HYBRID HEMOGLOBINS: MODELS FOR SINGLY AND DOUBLY LIGATED SPECIES

A. ^1H NMR Studies

Because of rapid dimer exchange, only the symmetric intermediates, such as $(\alpha^{+CN}\beta)_2$ and $(\alpha\beta^{+CN})_2$ or symmetric metal hybrid Hbs— two out of eight intermediate forms—have been isolated as stable tetramers. However, the characterization of other intermediate species (such as singly, doubly, and triply ligated species) is required for a full understanding of the structures and properties of partially ligated species.

The exchange of two dissociative dimers of Hb can be prevented by introduction of a cross-linking reagent, i.e., bis(3,5-dibromosalicyl) fumarate, which cross-links the Ly-82 residues of the two β subunits (Walder et al., 1979, 1980). This has enabled us to prepare various asymmetric cyanomet mixed-valency hybrid Hbs, i.e., with one cyanomet heme in either an α or β subunit, or with two cyanomet hemes, one in an α and one in a β subunit within a tetrameric Hb molecule (Miura and Ho, 1982). The following cross-linked mixed-valency hybrid Hbs have been prepared from derivatives of Hb A and Hb C (β6Glu $\rightarrow$ Lys): $(\alpha^{+CN}\beta)_A(\alpha\beta)_C XL$, $(\alpha\beta^{+CN})_A(\alpha\beta)_C XL$, $(\alpha^{+CN}\beta^{+Cn})_A(\alpha\beta)_C XL$, and $(\alpha\beta^{+CN})_A(\alpha^{+CN}\beta)_C XL$, where subscripts A and C denote that the $\alpha\beta$ dimers are from Hb A and mutant Hb C, respectively, and XL denotes cross-linked Hb. Hb C chains were used with Hb A chains as partners to prepare various hybrid Hb molecules in order to take advantage of the positive charge in the β chain of Hb C (β6Glu $\rightarrow$ Lys), which facilitates their separation by chromatography (Miura and Ho, 1982). Because it has been recognized that, as a first approximation, cyanomet heme is a good model for ligated heme, it is now possible to prepare stable Hb molecules containing various degrees of ligation at any specific subunit. $(\alpha^{+CN}\beta)_A(\alpha\beta)_C XL$ and $(\alpha\beta^{+CN})_A(\alpha\beta)_C XL$ can serve as models for singly ligated species, whereas the models for doubly ligated species, $(\alpha^{+CN}\beta^{+CN})_A(\alpha\beta)_C XL$ and $(\alpha\beta^{+CN})_A(\alpha^{+CN}\beta)_C XL$, can also provide valuable insights into the structures of partially ligated $\alpha_1\beta_1$ dimers and $\alpha_1\beta_2$ dimers within a tetrameric Hb molecule, respectively.

Figures 49 and 50 show the 600-MHz ^{1}H NMR spectra of cross-linked mixed-valency asymmetric hybrid Hbs with one cyanomet chain per Hb tetramer. Figures 49A and 50A give the proton resonances of $(\alpha^{+CN}\beta)_A(\alpha\beta)_C XL$ and $(\alpha\beta^{+CN})_A(\alpha\beta)_C XL$ in D_2O from $+7$ to $+20$ ppm from HDO. These hyperfine-shifted resonances arise from the protons in the heme groups and/or the protons of the amino acid residues in the immediate surroundings of the heme group. They are shifted from their usual diamagnetic positions by hyperfine interactions between the unpaired electrons from the low-spin ferric and high-spin ferrous heme iron atoms and the nearby protons. As discussed earlier, they are very sensitive to the environment of the heme pocket. The signals from the subunits with high-spin ferrous heme iron are much broader than those from cyanomet (low-spin

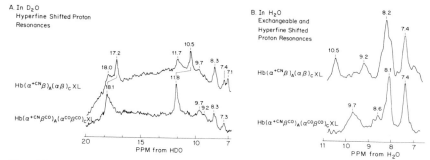

FIG. 49. 600-MHz ^{1}H NMR spectra of cross-linked mixed-valency asymmetric hybrid hemoglobin $(\alpha^{+CN}\beta)_A(\alpha\beta)_C$XL, in deoxy and CO forms, in the presence of 0.1 M phosphate at pH 6.8 and 21°C. (A) Hyperfine-shifted proton resonances in D$_2$O; (B) exchangeable and hyperfine-shifted proton resonances in H$_2$O. [From Miura and Ho (1982)].

ferric) chains. The linewidths of the high-spin ferrous hfs proton resonances at 600 MHz are expected to be >1200 Hz (Johnson et al., 1977). Thus, all sharp resonances in Figs. 49A and 50A come from the low-spin cyanomet chains. This conclusion is also supported by the proton resonances of $(\alpha^{+CN}\beta^{CO})_A(\alpha^{CO}\beta^{CO})_C$XL and $(\alpha^{CO}\beta^{+CN})_A(\alpha^{CO}\beta^{CO})_C$XL. The resonances that occur from +7 to +20 ppm from HDO come from the hyperfine interactions between the unpaired electrons from cyanomet heme iron atoms and the nearby protons. In the CO-ligated chains, the heme iron atoms are diamag-

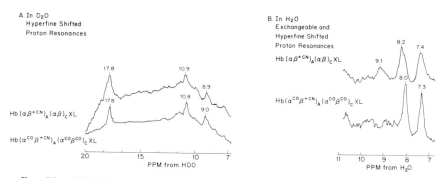

FIG. 50. 600-MHz ^{1}H NMR spectra of cross-linked mixed-valency asymmetric hybrid hemoglobin $(\alpha\beta^{+CN})_A(\alpha\beta)_C$XL, in deoxy and CO forms, in the presence of 0.1 M phosphate at pH 6.8 and 21°C. (A) Hyperfine-shifted proton resonances in D$_2$O; (B) exchangeable and hyperfine-shifted proton resonances in H$_2$O. [From Miura and Ho (1982)].

netic, and thus no hfs proton resonances are observable. The binding of CO to $(\alpha^{+CN}\beta)_A(\alpha\beta)_C XL$ produces significant spectral changes in the low-spin ferric hfs proton resonances of the α^{+CN} chain, as shown in Fig. 49A, especially in the spectral regions +10 to +12 and +17 to +18 ppm. On the other hand, the spectral changes of $(\alpha\beta^{+CN})_A(\alpha\beta)_C XL$ in going from the deoxy to the CO form are quite minor, as shown in Fig. 50A. These results on the low-spin ferric hfs proton resonances from the cyanomet chains in the asymmetric valency hybrid Hbs with one cyanomet chain are consistent with those from the symmetric valency hybrid Hbs with two cyanomet chains (Ogawa and Shulman, 1972). That is, the ^{1}H resonances from the cyanomet α chain change on ligation at neighboring ferrous subunits in 0.1 M phosphate buffer at pH 6.8 and 21°C, but those from the β chain do not (Figs. 49 and 50). Ogawa and Shulman (1972) found that only in the presence of IHP are the ^{1}H resonances from the cyanomet β chains in $(\alpha\beta^{+CN})_2$ observed to undergo alterations on ligation. Unfortunately, the cross-linked Hb is not sensitive to organic phosphates because the cross-linking reagent occupies one of the 2,3-DPG binding sites (Walder *et al.*, 1980). The hfs proton resonances from the cyanomet β chains of $(\alpha\beta^{+CN})_A(\alpha\beta)_C XL$ are not sensitive to the ligation state of the ferrous chains in this mixed-valency hybrid Hb (Miura and Ho, 1982).

Figures 49B and 50B give the proton resonances of $(\alpha^{+CN}\beta)_A(\alpha\beta)_C XL$ and $(\alpha\beta^{+CN})_A(\alpha\beta)_C XL$ in H_2O in both deoxy and CO forms over the spectral region from +7 to +11 ppm from H_2O. There are three different types of proton resonances shown in these two figures. The hfs proton resonances due to the cyanomet chains, such as resonances at +10.5 and +9.7 ppm in Fig. 49B, come from the α^{+CN} chain. The second type is due to the hfs proton resonances of the ferrous chains. As discussed earlier, these resonances are too broad to be detectable under our NMR measurement conditions. The third type is due to the exchangeable proton resonances. In going from the deoxy to the CO form (Figs. 49B and 50B), the most obvious change is the disappearance of the resonance at about +9.2 ppm. It should be noted that the intensity of the resonance at about +9.2 ppm in both $(\alpha^{+CN}\beta)_A(\alpha\beta)_C XL$ and $(\alpha\beta^{+CN})_A(\alpha\beta)_C XL$ appears to be approximately 50% that of the resonance at about +9.3 ppm in deoxy-Hb A or cross-linked Hb A (Miura and Ho, 1982).

Figures 51 and 52 show the hfs and exchangeable proton resonances of cross-linked valency hybrid Hbs with two cyanomet chains— $(\alpha^{+CN}\beta^{+CN})_A(\alpha\beta)_C XL$ and $(\alpha\beta^{+CN})_A(\alpha^{+CN}\beta)_C XL$—in both deoxy and CO forms from +7 to +20 ppm from H_2O. Overall, the 600-MHz

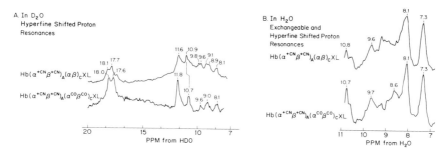

FIG. 51. 600-MHz ^{1}H NMR spectra of cross-linked mixed-valency symmetric hybrid hemoglobin $(\alpha^{+CN}\beta^{+CN})_A(\alpha\beta)_C$XL, in deoxy and CO forms, in the presence of 0.1 M phosphate at pH 6.8 and 21°C. (A) Hyperfine-shifted proton resonances in D_2O; (B) exchangeable and hyperfine-shifted proton resonances in H_2O. [Adapted from Miura and Ho (1982)].

^{1}H NMR spectra of these two valency hybrid Hbs are quite similar to each other. There are small spectral changes in these two hybrid Hbs in going from the deoxy to the CO form (Miura and Ho, 1982).

A comparison of the ^{1}H NMR spectra of the fully ligated, cross-linked hybrid Hbs with two cyanomet chains $[(\alpha^{+CN}\beta^{+CN})_A(\alpha^{CO}\beta^{CO})_C$XL and $(\alpha^{CO}\beta^{+CN})_A(\alpha^{+CN}\beta^{CO})_C$XL], as shown in Figs. 51A and 52A, with those with one cyanomet chain $[(\alpha^{+CN}\beta^{CO})_A(\alpha^{CO}\beta^{CO})_C$XL and $(\alpha^{CO}\beta^{+CN})_A(\alpha^{CO}\beta^{CO})_C$XL], as shown in Figs. 49A and 50A, clearly indicates that the former spectra are the spectral sum of the latter individual spectra. We have assigned the low-spin ferric hfs proton

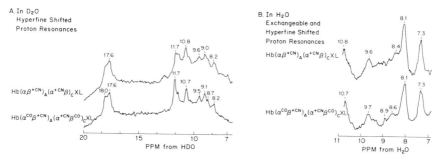

FIG. 52. 600-MHz ^{1}H NMR spectra of cross-linked mixed-valency asymmetric hybrid hemoglobin $(\alpha\beta^{+CN})_A(\alpha^{+CN}\beta)_C$XL, in deoxy and CO forms, in the presence of 0.1 M phosphate at pH 6.8 and 21°C. (A) Hyperfine-shifted proton resonances in D_2O; (B) exchangeable and hyperfine-shifted proton resonances in H_2O. [From Miura and Ho (1982)].

resonances at +18.1, +11.8, +9.7, +8.2, and +7.3 ppm from HDO to the cyanomet α chain and those at +17.7, +10.8, and +8.7 ppm to the cyanomet β chain (Miura and Ho, 1982).

The most obvious difference in the exchangeable proton resonances between the cross-linked hybrid Hbs with two cyanomet chains and those with one cyanomet chain is the absence of the characteristic deoxy quaternary spectral marker at about +9.2 ppm downfield from H_2O in the former hybrid Hbs, as shown in Figs. 51B and 52B. This suggests that the hybrid Hbs with two cyanomet chains per αβ tetramer do not have the usual deoxy quaternary structure as manifested by the intersubunit hydrogen bond between α42Tyr and β99Asp (Miura and Ho, 1982).

X-Ray crystallographic data suggest that, due to the dovetailed nature of the subunit interfaces, there are only two stable conformations of the $α_1β_2$ subunit interface, i.e., one for deoxy-Hb and the other for ligated Hb (Perutz, 1970, 1976, 1979; Baldwin and Chothia, 1979). Thus, the intersubunit hydrogen bond responsible for the resonance at about +9.2 ppm downfield from H_2O should be formed only in the deoxy quaternary structure. However, we have found that the intensity of this resonance for valency hybrid Hbs with one cyanomet chain is greatly reduced compared to that in fully deoxy-Hb A or cross-linked deoxy-Hb A (Figs. 49B and 50B; Miura and Ho, 1982). The reduction of the resonances at about +9.2 ppm suggests that the $α_1β_2$ subunit interface in these "partially ligated" hybrid Hbs is altered. One might speculate that the reduction of the signal intensity at about +9.2 ppm reflects the coexistence of two quaternary structures, i.e., T and R structures. This suggestion is not supported by the ^{1}H NMR results discussed above. For example, the hfs proton resonances from the cyanomet α chain in $(α^{+CN}β)_A(αβ)_CXL$ consist of two types of signals (Fig 49A): those that are shifted on the addition of CO (such as resonances at +10.5 and +17.2 ppm) and those that are essentially unchanged on ligation (such as resonances at +7.3 to +9.7, +11.7, and +18.0 ppm). The signals at +10.5 and +17.2 ppm have been suggested by Ogawa and Shulman (1972) as markers for the deoxy quaternary structure from their ^{1}H NMR investigation of symmetric valency hybrid Hbs. The hfs proton resonances from the cyanomet β chain in $(αβ^{+CN})_A(αβ)_CXL$ are essentially shifted to the positions found in the oxylike quaternary structure (Fig. 50A), whereas in this hybrid Hb, the exchangeable proton resonance is about half of the original intensity (Fig. 50B). Thus, these results suggest that at least one intermediate quaternary structure must be taken into account, other than the oxy and deoxy quaternary structures, in which almost half of the hydrogen bonds responsible for the resonance at about +9.2 ppm are altered. In other words, the structural changes during the ligation process are not concerted and two-structure allosteric models are not sufficient to describe fully the cooperative oxygenation of Hb. It should be mentioned that the spectral changes observed for the high-spin ferric hfs proton resonances of the β chains in Hb M Milwaukee on the ligation of the ferrous α chains are consistent with the results on cross-linked asymmetric mixed-valency hybrid Hbs with one cyanomet chain (Fung et al., 1976, 1977). As a first approximation, the following two possibilities for the intermediate quaternary structure should be considered: (1) an intermediate quaternary structure symmetric about the molecular diad axis in which both of the two hydrogen bonds responsible for the resonance at about +9.2 ppm are equally altered and (2) an intermediate quaternary structure asymmetric about the molecular diad axis in which one of the two hydrogen bonds remains intact and the other is broken. It is very difficult to assume that in the former case, the intensity from both hydrogen-bonded protons would be reduced to an equal extent without a change in the chemical shift. Furthermore, in the fully deoxygenated symmetric valency hybrid Hbs—$(\alpha^{+CN}\beta)_A(\alpha^{+CN}\beta)_A$ and $(\alpha\beta^{+CN})_A(\alpha\beta^{+CN})_A$—with IHP, the exchangeable proton resonance at about +9.2 ppm is missing (S. Miura and C. Ho, unpublished results). Thus, the present evidence supports the second possibility (Miura and Ho, 1982).

B. Equilibrium O_2 Binding Studies

The next step is to correlate the structural changes of these four cross-linked asymmetrical cyanomet valency hybrid Hbs to their respective functional properties. Oxygen equilibrium curves were obtained to determine the binding constant at each oxygenation step (K_i) (Miura et al., 1987). The O_2 equilibrium curves for the cross-linked Hb, $(\alpha\beta)_A(\alpha\beta)_C$XL, were obtained at various pH values. The curves approach closely positioned asymptotes at the high-saturation range and diverge at low-saturation range, as seen in Hb A (Imai and Yonetani, 1977). The cross-linking between two αβ dimers appears to cause a reduction in cooperativity as measured by the Hill coefficient (n = 2.3 for the cross-linked Hb compared to n = 2.9 for Hb A) and increases O_2 affinity as compared to Hb A. Essentially, the same Hill coefficient is obtained whether the two αβ dimers are derived from Hb A or one αβ dimer is derived from Hb A and the other from Hb C (Miura et al., 1987). The P_{50} value and Hill coefficient are found not to be significantly affected by organic phosphates (Miura et al., (1987), in agreement with the fact that one of the 2,3-DPG binding sites is occupied by the cross-linking reagent.

A comparison of each estimated K_i value with that of Hb A shows the effects of the cross-linking on the intrinisic binding constants. The K_4 value is found to be essentially unaffected by the reaction of Hb with bis(3,5-dibromosalicyl) fumarate. Thus, the increase in O_2 affinity must be attributed to an increase in K_i values other than K_4. The pH dependence of K_1 is smaller than that of K_2 and K_3 as is seen in Hb A (Imai and Yonetani, 1977). The relative magnitudes of the K_1, K_2, and K_3 are altered, however, compared to those of Hb A. The relative magnitudes of K_i of cross-linked Hb are found to be in the order $K_2 \leq K_3 < K_1 << K_4$, whereas those of Hb A are found to be in the order $K_1 \leq K_2 < K_3 << K_4$. The number of protons released during oxygenation from the cross-linked Hb (Bohr protons), which is derived from the pH dependence of P_{50}, is about 0.5 per heme (Miura et al., 1987), which is equal to that of Hb A.

Very recently, Shibayama et al. (1991) reported that cross-linked $(\alpha\beta)_A$ $(\alpha\beta)_A$XL isolated by our procedure (i.e., gel filtration in the presence of 1 M $MgCl_2$ followed by ion-exchange chromatography) was found to be contaminated with about 20% of an electrophoretically silent impurity that shows higher O_2 affinity. They developed a purification technique to separate the desired cross-linked Hb A from the impurity. They found that their purified $(\alpha\beta)_A(\alpha\beta)_A$XL in the absence of organic phosphate has the same O_2 affinity, cooperativity, and alkaline Bohr effect as native Hb A.

1. Asymmetric Hybrid Hemoglobins with One Cyanomet Heme

Figure 53A illustrates the Hill plot of the O_2 equilibrium curves of $(\alpha^{+CN}\beta)_A(\alpha\beta)_C$XL in 0.1 M phosphate buffer at various pH values. The O_2 equilibrium curves converge at a high-saturation range, as seen in $(\alpha\beta)_A(\alpha\beta)_C$XL, whereas the lower asymptotes diverge more than those of $(\alpha\beta)_A(\alpha\beta)_C$XL (see Miura et al. (1987). The O_2 binding curves show high cooperativity at low pH, with a Hill coefficient value of 1.8. The estimated K_i (i = 1, 2, or 3) values are plotted against pH as shown in Fig. 53B. K_3 of $(\alpha^{+CN}\beta)_A(\alpha\beta)_C$XL is observed to be less pH dependent than K_1 and K_2. Its value is very similar to those of K_4 of $(\alpha\beta)_A(\alpha\beta)_C$XL and Hb A, which are essentially pH independent. The K_1 value of $(\alpha^{+CN}\beta)_A(\alpha\beta)_C$XL shows a pH dependence similar to that of K_2 or K_3 of $(\alpha\beta)_A(\alpha\beta)_C$XL rather than that of K_1 of $(\alpha\beta)_A(\alpha\beta)_C$XL. The slope of the plot of P_{50} of $(\alpha^{+CN}\beta)_A(\alpha\beta)_C$XL versus pH is about 0.6 at pH 7.1, which is steeper than that of $(\alpha\beta)_A(\alpha\beta)_C$XL,

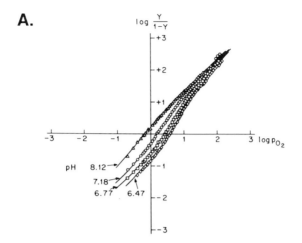

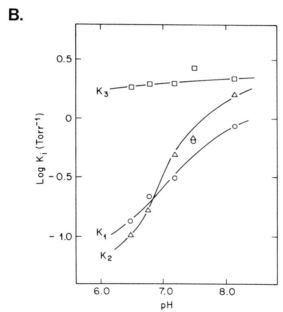

FIG. 53. Oxygen equilibrium curves of cross-linked mixed-valency hybrid hemoglobin $(\alpha^{+CN}\beta)_A(\alpha\beta)_C$XL, in 0.1 M phosphate plus 2 mM KCN at various pH values. (A) Hill plots; (B) O$_2$ binding constants. [From Miura et al. (1987)].

indicating that about 0.6 proton per ferrous heme and 1.8 protons per tetramer of $(\alpha^{+CN}\beta)_A(\alpha\beta)_C XL$ are released on oxygenation (Fig. 53B). The number 1.8 is comparable to that released by $(\alpha\beta)_A(\alpha\beta)_C XL$ after the initial ligation, suggesting that the same structural change as that of $(\alpha\beta)_A(\alpha\beta)_C XL$ would occur on full oxygenation of $(\alpha^{+CN}\beta)_A(\alpha\beta)_C XL$. Based on the recent findings of Shibayama *et al.* (1991), one would expect that the K_1 value of a "purified" sample of $(\alpha^{+CN}\beta)_A(\alpha\beta)_C XL$ would be lower than our measured value due to a 20% contamination of a high-affinity compound. Thus, the presence of this presumed high-affinity impurity would affect our K_1 values, but should not significantly alter our ^1H NMR results of this or other cross-linked mixed-valency hybrid Hbs studied. This is because a ^1H NMR spectrum is normally a linear combination of the components in the sample. On the other hand, the effect of a high-affinity impurity is weighted by its affinity in the O_2 binding measurement, e.g., the presence of a 10% impurity that has 10 times higher O_2 affinity than the parent molecule will give an apparent affinity twice its true value.

The O_2 equilibrium curves of $(\alpha\beta^{+CN})_A(\alpha\beta)_C XL$ at various pH values are shown in Fig. 54A. Introducing a cyanomet heme at the β subunit can cause a drastic change in the O_2 equilibrium curve. The affinity for O_2 of this cross-linked hybrid Hb is considerably higher than that of $(\alpha^{+CN}\beta)_A(\alpha\beta)_C XL$. K_1 and K_2 of $(\alpha\beta^{+CN})_A(\alpha\beta)_C XL$ are much larger than those of $(\alpha^{+CN}\beta)_A(\alpha\beta)_C XL$ at low pH. Each O_2 equilibrium constant increases on raising the pH (Fig. 54B). Consequently, the Hill coefficient remains constant over the pH range investigated (Fig. 55). A large Bohr effect is still present in this hybrid Hb, which has, in alkaline pH, a slope of P_{50} against pH similar to that of $(\alpha\beta)_A(\alpha\beta)_C XL$ (Fig. 55). Thus, the same number of protons per ferrous subunit as from $(\alpha\beta)_A(\alpha\beta)_C XL$ is released from $(\alpha\beta^{+CN})_A(\alpha\beta)_C XL$ during oxygenation. Because the K_1 value of $(\alpha\beta^{+CN})_A(\alpha\beta)_C XL$ is larger than that of $(\alpha\beta)_A(\alpha\beta)_C XL$ (see Miura *et al.*, 1987), the K_1 value would be less affected by the presence of the high-affinity component.

2. *Asymmetric Hybrid Hemoglobins with Two Cyanomet Hemes*

The O_2 equilibrium curves of asymmetric cyanomet hybrid Hbs with two cyanomet subunits—$(\alpha^{+CN}\beta^{+CN})_A(\alpha\beta)_C XL$ and $(\alpha\beta^{+CN})_A(\alpha^{+CN}\beta)_C XL$—were also measured. However, it was not feasible to estimate reliable values of K_1 and K_2 by curve fitting (Miura *et al.*, 1987). We found that it was difficult to fit the O_2 binding scheme with two binding constants over the entire saturation range of the

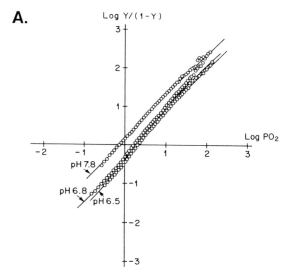

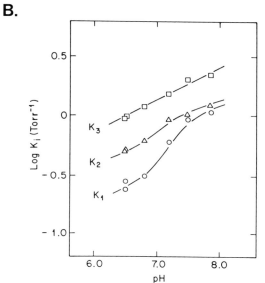

FIG. 54. Oxygen equilibrium curves of cross-linked mixed-valency hybrid hemoglobin $(\alpha\beta^{+CN})_A(\alpha\beta)_C XL$ in 0.1 M phosphate plus 2 mM KCN at various pH values. (A) Hill plots; (B) O_2 binding constants. [From Miura et al. (1987)].

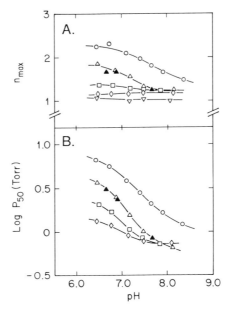

FIG. 55. Effects of pH on the parameters of oxygenation of cross-linked hybrid hemoglobins in 0.1 M phosphate. (A) The Hill coefficient (n); (B) P_{50}. ○, $(\alpha\beta)_A(\alpha\beta)_C$XL; △, $(\alpha^{+CN}\beta)_A(\alpha\beta)_C$XL; □, $(\alpha\beta^{+CN})_A(\alpha\beta)_C$XL; ▽, $(\alpha^{+CN}\beta^{+CN})_A(\alpha\beta)_C$XL; and ◇, $(\alpha^{+CN}\beta)_A(\alpha^{+CN}\beta)_C$XL. Filled symbol denotes containing 2 mM IHP. [From Miura et al. (1987)].

O_2 equilibrium curve. According to Miura et al. (1987), this difficulty might possibly come from the instability of the ferrous heme, which is readily oxidized to the ferric (or met) form. Thus, only the pH dependence of P_{50} and the Hill coefficient (n_{max}) are plotted in Fig. 55, together with those of other samples. Both of these asymmetric hybrid Hbs with two-cyanomet subunits show a low cooperativity, although the value of the Hill coefficient (n = 1.2) for $(\alpha\beta^{+CN})_A$-$(\alpha^{+CN}\beta)_C$XL is slightly higher than that of $(\alpha^{+CN}\beta^{+CN})_A(\alpha\beta)_C$XL ($n$ = 1.1). The O_2 affinity of these two hybrid Hbs is much lower than that of the isolated α or β subunits of Hb A. No essential difference in affinity for O_2 has been found for these two samples, so that only the data for $(\alpha\beta^{+CN})_A(\alpha^{+CN}\beta)_C$XL are plotted in Fig. 55. We have found that K_1 and K_2 of doubly ligated species fail to represent the characteristics of K_3 and K_4 of $(\alpha\beta)_A(\alpha\beta)_C$XL. The values of log P_{50} for these models of the doubly ligated species are much closer to an average of log K_3 and log K_4 of $(\alpha\beta)_A(\alpha\beta)_C$XL than to the log P_{50} of isolated α or β subunits over the pH range examined.

It is inferred that the affinity state of these doubly ligated species is different from that of fully ligated Hb, but is perhaps an intermediate affinity state. We do not understand why the cooperativity of these intermediate species is greatly reduced. The symmetric cyanomet valency hybrid Hbs [$(\alpha^{+CN}\beta)_2$) and $(\alpha\beta^{+CN})_2$] have also been found to have reduced cooperativity ($n = 1.1$–1.2) and increased affinity for O_2 compared to Hb A (Cassoly and Gibson, 1972; Maeda et al., 1972; Hofrichter et al., 1985). Thus, none of the four models for doubly ligated species can represent the features of K_3 and K_4 of cross-linked Hb or Hb A, as pointed out by Miura et al. (1987).

The O_2 equilibrium data indicate that the O_2 affinity increases with the number of cyanomet hemes carried by cross-linked mixed-valency hybrid Hbs. The O_2-binding property depends not only on the number of the subunits carrying cyanomet hemes, but also on the distribution of cyanomet hemes among the four subunits. The present results on the O_2-binding properties of cross-linked mixed-valency hybrid Hbs have provided additional support for the conclusion based on ^{1}H NMR studies of these hybrid Hbs—that there are at least three functional and energetically important structures of Hb in going from the deoxy to the ligated state.

VIII. INFLUENCE OF SALT BRIDGES ON TERTIARY AND QUATERNARY STRUCTURES OF HEMOGLOBIN

An important feature in the stereochemical mechanism of Perutz for the oxygenation of Hb is the critical roles played by the salt bridges (Perutz, 1970; Perutz and Ten Eyck, 1971). The salt bridges of the α chain involve the α-carboxyl of Arg($141\alpha_1$) with both the α-amino group of Val($1\alpha_2$) and the ε-amino group of Lys($127\alpha_1$) and the guanidinium group of Arg($141\alpha_2$) with the carboxyl group of Asp($126\alpha_1$). The β-chain salt bridges involve the α-carboxyl of His($146\beta_2$) with the ε-amino group of Lys($40\alpha_1$) and the imidazole group of His($146\beta_2$) with the carboxyl of Asp($94\beta_2$). In order to assess the roles of the salt bridges in the tertiary and quaternary structures of Hb, we have prepared the following cross-linked asymmetrically modified Hbs: [α(des-Arg)β]$_A$[$\alpha\beta$]$_C$XL; [α(des-Arg-Tyr)β]$_A$[$\alpha\beta$]$_C$XL; [α(des-Arg)β(NES)]$_A$[$\alpha\beta$]$_C$XL; and [α(des-Arg)β]$_A$[$\alpha\beta$(NES)]$_C$XL, where the subscript A or C denotes that the $\alpha\beta$ dimer is from Hb A or Hb C, respectively, and XL indicates a cross-linked Hb. We have used the same bifunctional reagent, bis(3,5-dibromosalicyl) fumarate, for cross-linking as was used in our studies on cross-linked mixed-valency hybrid Hbs in order to prevent exchange between dimers.

NES represents the product of a reaction between N-ethylmaleimide and the sulfhydryl group of β93 cysteine. Attaching NES to the sulfhydryl group of Cys(β93) inhibits the formation of the salt bridge between His(β146) and Asp(β94). ^{1}H NMR spectroscopy was used to investigate these modified Hbs and thus to assess the influence of the salt bridges located at the carboxy terminals of both the α and β chains on the tertiary and quaternary structures of Hb (for details, refer to Miura and Ho, 1984).

Figure 56 gives the ^{1}H NMR spectra of deoxy-des-Arg(α141)-Hb A in 0.1 M Bis–Tris, 0.1 M Tris, and 0.2 M chloride. In the spectral region from 5 to 10 ppm downfield from H_2O, the exchangeable proton resonance at ~ +9.4 ppm downfield from H_2O is an indicator

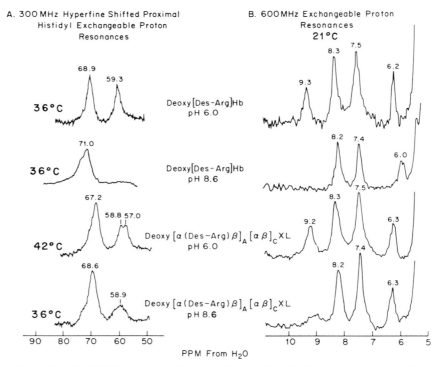

FIG. 56. ^{1}H NMR spectra of deoxy-des-Arg(α141)-Hb A and deoxy[α(des-Arg)β]$_A$[αβ]$_C$XL in 0.1 M Bis–Tris, 0.1 M Tris, and 0.2 M chloride in 95% H_2O and 5% D_2O. (A) 300-MHz ^{1}H NMR spectra of hyperfine-shifted exchangeable $N_δH$ of proximal histidyl residues; (B) 600-MHz ^{1}H NMR spectra of exchangeable proton resonances. [From Miura and Ho (1984)].

of the deoxy (T) quaternary structure, whereas the resonance at $\sim +6.4$ ppm from H_2O is a characteristic feature of a deoxy structural marker.

In the spectral region from 50 to 80 ppm downfield from H_2O, the hyperfine-shifted exchangeable proton resonance at $+59$ ppm is from the proximal histidine of the α chain and that at $+71$ ppm is from the β chain. It has been reported that the removal of the carboxy-terminal Arg(α141) can expose the carboxy group of Tyr(α140), which can then form a salt bridge with Val(α1) (Perutz and Ten Eyck, 1971). The newly formed salt bridge between the carboxy group of tyrosine on the α_1 subunit and the amino-terminal amino group of valine on the α_2 subunit is sensitive to pH in the physiological range. Thus, breaking of the newly formed salt bridge by increasing the pH can cause a downfield shift of the proximal histidyl $N_\delta H$ exchangeable proton resonance of the α subunit. Kilmartin *et al.* (1975) reported that at low pH, deoxy[des-Arg(α141)]-Hb A is in a T-like quaternary structure, whereas at high pH, it is converted into an R-like quaternary structure. The ^1H NMR spectra of deoxy[des-Arg(α141)]-Hb A show pH-dependent changes in both the proximal histidyl $N_\delta H$ exchangeable proton resonances (the α subunit resonance at $+59$ ppm is shifted downfield by about 12 ppm on raising the pH) and the hydrogen-bonded exchangeable proton resonances (the $+9.3$-ppm resonance is lost on raising the pH). Thus, the NMR results are in agreement with the finding of Kilmartin *et al.* (1975).

The ^1H NMR spectra of deoxy[α(des-Arg)β]$_A$[$\alpha\beta$]$_C$XL at pH 6.0 and 8.6 are also illustrated in Fig 56. At low pH, the exchangeable proton resonances in the region $+5$ to $+10$ ppm from H_2O are essentially the same for deoxy(des-Arg)-Hb A and deoxy[α(des-Arg)β]$_A$[$\alpha\beta$]$_C$XL. The resonances due to the α-subunit proximal histidyl $N_\delta H$ proton of deoxy[α(des-Arg)β]$_A$[$\alpha\beta$]$_C$XL split into two peaks at 42°C. The high-field peak shows the same chemical shift as that of the α-subunit proximal histidyl $N_\delta H$ exchangeable proton resonance of deoxy-Hb A (56.8 ppm at 42°C), presumably due to the normal α chain in the $\alpha\beta$ dimer from Hb C. The low-field peak shows the same chemical shift as that of deoxy[des-Arg(α141)]-Hb A, and is presumably due to the modified α chain in the α(des-Arg) dimer from Hb A. The integrated intensity of the resonances due to the α-subunit proximal histidyl $N_\delta H$ exchangeable protons is similar to that of the β-subunit proximal histidyl $N_\delta H$ exchangeable proton signal around $+70$ ppm downfield from H_2O. Raising the pH to 8.6 causes about half the intensity of the α-subunit proximal histidyl $N_\delta H$ proton resonance to shift by about +10 ppm downfield, as does that of deoxy[des-Arg(α141)]-Hb A on raising the pH. Thus, the tertiary structural change due to the destruction of the salt bridge is localized within the modified subunit, and the other intact α subunit in [α(des-Arg)β]$_A$[$\alpha\beta$]$_C$XL remains unaffected. The $\alpha_1\beta_2$ (or $\alpha_2\beta_1$) subunit interface is altered on raising the pH as a result of the tertiary structural alteration of the α(des-Arg) subunit. The intensity of the exchangeable proton resonance at +9.2 ppm is reduced on increasing the pH, but the resonance at +6.3 ppm remains unchanged.

According to Kilmartin *et al.* (1975), deoxy[des-Arg(α141)-Tyr(α140)]-Hb A is believed to exist in an R-like structure in the absence of IHP and can be converted to a T-like structure in the presence of IHP. The ^{1}H NMR spectra of deoxy[des-Arg(α141)-Tyr(α140)]-Hb A in the regions from +5 to +10 ppm and from +50 to +80 ppm from H$_2$O with and without IHP are shown in Fig 57. In the absence of IHP, the exchangeable proton resonances in the spectral region from +5 to +10 ppm from H$_2$O show the spectral features of an R-like structure, i.e., lacking the +9.3- and +6.3-ppm resonances. Corresponding to this, the hyperfine-shifted proximal histidyl N$_\delta$H exchangeable proton resonances from the α subunit are shifted downfield by more than 10 ppm from the position expected for deoxy-Hb A (Fig. 57A). Addition of IHP builds up the deoxy quaternary structure marker resonance at +9.2 ppm and makes the proximal histidyl N$_\delta$H exchangeable proton signals merge into one resonance at +68 ppm. It should be noted that despite the buildup of the +9.2-ppm resonance by addition of IHP, the exchangeable proton resonance expected at +6.3 ppm from H$_2$O, a deoxy structure marker, is absent.

As also shown in Fig 57A, for deoxy[α(des-Arg-Tyr)β]$_A$[$\alpha\beta$]$_C$XL at pH 6.0, about half the intensity of the hyperfine-shifted proximal histidyl N$_\delta$H exchangeable proton resonances has already shifted downfield by about 15 ppm, but the remaining resonance remains at a position close to that of the α-subunit proximal histidyl N$_\delta$H proton resonance in intact Hb A. It is reasonable to assign the resonance that remains unchanged to the normal intact α subunit of [$\alpha\beta$]$_C$ in deoxy[α(des-Arg-Tyr)β]$_A$[$\alpha\beta$]$_C$XL. In correspondence with the shifted N$_\delta$H proton resonance of α(des-Arg-Tyr), both resonances at +9.2 and +6.3 ppm from H$_2$O have lost about half of their respective intensities, and they disappear when the pH is raised to 8.6. The increase in pH also causes the α-subunit histidyl N$_\delta$H exchangeable proton resonance, normally occurring around +57

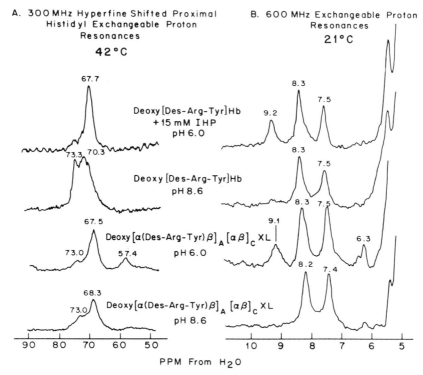

FIG. 57. ^{1}H NMR spectra of deoxy[des-Arg(α141)-Tyr(α140)]-Hb A and deoxy[α(des-Arg-Tyr)β]$_A$[αβ]$_C$XL in 0.1 M Bis–Tris, 0.1 M Tris, and 0.2 M chloride in 95% H$_2$O and 5% D$_2$O. (A) 300-MHz ^{1}H NMR spectra of hyperfine-shifted exchangeable N$_δ$H of proximal histidyl residues; (B) 600-MHz ^{1}H NMR spectra of exchangeable proton resonances. [From Miura and Ho (1984)].

ppm from H$_2$O at low pH, to shift downfield by about 14 ppm and the broad resonance at +73 ppm from H$_2$O to increase in intensity.

Figure 58 shows the ^{1}H NMR spectra of deoxy-NES-Hb A and deoxy-NES[des-Arg(α141)]-Hb A in the presence and absence of IHP. As previously reported, deoxy-NES[des-Arg(α141)]-Hb A in the absence of IHP shows a spectrum suggesting that the Hb molecule has an R-like structure, i.e., lacking a +9.4-ppm resonance (Fung and Ho, 1975) and exhibiting a downfield-shifted α(des-Arg) proximal histidyl N$_δ$H proton resonance (Nagai et al., 1982). Consequently, the spectrum of deoxyNES[des-Arg(α141)]-Hb A without IHP is

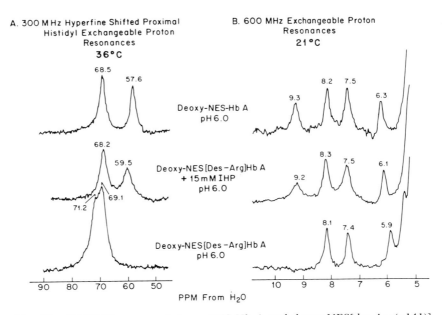

FIG. 58. ^{1}H NMR spectra of deoxy-NES-Hb A and deoxy-NES[des-Arg(α141)]-Hb A with and without inositol hexaphosphate in 0.1 M Bis–Tris, 0.1 M Tris, and 0.2 M chloride in 95% H_2O and 5% D_2O. (A) 300-MHz ^{1}H NMR spectra of hyperfine-shifted exchangeable $N_\delta H$ of proximal histidyl residues; (B) 600-MHz ^{1}H NMR spectra of exchangeable proton resonances. [From Miura and Ho (1984)].

quite similar to that of deoxy[des-Arg(α141)]-Hb A at high pH (Fig. 56). Addition of IHP converts this modified Hb molecule into a T-like structure with spectral features similar to those of deoxy-Hb A. These results are consistent with the O_2-binding properties of this modified Hb (Kilmartin et al., 1975). The intensity of the proximal histidyl $N_\delta H$ exchangeable proton resonance around +60 ppm from H_2O, due to the des-Arg subunits in deoxy-NES[des-Arg(α141)]-Hb A, however, is somewhat smaller than that of the resonance around +70 ppm from H_2O even in the presence of IHP. Incomplete recovery of the α(des-Arg) subunit resonance in deoxy-NES[des-Arg(α141)]-Hb A on the addition of IHP would correlate with the reduced intensity of the +9.2-ppm resonance.

Figure 59 shows the ^{1}H NMR spectra of deoxy[α(des-Arg)β]$_A$-[$\alpha\beta$(NES)]$_C$XL and deoxy[α(des-Arg)β(NES)]$_A$[$\alpha\beta$]$_C$XL. At low pH, both modified Hbs have hyperfine-shifted proximal histidyl $N_\delta H$ exchangeable proton resonances around +60 ppm from H_2O, presumably due to the unmodified α subunits, but have reduced intensity

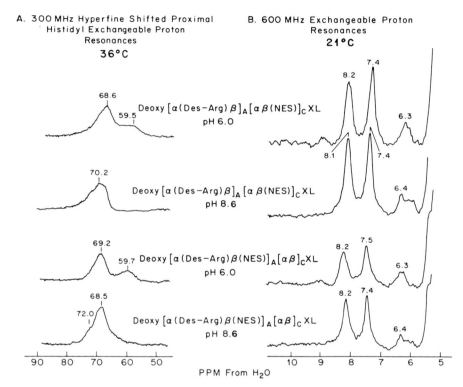

FIG. 59. ^{1}H NMR spectra of deoxy[α(des-Arg)β]$_A$[αβ(NES)]$_C$XL and deoxy[α(des-Arg)β(NES)]$_A$[αβ]$_C$XL in 0.1 M Bis–Tris, 0.1 M Tris, and 0.2 M chloride in 95% H$_2$O and 5% D$_2$O. (A) 300-MHz ^{1}H NMR spectra of hyperfine-shifted exchangeable N$_δ$H of proximal histidyl residues; (B) 600-MHz ^{1}H NMR spectra of exchangeable proton resonances. [From Miura and Ho (1984)].

due to the modified subunits. In contrast, both modified Hbs have completely lost the resonance at +9.3 ppm from H$_2$O, suggesting that the intersubunit interfaces between $α_1β_2$ and $α_2β_1$ are significantly altered in both Hbs even at low pH. It is noteworthy that the relative intensity of the proximal histidyl N$_δ$H exchangeable proton resonance located around +60 ppm from H$_2$O in deoxy[α(des-Arg)β(NES)]$_A$[αβ]$_C$XL is larger than that in deoxy[α(des-Arg)β]$_A$[αβ(NES)]$_C$XL. In both modified Hbs, raising the pH diminishes the resonance around +60 ppm and increases the intensity at +72 ppm instead. Consequently, both these modified Hbs show hyperfine-shifted proximal histidyl N$_δ$H exchangeable proton resonances of an R-like structure.

The ^{1}H NMR results indicate that the effects on the hyperfine-shifted proximal histidyl $N_\delta H$ exchangeable proton resonances at pH 6.0 of removing Arg(α141) or Arg(α141)-Tyr(α140) from one of the two α subunits are limited to within the α subunit from which the C-terminal amino acids are specifically removed. The two asymmetrical modified Hbs have the exchangeable proton resonance at +9.3 ppm from H_2O, which has been assigned to the hydrogen bond between α42Tyr and β99Asp located at the $\alpha_1\beta_2$ subunit interface. This suggests that these asymmetrically modified Hbs preserve the deoxylike quaternary structure in the $\alpha_1\beta_2$ subunit interface, as manifested by the presence of this intersubunit hydrogen bond. On the other hand, the ^{1}H NMR spectrum of deoxy[α(des-Arg-Tyr)β]$_A$-[$\alpha\beta$]$_C$XL at high pH shows no resonance at $\sim$+57 ppm from H_2O in the spectral region expected for the proximal histidyl $N_\delta H$ exchangeable proton resonance of the unmodified α subunit in the $\alpha\beta$ dimer from Hb C (Fig. 57A). Thus, these results clearly show that several features of the ^{1}H NMR spectra of these asymmetrically modified Hbs cannot be accounted for as a spectral sum of the intact deoxy-Hb C and chemically modified deoxy-Hb A. The ^{1}H NMR spectra of deoxy[α(des-Arg)β(NES)]$_A$[$\alpha\beta$]$_C$XL and deoxy[α(des-Arg)β]$_A$[$\alpha\beta$(NES)]$_C$XL at low pH cannot be explained simply as a sum of the spectral features specific for the deoxylike and the oxylike quaternary structures (Fig. 59). These results suggest that intermediate structures exist in which the tertiary and the quaternary structural transitions occur asymmetrically about the diad axis of the Hb molecule during the course of the successive removal of the salt bridges. The results presented in this section clearly show that within the tetrameric Hb molecule in the deoxy form, the conformation at the two intersubunit interfaces ($\alpha_1\beta_2$ and $\alpha_2\beta_1$) is different in these modified Hbs. This implies that there is at least one additional structure other than the T and R structures that can exist in the Hb molecule in going from the deoxy to the oxy state. The crystal structure of HbCO Ypsilanti (β99Asp → Tyr) recently reported by Smith *et al.* (1991) has provided support to our ^{1}H NMR studies on the structural properties exhibited by these cross-linked modified Hbs (see later, Fig. 61).

IX. Other Evidence for Existence of Ligation Intermediates of Hemoglobin

During the past decade, a number of kinetic, thermodynamic, and spectroscopic studies have provided strong support for our earlier

^{1}H NMR findings that two-structure allosteric models are not sufficient to describe the cooperative oxygenation of Hb A. In this section, we shall summarize some of these results.

Ackers and co-workers have carried out a systematic investigation of the free energies of the dimer–tetramer assembly for all of the partially ligated species of Hb using a combination of kinetic and equilibrium approaches (Smith and Ackers, 1985; Ackers and Smith, 1987; Smith et al., 1987; Perrella et al., 1990a; Ackers, 1990; Daugherty et al., 1991; Speros et al., 1991). Perrella, Rossi-Bernardi, and co-workers designed cryogenic techniques by which ligation intermediates of Hb are trapped by rapidly quenching an aqueous solution of Hb A into a cryosolvent containing ferricyanide at −25°C. The valency hybrids formed by oxidation of the unligated hemes of the intermediates are separated from met-Hb A and HbCO A by isoelectric focusing at −25°C and then are identified (Perrella et al., 1983, 1986, 1990b, 1991).

Using these several approaches and different model systems and ligands, Ackers, Perrella, and co-workers have determined the cooperative free energy levels of the various possible ligation species of Hb. In the cyanomet system, these ligation species are distributed into three discrete cooperative free energy levels, i.e., third allosteric state exists in addition to the unligated (T) and fully ligated (R) species. In the notation proposed by Ackers and Smith (1987) (see Fig. 60), three partially ligated species, singly ligated cyanomet α chain (species 11), singly ligated cyanomet β chain (species 12), and a doubly ligated cyanomet $\alpha_1\beta_1$ chain (species 21), have a cooperative free energy value intermediate to those of the fully unligated (species 01) and fully ligated species (species 41). The other three doubly ligated cyanomet species (i.e., species 22, 23, and 24) have a cooperative free energy the same as that for the triligated (i.e., species 31 and 32) and fully ligated (species 41) R-state species (Perrella et al., 1990a; Ackers, 1990; Daugherty et al., 1991).

Smith et al. (1987) have investigated Mn(II)/Fe(II)–CO and Fe(II)/Mn(III) mixed-metal Hb hybrid systems that mimic intermediate ligation states. In the first system, the hemes are replaced in some subunits by Mn(II) protoporphyrin IX, providing a functional analog of unligated subunits (Scheidt, 1977; Blough and Hoffman, 1984), whereas the remaining subunits have Fe(II) hemes that can be ligated with CO. In the second system, the ligated subunits that contain hemes are replaced by Mn(III) protoporphyrin IX, whereas the unligated subunits contain normal heme [i.e., Fe(II)]. These hybrids form the same distribution of cooperative free energies as re-

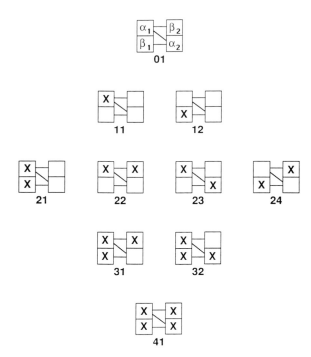

FIG. 60. Representation of the 10 ligation states of hemoglobin: X denotes ligation of a given chain or a metal substitution for Fe in a metal hybrid hemoglobin. Index numbers provide species designation. Subunit positions are shown in species 01. [Adapted from Ackers and Smith (1987)].

ported for the deoxy/cyanomet-Hb system discussed above, i.e., species 21 has a cooperative free energy intermediate between those of species 01 and 41 and different from those of species 22, 23, and 24 (Smith *et al.*, 1987; Ackers, 1990).

Ackers and co-workers pointed out that the doubly ligated Hb tetramers distribute into two separate sets of cooperative free energy values, which rules out even the most general model of the two-state class. The third or intermediate state differs from the fully unligated or fully ligated cyanomet-Hb by 3 kcal at pH 7.4 and 21.5°C (Perrella *et al.*, 1990a; Ackers, 1990). Furthermore, the existence of three molecular states with different free energy values strongly suggests that there are at least three molecular structures during the transition from the unligated to the ligated state of an Hb molecule.

What are the characteristic structural features of this intermediate molecular state? Based on their studies on the effects of pH, temperature, and single-site mutations on the free energy of quaternary assembly, in parallel with the corresponding data on the unligated (T) and fully ligated (R) species, Daugherty et al. (1991) have proposed that the intermediate allosteric tetramer has the deoxy (T) quaternary structure. They suggest that the switching from T to R occurs whenever a binding step results in a tetrameric Hb molecule with one or more ligated subunits on each side of the $\alpha_1\beta_2$ subunit interface. They further propose that significant cooperativity must exist within each $\alpha_1\beta_1$ dimer of the T-state tetramer, because they find that at pH 7.4, the ligand-induced tertiary free energy alters the binding affinity with the T structure by 170-fold prior to the quaternary switching. Thus, their findings support one of our earlier conclusions that some cooperativity must be present within the deoxy quaternary state during the oxygenation process (Viggiano and Ho, 1979).

A Co(II)/Fe(II)–CO system has been used as another model system to determine the cooperative free energy of ligation species. Cobaltous Hb, in which Fe(II) protoporphyrin IX is replaced by Co(II) protoporphyrin IX, binds O_2 cooperatively, but does not bind CO (Hoffman and Petering, 1970; Yonetani et al., 1974; Imai et al., 1977, 1980; Doyle et al., 1991). Thus, in the Co(II)/Fe(II)–CO system, the cobalt sites serve as the unligated sites whereas the Fe(II) sites contain CO molecules as their ligands. Speros et al. (1991) have investigated the cooperative free energies for the 10 ligation species of the Co(II)/Fe(II)–CO system. They find that the 10 ligation species distribute into five discrete cooperative free energy levels (i.e., for species 01, 11, 21, 23, and 41). A unique feature of this work lies in the resolution of cooperative energetics for species 21, in comparison with the other doubly ligated species. Thus, the experimental results from the Co(II)/Fe(II)–CO system rule out a two-state allosteric mechanism and support the combinatorial switching code proposed by Ackers and co-workers (Daugherty et al., 1991).

In a study using CO as the ligand, Perrella et al. (1990b) confirmed that the cooperative free energy value for the singly ligated species (species 11 and 12) is approximately one-half the total cooperative free energy, but, within the experimental accuracy, the cooperative free energy value for all of the doubly ligated species (species 21–24) is not significantly different from that of HbCO A. Thus, this work does not provide conclusive evidence for or against the third-state hypothesis, and they caution that cyanomet-Hb and metal-substituted Hb systems may not be satisfactory models for the ligation of Hb with O_2 or CO.

More recently, Perrella et al. (1992) have carried out a kinetic investigation on the binding of CO to Hb A. They find that the functional heterogeneity of the α and β chains of Hb A is dependent on the nature of allosteric effectors. In 0.1 M KCl at pH 7 and 20°C, the β chains react about 1.5 times faster than the α chains with the first CO molecule. In the presence of IHP, the α chains react about 1.5 times faster than the β chains, but the overall rate of the first CO binding is unchanged. These data emphasize the point seen in much of the NMR work that the nature of ligands and allosteric effectors plays an important role in the functional properties of Hb A.

A series of hemoglobin hybrids in which Fe has been replaced by Zn have been studied by Simolo et al. (1985). Zn(II) does not react with ligands, so that the Fe can be oxygenated without affecting the ligation state of the Zn. These Hb hybrids have been characterized by measurements of the CO dissociation kinetics [on the Fe(II)-containing species] and by various spectroscopic techniques such as circular dichroism, electron nuclear double resonance, and ^{1}H and ^{13}C NMR studies. These different spectroscopic techniques give different answers as to the T or R conformation of Zn/Fe hybrid Hb of species 23 and 24. Simolo et al. (1985) stated that the spectral changes as a function of ligation are not concerted and a two-state allosteric model appears inadequate to explain their results.

Yonetani and co-workers (Inubushi et al., 1986) reported their ^{1}H NMR studies of monoligated Fe–Co cross-linked asymmetric hybrid Hbs—[α(Fe–CO)β(Co)]$_A$[α(Co)β(Co)]$_C$XL (species 11) and [α(Co)β(Fe–CO)]$_A$[α(Co)β(Co)]$_C$XL (species 12). They found that the ligation of a single CO induces different alterations in the intersubunit hydrogen bond at $\sim +9.4$ ppm in these two mixed-metal hybrids Hbs. The T-state marker at $\sim +9.4$ ppm decreases in intensity by about 50% in the species-11 hybrid Hb and disappears completely in the species-12 hybrid Hb. They suggested that one of the two intersubunit hydrogen-bonds (i.e., the one between α42Tyr and β99Asp) in either $\alpha_1\beta_2$ or $\alpha_2\beta_1$ is broken in the [α(Fe–CO)β(Co)]$_A$ dimer induced by the ligation of CO, similar to our ^{1}H NMR results on Hb($\alpha^{+CN}\beta$)$_A$-(αβ)$_C$XL, and that both intersubunit hydrogen bonds (α42Tyr and β99Asp) in the $\alpha_1\beta_2$ and $\alpha_2\beta_1$ subunit interfaces are broken in [α(Co)β(Fe–CO)]$_A$[α(Co)β(Co)]$_C$XL. This behavior differs from that of Hb($\alpha\beta^{+CN}$)$_A$(αβ)$_C$XL, as shown by our ^{1}H NMR results. They also found that there are differences between these two monoligated asymmetric Fe–Co hybrid Hbs in the hyperfine-shifted resonance region of the proximal histidyl N$_\delta$H proton resonances. The ligation of a single CO molecule to the β(Fe) subunit in [α(Co)β(Fe–CO)]$_A$-

[α(Co)β(Co)]$_C$XL shifts the proximal histidyl N$_δ$H proton resonances of the α(Co) and β(Co) subunits 7.6 and 3.4 ppm downfield, respectively, whereas no appreciable change is observed in the complementary hybrid. Based on their ^{1}H NMR results on Fe–Co cross-linked asymmetric hybrids Hbs, Inubushi *et al.* (1986) made two conclusions: (1) there are more than two quaternary structures of Hb during the ligation of Hb, which is inconsistent with the two-state allosteric mechanism, and (2) the α chain of native Hb A has a higher affinity than the β chain. These two conclusions are consistent with our ^{1}H NMR studies on the oxygenation of Hb A and on the cross-linked mixed-valency asymmetric hybrid Hbs. The different ^{1}H NMR spectral properties of these two Fe–Co hybrid Hbs imply that the energetic properties for species 11 and 12 may not be the same. If so, this is a deviation from the scheme based on free energy of cooperativity proposed by Ackers and co-workers (Smith and Ackers, 1985; Ackers and Smith, 1987; Smith *et al.*, 1987; Daugherty *et al.*, 1991; Ackers *et al.*, 1992).

X. X-Ray Crystallographic Investigations of Structural Characteristics of Partially Ligated Species of Hemoglobin

Because O_2 rearranges rapidly among different heme sites in a tetrameric Hb molecule, and partially oxygenated Hb species are susceptible to oxidation, it has been a real challenge for researchers to determine the structures of singly, doubly, and triply oxygenated Hb species under ordinary experimental conditions. During the past decade, a number of attempts (including using metal-substituted Hbs) have been made to gain insights into the structures of partially ligated species of Hb [for a discussion on this topic, see Perutz *et al.* (1987)].

The crystal structures of the following six Hb derivatives, which have bearing on the structures of partially ligated species of Hb, have been determined: (1) Hb A ($\alpha_2^{O_2}\beta_2$), (2) {α[Fe(II)–CO]β[Mn(II)]}$_2$, (3) {α[Fe(II)–CO]β[Co(II)]}$_2$, (4) {α[Fe(II)–O$_2$]β[Ni(II)]}$_2$, (5) {α[Ni(II)]β[Fe(II)–CO]}$_2$, and (6) HbCO Ypsilanti (β99Asp → Tyr). Crystals of (1) and (5) were grown from poly(ethylene glycol) (PEG) solutions according to the procedure of Ward *et al.* (1975), and crystals of (2), (3), (4), and (6) were grown from high concentrations of ammonium sulfate/phosphate according to the procedures of Perutz (1968). We shall give a brief summary of the main structural features of these Hb derivatives and discuss the relationship of these structural features to the functional properties of Hb A during the ligation process.

Dodson and co-workers (Brzozowski *et al.*, 1984; Liddington *et al.*, 1988) have carried out a structural determination of partially oxygenated Hb A, which contains two oxygenated α chains and two β chains in the deoxy or unligated form [($\alpha_2^{O_2}\beta_2$), i.e., species 23]. The crystals were grown from PEG by the procedure of Ward *et al.* (1975), except that air was allowed entry into the crystallization vessels and the crystals were transferred to more concentrated solutions of PEG before X-ray diffraction measurements. At 2.1 Å resolution, Dodson and co-workers have compared the crystal structures of deoxy-Hb A, met-Hb A, and ($\alpha_2^{O_2}\beta_2$), all grown from PEG, and they all exhibit the T-type quaternary structure (Liddington *et al.*, 1988). In the crystals of ($\alpha_2^{O_2}\beta_2$), the ligated or oxygenated α hemes are buckled in a manner identical to those in the α hemes of fully deoxygenated Hb A (i.e., about 0.2 Å out of the plane defined by the four pyrrole nitrogens) and the five-coordinated Fe(II) atoms of the β hemes are about 0.3 Å out of the heme plane (Brzozowski *et al.*, 1984; Perutz *et al.*, 1987; Liddington *et al.*, 1988). Liddington *et al.* (1988) concluded from their analysis of ($\alpha_2^{O_2}\beta_2$) and T-state met-Hb A that the intersubunit contacts prevent the F helix and FG corner from assuming a conformation that would permit the heme to become planar and the Fe atom to have an R-like conformation. They further observed larger changes in the α chains of the T-state met-Hb than of the T-state ($\alpha_2^{O_2}\beta_2$). Thus, they believe that there can be communication between the subunits across the interface within the T structure, similar to that suggested by our ^{1}H NMR studies on the oxygenation of Hb A (Viggiano and Ho, 1979; Viggiano *et al.*, 1979; Ho *et al.*, 1982b). More recently, Waller and Liddington (1990) reported their 1.5 Å-resolution structure of the T-state semioxygenated Hb A crystals. They find that when twofold excess of IHP is added to the crystallization medium containing 18–25% (v/v) PEG-1000 at pH 7.2–7.4, the crystals of Hb contain only the α chains oxygenated (in contrast with their early semioxygenated Hb crystals, which have α chains fully occupied by O_2 molecules and a low level of 25% ligation of the β chains in the absence of IHP). This observation is again consistent with ^{1}H NMR studies on the oxygenation of Hb A in the presence of IHP, i.e., at low O_2 saturations, O_2 molecules are bound to the α chains before the β chains (Ho *et al.*, 1982b).

Arnone *et al.* (1986) determined the crystal structure of a Mn–Fe hybrid Hb, {α[Fe(II)–CO]β[Mn(II)]}$_2$, corresponding to species 23 according to the notation given in Fig. 60. Crystals were grown in 2.3 M ammonium sulfate plus 0.3 M ammonium phosphate at pH 6.5 following the procedure of Perutz (1968) for deoxy-Hb A. At 3.0

Å resolution, they found that (1) the Mn(II)-substituted β chains are structurally isomorphous with normal deoxy β chains in Hb A, and (2) CO binding to the α chains induces only small, localized changes (such as the iron atoms in the α chains being drawn toward the heme planes) and does not produce the larger structural changes in subunit tertiary structures that normally accompany complete ligand binding to all four chains. Thus, $\{α[Fe(II)–CO]β[Mn(II)]\}_2$ maintains a T-type quaternary structure.

Luisi and Shibayama (1989) determined the crystal structures of the following three mixed-metal hybrid Hbs, $\{α[Fe(II)]β[CO(II)]\}_2$, $\{α[Fe(II)–CO]β[Co(II)]\}_2$, and $\{α[Fe(II)–O_2]\ β[Ni(II)]\}_2$ at 2.8, 2.9, and 3.5 Å resolution, respectively. The crystals were grown from high concentrations of ammonium sulfate/phosphate according to Perutz (1968). Crystals of $\{α[Fe(II)]β[Co(II)]\}_2$ and $\{α[Fe(II)]β[Ni(II)]\}_2$ were found to be isomorphous with crystals of native deoxy-Hb. Crystals of $\{α[Fe(II)–CO]β[Co(II)]\}_2$ were obtained by exposing the crystals of $\{α[Fe(II)]β[Co(II)]\}_2$ to CO and crystals of $\{α[Fe(II)–O_2]β[Ni(II)]\}_2$ were obtained by exposing crystals of $\{α[Fe(II)]β[Ni(II)]\}_2$ to O_2. The crystals of $\{α[Fe(II)–CO]β[Co(II)]\}_2$ and $\{α[Fe(II)–O_2]β[Ni(II)]\}_2$ were also crystallized in the T state. Luisi and Shibayama (1989) found that the structural differences between $\{α[Fe(II)–CO]β[Co(II)]\}_2$ or $\{α[Fe(II)–O_2]β[Ni(II)]\}_2$ and deoxy-Hb A are small and most of the changes occur in the immediate environment of the so-called "allosteric core," similar to those changes observed in $(α_2^{O_2}β_2)$ and $\{α[Fe(II)–CO]β[Mn(II)]\}_2$ discussed above. These four Hb derivatives all belong to species 23 according to the notation given in Fig. 60.

We turn to the crystal structure of deoxy-Hb with ligated β chains, using as a model the metal-substituted Hb $\{α[Ni(II)]β[Fe(II)]\}_2$. Shibayama et al. (1986a) reported that the oxygenation of $\{α[Ni(II)]β[Fe(II)]\}_2$ is noncooperative (with a Hill coefficient, $n = 1.0$) over the pH range 6.5 to 8.5 in the absence and presence of IHP. They further reported that the bond between Ni(II) and the proximal His(F8) in the α chains is broken on the change of structure from the oxylike to the deoxylike. Luisi et al. (1990) determined the structure of $\{α[Ni(II)]β[Fe(II)–CO]\}_2$ at 2.6 Å resolution. These crystals were grown under a CO atmosphere from PEG in the presence of 5.5 mM potassium phosphate at pH 6 and 1 mM IHP. They found that $\{α[Ni(II)]β[Fe(II)–CO]\}_2$ (species 24) is in the T state as defined by the packing of the $α_1β_2$ and $α_2β_1$ subunit interfaces. They further reported that the bond between the Ni and the proximal His(F8) residue in the α chain is broken, in agreement with the spectroscopic results (Shibayama et al., 1986b, 1987). Thus, this broken linkage results in small perturbations in the α-heme pocket and can alter the tertiary structure accompanying the ligation in the normal T state (Luisi et al., 1990).

Smith et al. (1991) reported the crystal structure of Hb Ypsilanti (β99Asp → Tyr) in the CO form. The HbCO Ypsilanti crystals were grown from solutions containing 2.25–2.30 M NaH$_2$PO$_4$/K$_2$HPO$_4$ at pH 6.7 and were stabilized by equilibration with 3.5 M Na-H$_2$PO$_4$/K$_2$HPO$_4$ prior to X-ray measurements. At 3.0 Å resolution, they found a new structure for HbCO Ypsilanti, which is denoted Y. This mutant hemoglobin, they proposed, exhibits a new quaternary structure distinctly different from those of unligated and ligated Hbs as described by Baldwin and Chothia (1979). The hydrogen-bonding interactions in the "switch" region and packing interactions in the "flexible joint" region of the new quaternary structure show non-covalent interactions characteristic of the $\alpha_1\beta_2$ subunit interfaces of both unligated and ligated forms of Hb A (Smith et al., 1991). Figure 61 gives a rigid-body representation of the structure of HbCO Ypsilanti, deoxy-, and oxy-Hb A, showing that there are distinct differences in the quaternary structures among these three forms of Hb. Smith et al. (1991) further proposed that this new quaternary structure of HbCO Ypsilanti may mimic the quaternary transitions that could occur in the Hb molecule in going from the deoxy T-type to the oxy R-type quaternary structure.

With the exception of this new structure HbCO Ypsilanti (β99Asp → Tyr), all crystal structures of hemoglobins that have been determined so far are either in the T or R quaternary structure [including partially ligated Hb species such as ($\alpha_2^{O_2}\beta_2$) and metal hybrid Hb systems]. The structures of ($\alpha_2^{O_2}\beta_2$), {α[Fe(II)–CO]β[Mn(II)]}$_2$, {α[Fe(II)–Co]β[Co(II)]}$_2$, {α[Fe(II)–O$_2$]β[Ni(II)]}$_2$, and {α[Ni(II)]β[Fe(II)–CO]}$_2$ all have the T-type structure. Because the first three belong to species 23 and the last one belongs to species 24, they should have an R-type structure according to the cooperative free energy scheme proposed by Ackers and co-workers (Smith and Ackers, 1985; Smith et al., 1987; Ackers and Smith, 1987; Daugherty et al., 1991; Ackers et al., 1992) and other biochemical results published by various investigators.

The original stereochemical mechanism of Perutz for the cooperative oxygenation of Hb suggests that the α chains should have a higher affinity for O$_2$ than the β chains (Perutz, 1970). More recent high-resolution X-ray crystallographic results show that there are substantial differences between the distal portions of the α- and β-heme pockets in Hb A (Shaanan, 1983; Derewenda et al., 1990).

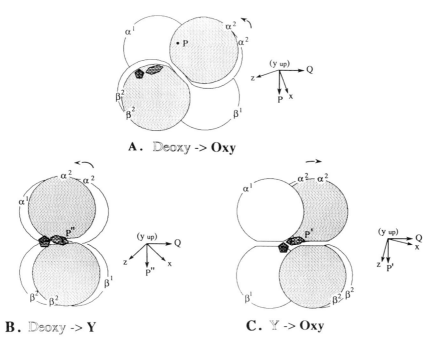

FIG. 61. Rigid-body representation of the quaternary structural differences among HbCO Ypsilanti (β99Asp → Tyr), deoxy-Hb A, and oxy-Hb A. Approximate locations of the switch and flexible joint regions are indicated by the symbols, pentagonal and hexagonal respectively. (A) The quaternary structural change in going from the deoxy (unshaded) to the oxy (shaded) form of Hb A. The rotation axis P is 20° from the x axis and 70° from the z axis, and intersects the molecular dyad of deoxy-Hb A at a point 12 Å from the molecular center and between the α subunits. (B) The quaternary structural change in going from the deoxy-Hb A (unshaded) to the Y state (shaded). The rotation axis P″ is 45° from the x and z axes, and intersects the molecular dyad of deoxy-Hb A at a point near the flexible joint region. (C) The quaternary structural change in going from the Y state (unshaded) to the oxy-Hb A (shaded). The rotation axis P′ is 76° from the x axis and 14° from the z axis, and intersects the molecular dyad of HbCO Ypsilanti in the flexible joint region of the molecule. [From Smith *et al.* (1991). Figure provided through the courtesy of Dr. F. R. Smith].

Val(E11) sterically restricts ligands in the β chains but not in the α chains of deoxy-Hb A. Thus, the conformation of the distal heme pocket of the β chain [such as Val(E11)] needs to be altered before a ligand can bind to the β chain. However, the crystal structure of $\{α[Ni(II)]β[Fe(II)-CO]\}_2$ shows that this T-type structure can have a ligand bind to the β chain (Luisi *et al.*, 1990). Most published data on the function of native, valency, and mixed-metal hybrid Hbs suggest that the native T-state α and β chains have similar affinities for ligands in the absence of organic phosphate, whereas the α chain preferentially binds O_2 in the presence of organic phosphate (for example, see Johnson and Ho, 1974; Olson, 1981; Mathews *et al.*, 1991, and references therein).

There is another serious contradiction between the X-ray crystallographic and optical results on Hb A crystals. Very recently, Eaton and co-workers (Mozzarelli *et al.*, 1991) reported their O_2 binding studies on single crystals of Hb A grown from solutions of PEG (relative molecular weight 8000, 62% (w/v) in 10 mM potassium phosphate) at a pH of 6.0–8.5 at 25°C. They found that (1) the deoxy-Hb A crystals remain intact on oxygenation, (2) the oxygenation of these Hb crystals is noncooperative (with a Hill coefficient, $n = 1$), (3) the O_2 affinity is independent of pH, and (4) the α and β chains of Hb A crystals have about the same affinity for O_2 (the difference in the affinity between α and β chains is no more than a factor or two). This last point directly contradicts the crystal structural results on ($\alpha_2^{O_2}\beta_2$) reported by Brzozowski *et al.* (1984), Liddington *et al.* (1988), and Waller and Liddington (1990), especially the latter. The major difference in experimental conditions of Mozzarelli *et al.* (1991) and Waller and Liddington (1990) is the presence of IHP in the PEG solution used by the latter workers. What are the causes for this contradiction?

The ($\alpha_2^{O_2}\beta_2$) crystals exhibit some unique properties (Brzozowski *et al.*, 1984; Liddington *et al.*, 1988). The original crystals were grown from PEG in the absence of IHP and were found to be in the T-type quaternary structure as determined by X-ray diffraction measurements. Under these conditions, the oxygenation occurs primarily in the α chains. When crystals are grown from PEG in the presence of IHP, only the α chains are oxygenated (Waller and Liddington, 1990). The results suggest that the crystals of deoxy-Hb A grown from PEG are biased toward the T-type quaternary structure. We have found that on the addition of an HbO_2 A solution to PEG, crystals of deoxy-Hb A are formed and the gradual formation of deoxy-Hb A is observed spectrophotometrically, i.e., PEG lowers the oxygen affinity of Hb A (N. T. Ho and C. Ho, unpublished results, 1987). It should be mentioned that Haire *et al.* (1981) investigated the effect of 17.5% PEG (relative molecular weight ~6000) on the O_2-binding property of Hb A in 0.1 M potassium phosphate at pH 7.02 and 25°C. They observed only small decreases in the O_2 binding curve (above 30% O_2 saturations) when PEG was added to Hb A. According to Waller and Liddington (1990), a solution of HbO_2 A was found to be deoxygenated in a few days following the addition of PEG. It appears that the addition of IHP to the PEG solution further pushes the Hb A to the T-type quaternary structure, even in the presence of O_2. This finding is consistent with our early ^1H NMR studies on the oxygenation of Hb A in IHP, i.e., both the singly and doubly oxygenated species all have O_2 molecules bound to the α chains of Hb A (Ho *et al.*, 1982b). Additional experiments are needed to clarify the somewhat conflicting results on the oxygenation of Hb A crystals grown from PEG in the absence and presence of IHP. It will be of great interest to determine whether optical data on O_2 binding to Hb A crystals grown from PEG with IHP added will agree with the X-ray crystallographic results of Waller and Liddington (1990) in showing preferential binding of O_2 to the α chains. One also needs to know the state of the Hb crystals, i.e., is there any oxidation in the β chains?

A number of functional studies have shown that the partially ligated species and valency hybrid Hbs exhibit slowly interconverting conformation states. Some of them have properties that are intermediate between those associated with the T- and R-quaternary conformations (Cassoly and Gibson, 1972; Samaja *et al.*, 1987; Sharma, 1989; Berjis *et al.*, 1990). Thus, one needs to be careful in making correlations between a given crystal structure of a specific Hb species and its functional properties. It should be kept in mind that crystallization is a selective procedure, i.e., it selects those proteins with specific structures that are crystallizable under a given set of crystallization conditions. The structures of these crystallized proteins may not be the dominant ones when function is measured under solution conditions different from those used in crystallizing the proteins.

XI. Possible Pathways for Heme–Heme Communication

Based on a comparison of the structures of deoxy-Hb and ligated forms of Hb (Perutz, 1970; Baldwin and Chothia, 1979; Brzozowski *et al.*, 1984; Perutz *et al.*, 1987; Liddington *et al.*, 1988; Derewenda *et al.*, 1990), the most likely pathway for the transmission of ligand-induced conformational changes from the heme of the α_1 subunit to the heme of the β_2 subunit or from the heme of the β_1 subunit to the heme of the α_2 subunit is from the proximal histidyl residue (F8) through the so-called switch region (Fig. 3) $\alpha_1FG-\beta_2C$ (or $\alpha_2FG-\beta_1C$) to the $\alpha_1\beta_2$ (or $\alpha_2\beta_1$) subunit interface and then again through the corresponding switch region of the unligated chain to its heme group.

The packing in the switch region of the Hb molecule is quite tight according to the crystallographic results. Figure 6 illustrates the packing of amino acid residues on the proximal histidyl (F8) side of the heme group in the β chain. The proximal histidyl residue (F8) is surrounded by a cage of hydrophobic amino acid residues [such as Leu (F4), Leu (F7), Leu (FG3), Val (FG5), and Leu (H19)] (Fig. 6). Figures 5A and 5B illustrate the conformational differences between deoxy- and oxy-Hb A as a result of the movement of the proximal histidyl residue (F8) of the β chain upon the binding of O_2 to the heme iron atom. Figure 7 illustrates the helix and heme motion in the α chain during the transition between deoxy-Hb A and HbCO A (for details, see Dickerson and Geis, 1983). Thus, even if there is a small perturbation [such as a small movement of the iron atom toward the proximal histidyl residue (F8)], one would expect that this conformational change could be sensed by the $\alpha_1\beta_2$ subunit interface. The shortest distance between two heme groups in Hb is 23 Å [i.e., the distance between the two hemes in the α_1 and β_2 chains; the other interheme distances are larger, 31 Å (α_1 to α_2), 37 Å (α_1 to β_1), and 39 Å (β_1 to β_2) (Baldwin and Chothia, 1979)].

It appears that there are at least three possible pathways for communication (or transmission of conformational changes) from the heme group of the α_1 chain to the heme group of the β_2 chain: (1) In the first pathway (Fig. 62A), the ligation of the α_1 heme will initiate the movement of proximal histidyl residue α87(F8), which can alter the conformation of Valα93(FG5). This conformational alteration or perturbation can be transmitted to Argα92(FG4), which can affect Argβ40(C6), i.e, across the $\alpha_1\beta_2$ interface through nonbonded interactions. The conformational alterations at Argβ40(C6) can be transmitted back to Tyrα42(C7), i.e., again across the $\alpha_1\beta_2$ interface, then back to the β_2 chain at Aspβ99(G1), to Valβ98(FG5), which is in contact with the heme of the β_2 chain, so as to alter its electronic environment and thus its ligand affinity. (2) The second pathway is illustrated in Fig. 62B. It uses an alternate route from Valα93(FG5) to Argβ40(C6). Valα93(FG5) forms an intramolecular hydrogen bond with Tyrα140(HC2) in the deoxy state. This hydrogen bond is broken on ligation of the heme (Perutz, 1970; Baldwin and Chothia, 1979). The altered conformation of Tyrα140(HC2) can be transmitted across the $\alpha_1\beta_2$ interface to Trpβ37(C3), then to Argβ40(C6) by way of Thrβ38(C4) and Glnβ39(C5). This conformational perturbation at Argβ40(C6) can be transmitted to Tyrα42(C7) through the $\alpha_1\beta_2$ subunit interface, and then back to Aspβ99(G1), which can alter the electronic structure and thus the ligand affinity of the β_2 heme by

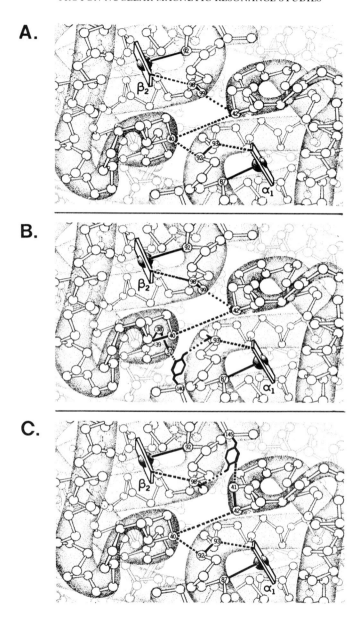

FIG. 62. Three possible pathways (A, B, and C) for "heme–heme" communication between α_1 and β_2 chains. Hydrogen bonds (· · ·) or salt bridges and nonbonded packing contacts (---) are indicated. For details, see the text. [Illustrations prepared and copyrighted by I. Geis].

way of Valβ98(FG5). (3) The third pathway is illustrated in Fig. 62C. Here the alternate route is from Tyrα42(C7) to Valβ98(FG5). The conformational alteration at Tyrα42(C7) can be sensed by Thrα41(C6) due to close packing in this region. Thrα41(C6) has nonbonded packing contacts with Tyrβ145(HC2) through the $\alpha_1\beta_2$ interface. The conformational alteration in Tyrβ145(HC2) can be sensed by Valβ98(FG5), which, in turn, can influence the electronic structure and properties of the β_2 heme. The same pathways can be operative if the first ligand binds to the heme of a β chain by just reversing the routes shown in Fig. 62. It should be emphasized that these three possible pathways for heme–heme communication do not have to operate independently or exclusively of each other. In fact, it is likely that they operate synchronously or synergistically to transmit ligand-induced conformational change from the ligated subunit to a neighboring unligated subunit. These three pathways for heme–heme communication provide a logical explanation for (1) the observed simultaneous alterations of the ferrous hfs proton resonances in both the abnormal and normal chains in mutant Hbs having amino acid substitutions in the $\alpha_1\beta_2$ subunit interface, such as Hb Chesapeake, Hb J Capetown, Hb Yakima, and Hb Kempsey (Fig. 39), and (2) the sum of deoxy α and β chains of Hb A giving rise to +12- and +18-ppm resonances is smaller than that predicted by the total O_2 saturation (Fig. 40).

These results can provide at least a qualitative picture for the molecular basis of the cooperative oxygenation of Hb A. Hemoglobin possesses two major quaternary structures (T and R), as amply demonstrated by X-ray crystallographic results. In the transition from the deoxy to the oxy state, Hb must go through various partially ligated species. Our ^{1}H NMR results provide evidence that ligand-induced conformational changes (tertiary structural changes) can be propagated from the proximal histidyl residue of the α_1 heme through the $\alpha_1\beta_2$ subunit interface and then affect the heme environment of the β_2 subunit. Thus, by implication, they can alter the ligand affinity of the β_2 heme (a similar argument also applies to the β heme if it binds ligand first). It is most likely that the tertiary structural change initiated from the first ligated subunit can affect or alter the subunit arrangement in the $\alpha_1\beta_2$ (or $\alpha_2\beta_1$) interface. Unfortunately, no direct information about the structure of a singly ligated Hb species is available at the moment. This is not surprising because (1) the highly cooperative oxygenation process strongly favors either the fully deoxy or the fully oxy state and (2) the oxygenation process is quite fast [the half-times for O_2 binding to the T state of Hb A are approximately 1×10^{-3} sec (Olsen, 1981)] and the rate of the R→T transition is also very fast, in the range 10^3 to 4×10^4 sec^{-1}, depending on the ligation state of Hb and other experimental conditions (for example, see Eaton *et al.*, 1991), which make a direct structural determination of singly oxygenated Hb species quite challenging. Even though the partially oxygenated Hb species only exist transiently, we believe that they play an indispensable role in facilitating the transition from the deoxy to the oxy state. The nature and the concentration of these partially ligated Hb species depend on the experimental conditions, such as the presence of allosteric effectors like inositol hexaphosphate. Under certain conditions, ligation intermediates may play a much less important role than under other experimental conditions.

The above discussion is not new. Several variations of these ideas have been described and debated by a large number of researchers of hemoglobin during the past six decades (for example, refer to Pauling, 1935; Coryell, 1939; Wyman, 1964; Antonini *et al.*, 1964, 1966; Monod *et al.*, 1965; Koshland *et al.*, 1966; Nagel *et al.*, 1967; Ogawa *et al.*, 1968; Antonini and Brunori, 1969, 1970; Haber and Koshland, 1969, 1971; Hayashi *et al.*, 1969; Brunori *et al.*, 1970; Perutz, 1970, 1989, 1990; Ho *et al.*, 1970, 1973, 1982b; Baldassare *et al.*, 1970; Edelstein, 1971; Davis *et al.*, 1971; Maeda and Ohnishi, 1971; Ogata and McConnell, 1971, 1972a; Ogawa and Shulman, 1971, 1972; McConnell, 1971; Szabo and Karplus, 1972; Shulman *et al.*, 1972, 1975; Saroff, 1972; Saroff and Minton 1972; Saroff and Yap, 1972; Huestis and Raftery, 1972d, 1973; Imai, 1973; Anderson, 1973; Lindstrom *et al.*, 1973; Hopfield, 1973; Herzfeld and Stanley, 1974; Minton, 1974; Minton and Imai, 1974; Fung *et al.*, 1976,1977; Gelin and Karplus, 1977; Vigiano and Ho, 1979; Viggiano *et al.*, 1979; Miura and Ho, 1982, 1984; Weber, 1982; Peller, 1982; Gelin et al., 1983; Lee and Karplus, 1983; Johnson *et al.*, 1984; Smith and Ackers, 1985; Edelstein and Edsall, 1986; Marden *et al.*, 1986; Gibson and Edelstein, 1987; Miura *et al.*, 1987; Karplus *et al.*, 1989; Daugherty *et al.*, 1991; Ackers *et al.*, 1992; and references therein). We hope that we have provided convincing experimental results to support the concept that ligand-induced tertiary structural changes can be propagated to the neighboring heme groups of the unligated subunits through the $\alpha_1\beta_2$ and/or $\alpha_2\beta_1$ subunit interface. This pathway provides a mechanism for the ligation of one chain to result in alteration of the ligand affinity of a neighboring unligated chain, a feature central to the conceptual framework of a sequential mechanism.

XII. CONCLUDING REMARKS

In this article, we have summarized and discussed various experimental results relating to our current understanding of the molecular basis for the cooperative oxygenation of Hb A, as well as the structural and functional properties of the ligation intermediates formed during the transition from the deoxy to the oxy state of a Hb molecule. The two-state allosteric model is based on the idea that there can be only two types of quaternary structure of the Hb molecule, the deoxy (T) and the oxy (R), with a single quaternary structural transition (T $\rightleftharpoons$ R) being responsible for the cooperativity of the oxygenation process (for example, see Monod *et al.*, 1965; Perutz, 1970). During the past two decades, a number of published results on hemoglobin have been found to be inconsistent with two-structure types of allosteric mechanisms for the oxygenation of Hb A. What is the likely mechanism for the cooperative oxygenation of Hb A?

We have taken three separate approaches to investigate the structural features of Hb A associated with the cooperative oxygenation process. First, we have investigated the structural changes during the oxygenation process of Hb A by monitoring the ^{1}H NMR spectral changes of several resonances, which are sensitive to tertiary and quaternary structural changes on the oxygenation of Hb A. The main advantage of these NMR signals is that they serve as intrinsic reporters to monitor the binding of O_2 to Hb A and the structural changes associated with the oxygenation process. Because they are intrinsic markers, they do not produce any perturbations in the Hb molecule (in contrast to some extrinsic reporter groups). The main disadvantage of this approach is that these intrinsic reporters do not provide definite information about the structures of singly, doubly, and triply oxygenated species, i.e., $Hb(O_2)_1$, $Hb(O_2)_2$, and $Hb(O_2)_3$. The results of our ^{1}H NMR findings can be summarized as follows (for details, see Lindstrom and Ho, 1972; Johnson and Ho, 1974; Viggiano and Ho, 1979; Viggiano *et al.*, 1979; Ho *et al.*, 1982b):

1. There is no preferential O_2 binding to the α or β chain of Hb A in the absence of organic phosphates.

2. In the presence of 2,3-DPG or IHP (especially IHP), the α chains have a higher affinity of O_2 than do the β chains.

3. The ligand-induced structural changes in the Hb molecule are not concerted.

4. Within the deoxy quaternary (or T) state, some cooperativity must be present during the oxygenation process.

Second, we have used ^{1}H NMR spectroscopy to investigate the ligation process in naturally occurring valency hybrid hemoglobins, such as Hb M Milwaukee (β67E11Val $\rightarrow$ Glu, $\alpha_2\beta_2^+$), and in synthetic cross-linked mixed-valency hybrid hemoglobins as well as cross-linked asymmetrically modified hemoglobins. These results can be summarized as follows:

1. The ^{1}H NMR spectral changes of the abnormal ferric β chains of Hb M Milwaukee on oxygenation of the normal deoxy α chains are not compatible with a two-structure model for cooperativity and support the concept of direct ligand-linked interactions between subunits embodied in a sequential-type model (Fung et al., 1976, 1977).

2. Cross-linked mixed-valency hybrid Hbs—$(\alpha^{+CN}\beta)_A(\alpha\beta)_C$XL and $(\alpha\beta^{+CN})_A(\alpha\beta)_C$XL—have been used as models for singly ligated species 11 and 12 respectively. Our ^{1}H NMR results suggest that the $\alpha_1\beta_2$ subunit interface is altered in these two singly ligated species and that an intermediate quaternary structure must exist during the oxygenation of Hb A.

3. For the doubly ligated Hb species—$(\alpha^{+CN}\beta^{+CN})_A(\alpha\beta)_C$XL and $(\alpha\beta^{+CN})_A(\alpha^{+CN}\beta)_C$XL (species 21 and 22, respectively)—the ^{1}H NMR results suggest that these two doubly ligated species do not have the usual deoxy quaternary structure, as manifested by the intersubunit hydrogen bond between α42Tyr and β99Asp (Miura and Ho, 1982)

4. Our O_2 equilibrium curves for the four cross-linked mixed-valency hybrid Hbs indicate that there are significant functional differences between $(\alpha^{+CN}\beta)_A(\alpha\beta)_C$XL and $(\alpha\beta^{+CN})_A(\alpha\beta)_C$XL. The two asymmetrical hybrid Hbs with two cyanomet subunits show a low cooperativity and the O_2 affinity of these two hybrids is much lower than that of the isolated α or β subunit of Hb A (Miura et al., 1987). The overall O_2-binding property as measured by P_{50} is on the order of $(\alpha\beta)_A(\alpha\beta)_C$XL > $(\alpha^{+CN}\beta)_A(\alpha\beta)_C$XL > $(\alpha\beta^{+CN})_A(\alpha\beta)_C$XL > $(\alpha^{+CN}\beta^{+CN})_A(\alpha\beta)_C$XL $\approx$ $(\alpha\beta^{+CN})_A(\alpha^{+CN}\beta)_C$XL.

These results are consistent with the assumption that the subunit carrying the cyanomet heme behaves like a model of an oxygenated subunit. The disadvantages of this approach are twofold: Even though the cyanomet chain can mimic certain properties of an oxygenated chain, it is not a perfect analog for an oxygenated chain. Cross-linked mixed-valency hybrid Hbs are not as stable as native Hb A and also require special purification procedures (Miura and Ho, 1982; Miura et al., 1987; Shibayama et al., 1991). It should be mentioned that these two limitations do not apply to naturally occurring mixed valency hybrids Hbs such as Hb M Milwaukee.

Third, we have used ^{1}H NMR spectroscopy to investigate cross-linked asymmetrically modifed Hbs—[α(des-Arg)β]$_A$[αβ]$_C$XL, [α(des-Arg-Tyr)β]$_A$[αβ]$_C$XL, [α(des-Arg)β(NES)]$_A$[αβ]$_C$XL, and [α(des-Arg)β]$_A$[αβ(NES)$_C$XL (Miura and Ho, 1984). These modified Hbs have provided additional insights into the roles played by the salt bridges in the stereochemical mechanism of Perutz for the cooperative oxygenation of Hb (Perutz, 1970; Perutz and Ten Eyck, 1971) and into the structural transitions experienced in the subunit interfaces during the oxygenation process. Our results suggest that there are intermediate structural states in which one of two subunits is converted to the oxylike tertiary structure, but the other subunit can remain in a deoxylike tertiary structure. In accordance with this asymmetric tertiary structural transition from the deoxylike to oxylike structure, the deoxylike quaternary structure is altered. Thus, these results indicate that there are more than two arrangements of the subunits (quaternary structures) in the course of successive removal of specific salt bridges by means of enzymatic and chemical modifications.

There are two key questions that need to be addressed regarding the ^{1}H NMR and biochemical results discussed in this article in relation to the structural and functional properties of partially ligated species formed during the oxygenation of Hb A. First, are those partially ligated or intermediate Hb species detected by our ^{1}H NMR spectroscopy energetically significant in the oxygenation process? Second, what are the structural characteristics of these partially ligated or intermediate Hbs (especially the singly ligated species)? In other words, do they represent new quaternary structures or just minor modifications of the T- and/or R-type quaternary structure? Recent systematic investigations on the cooperative free energies of Hb systems carried out by Ackers and co-workers (Smith and Ackers, 1985; Ackers and Smith, 1987; Smith *et al.*, 1987; Perrella *et al.*, 1990a; Ackers, 1990) clearly demonstrate that there are at least three distinct free energy levels. It is highly likely that the intermediate species detected by our ^{1}H NMR measurements during the transition from the deoxy to the oxy state belong to one of the three free energy states proposed by Ackers and co-workers. Thus, the partially ligated Hb species observed by our ^{1}H NMR spectroscopy are both energetically and functionally important in regard to the cooperative oxygenation of Hb A. The traditional viewpoint regarding the quaternary structure in hemoglobin results from a comparison of the crystal structures of deoxy- and oxy-Hb A (Perutz, 1970; Baldwin and Chothia,

1979). Based on such a comparison, it was concluded that there can be only two quaternary structures, one for deoxy-Hb (T type) and the other for oxy-Hb (R type). In fact, all published crystal structures for various forms of Hb (including metal hybrid Hbs) have been found to be either in the T- or R-type quaternary structure. However, Smith *et al.* (1991) have recently discovered a third crystal structure in HbCO Ypsilanti. This structure is distinctly different from those of the classical T- and R-type structures and Smith *et al.* (1991) have called it the Y structure (Fig. 61). This is the first direct evidence that more than two types of quaternary structure can exist in Hb crystals. It should be noted that very recently Arnone and co-workers (Silva *et al.*, 1992) reported a new ligated quaternary structure for HbCO A crystals grown in low salt conditions. This demonstrates that there are more than two quaternary structures for Hb A (for details, see Note Added in Proof). Unfortunately, we still do not have a clear picture regarding the structural characteristics of the singly, doubly, and triply oxygenated species of Hb A. The major difficulty in this regard is due to our inability to determine the complete three-dimensional structure of a short-lived protein species. Thus, at this time we cannot state definitely the quaternary structure type of the partially ligated Hb intermediates.

Ackers and co-workers (Daugherty *et al.*, 1991; Ackers *et al.*, 1992) have proposed a molecular code for cooperativity in hemoglobin. They suggest that the switching from the deoxy (or T) quaternary structure to the oxy (or R) quaternary structure occurs whenever a binding step creates a Hb tetramer with one or more ligated subunits on each side of the subunit interface between identical $\alpha_1\beta_1$ and $\alpha_2\beta_2$ dimers. They believe that the $\alpha\beta$ dimers are cooperative within a tetrameric Hb A molecule even though they are noncooperative when dissociated. They further believe that hemoglobin cooperativity arises, in part, from concerted quaternary switching and that the Hb molecule also operates by sequential cooperativity within each quaternary structure, i.e., T and R. They propose that the singly ligated species 11 and 12 should belong to the T quaternary structure, but with specific and substantial alterations in the tertiary structure in each case. This model has attempted to provide a logical framework to reconcile a number of structural and functional results reported by various researchers during the past 25 years.

However, there are serious discrepancies between the functional properties of ligation intermediates and the crystal structure features obtained for various partially ligated Hb species [such as $(\alpha_2^{O_2}\beta_2)$ and metal hybrid Hbs] as described in this article. We do not know whether these discrepancies are simply due to the fact that the Hb structures that are determined from crystals (grown under various crystallizing conditions) may not be the same ones as when the functional properties are determined under different solution conditions. Also, as mentioned in Section X, there is a serious inconsistency between the properties derived from X-ray crystallography and from optical spectroscopy on single crystals of Hb A as a function of oxygenation (Brzozowski et al., 1984; Liddington et al., 1988; Waller and Liddington, 1990; Mozzarelli et al., 1991). All these discrepancies need to be resolved and offer challenges for researchers of hemoglobin.

There is no doubt that researchers have made great progress in understanding the molecular basis for the cooperative oxygenation of Hb A. X-Ray crystallographic results clearly show that there are two well-defined structures for Hb A, i.e., the deoxy (T type) and the oxy (R type) quaternary structures. On the other hand, various published solution studies of Hb A also clearly indicate that these results are not consistent with a two-structure type of allosteric mechanism for the cooperative oxygenation process. Based on the results presented in this article, we would like to suggest that the molecular basis for the cooperative oxygenation of Hb is more consistent with a sequential-type model as proposed by Koshland et al. (1966) than a two-structure allosteric model such as the one proposed by Monod et al. (1965).

If a sequential-type mechanism is to operate in the cooperative oxygenation of Hb A, an essential requirement is that the ligand-induced conformational change of the ligated chain has to be able to propagate to the heme environments of its neighboring unligated chains within a tetrameric Hb molecule. This in turn enhances the ligand affinity of the unligated chains for the subsequent binding of the second, third, and fourth ligands. A two-state allosteric mechanism does not require that the ligand-induced conformational change needs to be transmitted to the neighboring unligated chains and alter their ligand affinities. On the contrary, it states that within a given quaternary structure, the ligation of one subunit will not affect the ligand affinity of its neighboring unligated subunits. It is the equilibrium between two quaternary structures [i.e., deoxy (or T) and oxy (or R)] that governs the ligand affinity of the Hb molecule. We would like to summarize seven separate sets of experimental results (five from our ^{1}H NMR results and two from other published results) that support the proposal that the cooperative oxygenation of Hb A follows a sequential-type mechanism. In the last section, we discussed the possible pathways for transmitting conformational change from one chain to another of the Hb molecules.

First, comparing the ferrous hyperfine-shifted proton resonances of four mutant deoxyhemoglobins [Hb Chesapeake (α92Arg → Leu), Hb J Capetown (α92Arg → Gln), Hb Yakima (β99Asp → His), and Hb Kempsey (β99Asp → Asn)] with single amino acid substitutions in the $\alpha_1\beta_2$ subunit interface with those of deoxy-Hb A, we have observed simultaneous alterations in these ferrous hfs protons (which have been assigned to either the α or the β chains), regardless of which chain is abnormal, as shown in Fig. 39 (Davis *et al.*, 1971). We have interpreted these results as suggesting that in Hb A, the unligated hemes can be influenced by the state of ligation of the hemes in the neighboring ligated chains of a tetrameric Hb molecule. This is an essential element for a sequential-type model such as the one proposed by Koshland *et al.* (1966). The structural features in the proximal histidyl residue, the switch region, and the $\alpha_1\beta_2$ subunit interface lend support to this idea (see Figs. 3–7).

Second, by monitoring the ferric hfs proton resonances of the abnormal β chains of Hb M Milwaukee (β67Val → Glu, $\alpha_2\beta_2^+$) as a function of oxygenation, we have found that the variations of these ferric hfs proton resonances of the β chains cannot be described as an average of the resonances of the fully deoxy and oxy species (Fung *et al.*, 1976, 1977). We have concluded that these ^{1}H NMR results are not consistent with a two-structure model for the oxygenation of Hb M Milwaukee and that by implication a two-structure concerted allosteric model cannot provide an adequate description for the oxygenation of Hb A.

Third, by monitoring the intensities of the two ferrous hfs proton resonances (one at $\sim+18$ ppm assigned to the β chain and one at $\sim+12$ ppm assigned to the α chain) as a function of oxygenation, we have found that their intensities decrease more than the total number of deoxy chains available as measured by the degree of O_2 saturation of Hb A (Figs. 40 and 41). We have interpreted this to mean that these two ferrous hfs proton resonances are sensitive to structural changes that are believed to occur in the unligated subunits on the ligation of their neighbors in an intact Hb molecule (Viggiano and Ho, 1979; Viggiano *et al.*, 1979; Ho *et al.*, 1982b). These results support a sequential-type model for the cooperative oxygenation of Hb A.

Fourth, our ^{1}H NMR studies on the singly ligated species using cross-linked asymmetric hybrid hemoglobins—$(\alpha^{+CN}\beta)_A(\alpha\beta)_C$XL and $(\alpha\beta^{+CN})_A(\alpha\beta)_C$XL—suggest that the $\alpha_1\beta_2$ or $\alpha_2\beta_1$ subunit interface of these singly ligated species is different from the interface in the deoxy (or T-like) or oxy (or R-like) quaternary structure. These results imply that (1) the ligand-induced tertiary structural changes can be propagated from the ligated subunit through the switch region to the $\alpha_1\beta_2$

subunit interface, and can thus affect the ligand affinity of the neighboring unligated chain and (2) there are more than two quaternary structures of Hb A in going from the deoxy to the oxy state (Miura and Ho, 1982). These findings provide support for a sequential-type model for the oxygenation of Hb A.

Fifth, we have prepared cross-linked modified hemoglobins—[α(des-Arg)β]$_A$[αβ]$_C$XL, [α(des-Arg-Tyr)β]$_A$[αβ]$_C$XL, [α(des-Arg)β(NES)]$_A$[αβ]$_C$XL, and [α(des-Arg)β]$_A$[αβ(NES)]$_C$XL—and used them to investigate the influence of salt bridges on tertiary and quaternary structures of Hb by ^{1}H NMR spectroscopy (Miura and Ho, 1984). We have found that several features of the ^{1}H NMR spectra of asymmetrically modified Hbs cannot be accounted for as a spectral sum of the intact dexoy-Hb C and chemically modified Hb A. The ^{1}H NMR spectra of deoxy [α(des-Arg)β(NES)]$_A$[αβ]$_C$XL and deoxy[α(des-ARG)β]$_A$[αβ(NES)]$_C$XL at low pH (pH ~6.0) cannot be explained simply as a sum of the spectral features specific for the deoxylike and the oxylike quaternary structures. Thus, our results suggest that there exist intermediate structures in which the tertiary and quaternary structural transitions occur asymmetrically about the diad axis of the Hb molecule during the course of successive removal of the salt bridges (Miura and Ho, 1984).

Sixth, the extensive thermodynamic and kinetic studies on dimer–tetramer assembly using various hybrid and mutant hemoglobins carried out by Ackers and co-workers clearly indicate that there are at least three molecular functional states for Hb A during the transition from the deoxy to the oxy state (Smith and Ackers, 1985; Ackers and Smith, 1987; Smith *et al.*, 1987; Daugherty *et al.*, 1991; Ackers *et al.*, 1992). These results are not consistent with a two-structure allosteric description for the oxygenation of Hb A.

Seventh, Smith *et al.* (1991) reported the structure of a mutant hemoglobin, HbCO Ypsilanti (β99Asp → Tyr). They have found that this mutant Hb has a new quaternary structure that is distinctly different from those of deoxy (or T) or oxy (or R) quaternary structures and that the hydrogen-bonding interactions in the $α_1β_2$ subunit interface (i.e., the so-called switch and flexible joint regions) have features characteristic of both unligated and ligated forms of Hb A (see Fig. 61). These findings also provide independent support to our conclusion based on our ^{1}H NMR studies of cross-linked modified Hbs regarding the existence of intermediate structures.

The hemoglobin problem is certainly not yet solved and there are more exciting discoveries to be made by researchers in the years ahead. A complete description of ligand binding to Hb A must include detailed analyses of both intermediate structural states of Hb and their functional properties. A challenge to hemoglobin researchers is to elucidate the structures of singly, doubly, and triply oxygenated Hb species and to determine their respective functional properties. We can then compare these structures to those of the deoxy (or T) quaternary and oxy (or R) quaternary structures and correlate these structural properties to their energetic and functional properties. These partially ligated Hb species may only exist transiently, but they play an essential role in making possible the transition from the deoxy to the oxy state of Hb A. These new findings will lead us not only to an understanding of hemoglobin cooperativity, but also to new knowledge of the molecular basis for the regulation of metabolic reactions by allosteric enzymes and multimeric proteins. Thus, we will come an important step closer to our eventual understanding of biological information transfer and signaling phenomena at the atomic level.

Acknowledgments

The ^{1}H NMR and biochemical results described in this article were carried out by present and former colleagues (undergraduates, graduate students, postdoctoral research associates, visiting scientists, research assistants, and collaborators from other institutions). Without their results and various mutant hemoglobins, this article could not have been written. I would like to make use of this opportunity to thank Dr. Mark R. Busch, Dr. Susan R. Dowd, Dr. John T. Edsall, Dr. David S. Eisenberg, Mr. Irving Geis, Dr. David D. Hackney, Dr. Alan P. Koretsky, Dr. Shigetoshi Miura, Dr. John S. Olson, Dr. E. Ann Pratt, and Dr. Irina M. Russu for helpful comments, to Mr. Irving Geis for providing Figs. 1–7, 11, and 62, and to Dr. Francine R. Smith for providing Fig. 61. I also want to express my gratitude to Dr. E. Ann Pratt for her constant advice to improve the text and to Ms. Cynthia J. Davis for her valuable assistance in manuscript preparation.

This research on hemoglobin has been supported by research grants from the National Institutes of Health and the National Science Foundation. During the past 15 years, our hemoglobin research has been supported by Grant HL-24525 from the National Institutes of Health.

References

Abragam, A. (1961). "The Principles of Nuclear Magnetism." Oxford Univ. Press (Clarendon) London and New York.
Ackers, G. K. (1990). *Biophys. Chem.* **37,** 371–382.
Ackers, G. K., and Smith, F. R. (1987). *Annu. Rev. Biophys. Biophys. Chem.* **16,** 583–609.
Ackers, G. K., Doyle, M. L., Myers, D., and Daugherty, M. A. (1992). *Science* **255,** 54–63.

Allerhand, A., and Gutowsky, H. G. (1964). *J. Chem. Phys.* **41,** 2115–2126.
Allerhand, A., and Gutowsky, H. G. (1965). *J. Chem. Phys.* **42,** 1587–1599.
Anderson, L. (1973). *J. Mol. Biol.* **79,** 495–506.
Antonini, E., and Brunori, M. (1969). *J. Biol. Chem.* **244,** 3909–3912.
Antonini, E., and Brunori, M. (1970). *Annu. Rev. Biochem.* **39,** 977–1042.
Antonini, E., and Brunori, M. (1971). "Hemoglobin and Myoglobin in their Reactions with Ligands." North-Holland Publ., Amsterdam.
Antonini, E., Wyman, J., Brunori, B., Taylor, J. F., Rossi-Fanelli, A., and Caputo, A. (1964). *J. Biol. Chem.* **239,** 907–912.
Antonini, E., Brunori, M., Wyman, J., and Noble, R. W. (1966). *J. Biol. Chem.* **241,** 3236–3238.
Arnone, A. (1972). *Nature (London)* **237,** 146–149.
Arnone, A., and Perutz, M. F. (1974). *Nature (London)* **249,** 34–36.
Arnone, A., Rogers, P., Blough, N. V., McGourty, J. L., and Hoffman, B. M. (1986). *J. Mol. Biol.* **188,** 693–706.
Asakura, T., Adachi, K., Wiley, J. S., Fung, L. W.-M., Ho, C., Kilmartin, J. V., and Perutz, M. F. (1976). *J. Mol. Biol.* **104,** 185–195.
Baldassare, J. J., Charache, S., Jones, R. T., and Ho, C. (1970). *Biochemistry* **9,** 4707–4713.
Baldwin, J. M. (1980). *J. Mol. Biol.* **136,** 103–128.
Baldwin, J. M., and Chothia, C. (1979). *J. Mol. Biol.* **129,** 175–220.
Banerjee, R., Statzowski, F., and Henry, Y. (1973). *J. Mol. Biol.* **73,** 455–467.
Bax, A. (1989). *Annu. Rev. Biochem.* **58,** 223–256.
Becker, E. D. (1980). "High Resolution NMR: Theory and Chemical Applications," 2nd ed. Academic Press, New York.
Benesch, R., and Benesch, R. E. (1964). *Nature (London)* **202,** 773–775.
Benesch, R., and Benesch, R. E. (1967). *Biochem. Biophys. Res. Commun.* **26,** 162–167.
Benesch, R. E., Edalji, R., and Benesch, R. (1977). *Biochemistry* **16,** 2594–2597.
Berjis, M., Bandyopadhyay, D., and Sharma, V. S. (1990). *Biochemistry* **29,** 10106–10113.
Bernal, J. D. (1958). *Discuss. Faraday Soc.* **25,** 7–18.
Bloembergen N. (1957). *J. Chem. Phys.* **27,** 595–596.
Blough, N. V., Zemel, H., and Hoffman, B. M. (1984). *Biochemistry* **23,** 2883–2891.
Braunitzer, G., Hilse, K., Rudloff, V., and Hilschmann, N. (1964). *Adv. Protein Chem.* **19,** 1–71.
Brunori, M., Amiconi, G., Antonini, E., Wyman, J., and Winterhalter, K. (1970). *J. Mol. Biol.* **49,** 461–471.
Brzozowski, A., Derewenda, Z., Dodson, E., Dodson, G., Grabowski, M., Liddington, R., Skarzynski, T., and Valley, D. (1984). *Nature (London)* **307,** 74–76.
Bunn, H. F., and Forget, B. G. (1986). "Hemoglobin: Molecular, Genetic, and Clinical Aspects." Saunders, Philadelphia.
Busch, M. R., and Ho, C. (1990). *Biophys. Chem.* **37,** 313–322.
Busch, M. R., Mace, J. E., Ho, N. T., and Ho, C. (1991). *Biochemistry* **30,** 1865–1877.
Carrington, A., and McLachlan, A. D. (1967). "Introduction to Magnetic Resonance." Harper & Row, New York.
Cassoly, R. (1978). *J. Biol. Chem.* **253,** 3602–3606.
Cassoly, R., and Gibson, Q. H. (1972). *J. Biol. Chem.* **247,** 7332–7341.
Cassoly, R., Gibson, Q. H., Ogawa, S., and Shulman, R. G. (1971). *Biochem. Biophys. Res. Commun.* **44,** 1015–1021.
Chada, K., Magram, J., Rapael, K., Radice, G., Lacy, E., and Constantini, F. (1985). *Nature (London)* **314,** 377–380.

Chanutin, A., and Curnish, R. R. (1967). *Arch. Biochem. Biophys.* **121,** 96–102.
Clore, G. M., and Gronenborn, A. M. (1991). *Science* **252,** 1390–1399.
Cole, F. X., and Gibson, Q. H. (1973). *J. Biol. Chem.* **248,** 4998–5004.
Coyrell, C. D. (1939). *J. Phys. Chem.* **43,** 841–852.
Craescu, C. T., and Mispelter, J. (1988). *Eur. J. Biochem.* **176,** 171–178.
Craescu, C. T., and Mispelter, J. (1989). *Eur. J. Biochem.* **181,** 87–96.
Craescu, C. T., Mispelter, J., and Blouquit, Y. (1990). *Biochemistry* **29,** 3953–3958.
Dalvit, C., and Ho, C. (1985). *Biochemistry* **24,** 3398–3407.
Dalvit, C., and Wright, P. E. (1987). *J. Mol. Biol.* **194,** 3398–3407.
Dalziel, K., and O'Brien, J. R. P. (1957a). *Biochem. J.* **67,** 199–124.
Dalziel, K., and O'Brien, J. R. P. (1957b). *Biochem. J.* **67,** 124–136.
Daugherty, M. A., Shea, M. A., Johnson, J. A., LiCata, V. J., Turner, G. J., and Ackers, G. K. (1991). *Proc. Natl. Acad. Sci. U.S.A.* **88,** 1110–1114.
Davis, D. G., Mock, N. L., Laman, V. R., and Ho, C. (1969a). *J. Mol. Biol.* **40,** 311–313.
Davis, D. G., Charache, S., and Ho, C. (1969b). *Proc. Natl. Acad. Sci. U.S.A.* **63,** 1403–1409.
Davis, D. G., Lindstrom, T. R., Mock, N. H., Baldassare, J. J., Charache, S., Jones, R. T., and Ho, C. (1971). *J. Mol. Biol.* **60,** 101–111.
Derewenda, Z., Dodson, G., Emsley, P., Harris, D., Nagai, K., Perutz, M., and Renaud, J.-P. (1990). *J. Mol. Biol.* **211,** 515–519.
Di Cera, E., Doyle, M. L., Connelly, P. R., and Gill, S. J. (1987). *Biochemistry* **26,** 6494–6502.
Dickerson, R. E., and Geis, I. (1969). "The Structure and Action of Proteins." Harper & Row, New York.
Dickerson, R. E., and Geis, I. (1983). "Hemoglobin: Structure, Function, Evolution, and Pathology." Benjamin/Cummings, Menlo Park, California.
Doyle, M. L., Speros, P. C., LiCata, V. J., Gingrich, D., Hoffman, B. M., and Ackers, G. K. (1991). *Biochemistry* **30,** 7263–7271.
Eaton, W. A., Henry, E. R., and Hofrichter, J. (1991). *Proc. Natl. Acad. Sci. U.S.A.* **88,** 4472–4475.
Edelstein, S. J. (1971). *Nature (London)* **230,** 224–227.
Edelstein, S. J., and Edsall, J. T. (1986). *Proc. Natl. Acad. Sci. U.S.A.* **83,** 3796–3800.
Edsall, J. T. (1972). *J. Hist. Biol.* **5,** 205–257.
Enoki, Y., and Tomita, S. (1968). *J. Mol. Biol.* **32,** 121–134.
Fermi, G. (1975). *J. Mol. Biol.* **97,** 237–256.
Fermi, G., and Perutz, M. F. (1981). In "Hemoglobin and Myoglobin: Atlas in Molecular Structures in Biology" (D. C. Phillips and F. M. Richards, eds.). Oxford Univ. Press (Clarendon), London and New York.
Fermi, G., Perutz, M. F., Shaanan, B., and Fomme, R. (1984). *J. Mol. Biol.* **175,** 159–174.
Ferrige, A. G., Lindon, J. C., and Paterson, R. A. (1979). *J. Chem. Soc., Faraday Trans.* **75,** 2851–2864.
Fung, L. W.-M. and Ho, C. (1975). *Biochemistry* **14,** 2526–2535.
Fung, L. W.-M., Ho, C., Roth, E. F., Jr., and Nagel, R. L. (1975). *J. Biol. Chem.* **250,** 4786–4789.
Fung, L. W.-M., Minton, A. P., and Ho, C. (1976). *Proc. Natl. Acad. Sci. U.S.A.* **73,** 1581–1585.
Fung, L. W.-M., Minton, A. P., Lindstrom, T. R., Pisciotta, A. V., and Ho, C. (1977). *Biochemistry* **16,** 1452–1462.
Gelin, B. R., and Karplus, M. (1977). *Proc. Natl. Acad. Sci. U.S.A.* **74,** 801–805.

Gelin, B. R., Lee, A. W.-M., and Karplus, M. (1983). *J. Mol. Biol.* **171,** 489–559.
German, B., and Wyman, J., Jr. (1937). *J. Biol. Chem.* **117,** 533–550.
Gibson, Q. H. (1973). *Proc. Natl. Acad. Sci. U.S.A.* **70,** 1–4.
Gibson, Q. H., and Edelstein, S. J. (1987). *J. Biol. Chem.* **262,** 516–519.
Glasoe, P. K., and Long, F. A. (1960). *J. Phys. Chem.* **64,** 188–191.
Gray, R. D., and Gibson, Q. H. (1971a). *J. Biol. Chem.* **246,** 5176–5178.
Gray, R. D., and Gibson, Q. H. (1971b). *J. Biol. Chem.* **246,** 7168–7174.
Groebe, D. R., Chung, A. E., and Ho, C. (1990). *Nucleic Acids Res.* **18,** 4033.
Groebe, D. R., Busch, M. R., Tsao, T. Y. M., Luh, F.-Y., Tam, M. F., Chung, A. E., Gaskell, M., Liebhaber, S. A., and Ho, C. (1992). *Protein Express. Purif.* **3,** 134–141.
Grosveld, F., van Assendelft, G. B., Graves, D. R., and Kollias, G. (1987). *Cell (Cambridge, Mass.)* **51,** 975–985.
Guéron, M. (1975). *J. Magn. Reson.* **19,** 58–66.
Gupta, R. K., Benovic, J. L., and Rose, Z. B. (1979). *J. Biol. Chem.* **254,** 8250–8255.
Haber, J. E., and Koshland, D. E., Jr. (1969). *Biochim. Biophys. Acta* **194,** 339–341.
Haber, J. E., and Koshland, D. E., Jr. (1971). *J. Biol. Chem.* **246,** 7790–7793.
Haigh, C. W., and Mallion, R. B. (1971). *Mol. Phys.* **22,** 955–970.
Haire, R. N., Tisel, W. A., Niazi, G., Rosenberg, A., Gill, S. J., and Richey, B. (1981). *Biochem. Biophys. Res. Commun.* **101,** 177–182.
Hanscombe, O., Vidal, M., Kaeda, J., Luzzatto, L., Graves, D. R., and Grosveld, F. (1989). *Genes Dev.* **3,** 1572–1581.
Hayashi, A., Motokama, Y., and Kikuchi, G. (1966). *J. Biol. Chem.* **241,** 79–84.
Hayashi, A., Suzuki, T., Imai, K., Morimoto, H., and Watari, H. (1969). *Biochim. Biophys. Acta* **194,** 6–15.
Hayashi, A., Suzuki, T., and Shin, M. (1973). *Biochim. Biophys. Acta* **310,** 309–316.
Heidner, E. J., Ladner, R. C., and Perutz, M. F. (1976). *J. Mol. Biol.* **104,** 707–722.
Herzfeld, J., and Stanley, E. (1974). *J. Mol. Biol.* **82,** 231–265.
Hille, R., Palmer, G., and Olson, J. S. (1977). *J. Biol. Chem.* **252,** 403–405.
Hille, R., Olson, J. S., and Palmer, G. (1979). *J. Biol. Chem.* **254,** 12100–12120.
Ho, C., and Lindstrom, T. R. (1972). *Adv. Exp. Med. Biol.* **28,** 65–76.
Ho, C., and Russu, I. M. (1981). *In* "Methods in Enzymology" (E. Antonini, L. Russi-Bernardis and E. Chiancone, eds.) vol 76, pp. 275–312. Academic Press, New York.
Ho, C., and Russu, I. M. (1985). "New Methodologies in Studies of Protein Configuration," pp. 1–35. Van Nostrand-Reinhold, New York.
Ho, C., and Russu, I. M. (1987). *Biochemistry* **26,** 6299–6305.
Ho, C., Baldassare, J. J., and Charache, S. (1970). *Proc. Natl. Acad. Sci. U.S.A.* **66,** 722–729.
Ho, C., Lindstrom, T. R., Baldassare, J. J., and Breen, J. J. (1973). *Ann. N. Y. Acad. Sci.* **222,** 21–39.
Ho, C., Fung, L. W.-M., Wiechelman, K. J., Pifat, G., and Johnson, M. E. (1975). *In* "Erythrocyte Structure and Function" (G. J. Brewer,ed.), pp. 43–64. Alan R. Liss, New York.
Ho, C., Fung, L. W.-M., and Wiechelman, K. J. (1978). *In* "Methods in Enzymology" (S. Fleischer and L. Packer, eds.), Vol. 54, pp. 192–223. Academic Press, New York.
Ho, C., Eaton, W. A., Collman, J. P., Gibson, Q. H., Leigh, J. S., Jr., Margoliash, E., Moffat, K., and Scheidt, W. R., eds. (1982a). "Hemoglobin and Oxygen Binding." Elsevier/North-Holland, New York.
Ho, C., Lam, C.-H. J., Takahashi, S., and Viggiano, G. (1982b). *In* "Hemoglobin and Oxygen Binding" (C. Ho, W. A. Eaton, J. P. Collman, Q. H. Gibson, L. S. Leigh,

Jr., E. Margoliash, K. Moffat, and W. R. Scheidt, eds.) pp. 141–149. Elsevier/North-Holland, New York.
Hochmann, J., and Kellerhals, H. (1980). *J. Magn. Reson.* **38,** 23–39.
Hoffman, B. M., and Petering, D. H. (1970). *Proc. Natl. Acad. Sci. U.S.A.* **67,** 637–641.
Hoffman, S. J., Locker, D. L., Roehrich, J. M., Cozart, P. E., Durfee, S. L., Tedesco, J. L., and Stettler, G. L. (1990). *Proc. Natl. Acad. Sci. U.S.A.* **87,** 8521–8525.
Hofrichter, J., Henry, E. R., Sommer, J. H., Deutsch, R., Ikeda-Saito, M., Yonetani, T., and Eaton, W. A. (1985). *Biochemistry* **24,** 2667–2679.
Hopfield, J. J. (1973). *J. Mol. Biol.* **77,** 207–222.
Huang, T.-H., and Redfield, A. G. (1976). *J. Biol. Chem.* **251,** 7114–7119.
Huestis, W. H., and Raftery, M. A. (1972a). *Biochem. Biophys. Res. Commun.* **48,** 678–683.
Huestis, W. H., and Raftery, M. A. (1972b). *Biochem. Biophys. Res. Commun.* **49,** 428–433.
Huestis, W. H., and Raftery, M. A. (1972c). *Biochem. Biophys. Res. Commun.* **49,** 1358–1365.
Huestis, W. H., and Raftery, M. A. (1972d). *Biochemistry* **11,** 1648–1654.
Huestis, W. H., and Raftery, M. A. (1973). *Biochemistry* **12,** 2531–2535.
Imai, K. (1982). "Allosteric Effects in Haemoglobin." Cambridge Univ. Press, Cambridge, England.
Imai, K., and Yonetani, T. (1977). *Biochim. Biophys. Acta* **490,** 164–170.
Imai, K., Yonetani, T., and Ikeda-Saito, M. (1977). *J. Mol. Biol.* **109,** 83–97.
Imai, K., Ikeda-Saito, M., Yamamoto, H., and Yonetani, T. (1980). *J. Mol. Biol.* **138,** 635–648.
Imai, K., Fushitani, K., Miyazaki, G., Ishimori, K., Kitagawa, T., Wada, Y., Morimoto, H., Morishima, I., Shih, D. T.-B. and Tame, J. (1991). *J. Mol. Biol.* **218,** 769–778.
Inubushi, T., D'Ambrosio, C., Ikeda-Saito, M., and Yonetani, T. (1986). *J. Am. Chem. Soc.* **108,** 3799–3083.
Ishimori, K., Imai, K., Miyazaki, G., Kitagawa, T., Wada, Y., and Morimoto, I. (1992). *Biochemistry* **31,** 3256–3264.
Jesson, J. P. (1967). *J. Chem. Phys.* **47,** 579–581.
Johnson, C. E., Jr., and Bovey, F. A. (1958). *J. Chem. Phys.* **29,** 1012–1014.
Johnson, M. E., and Ho, C. (1974). *Biochemistry* **13,** 3653–3661.
Johnson, M. E., Fung, L. W.-M., and Ho, C. (1977). *J. Am. Chem. Soc.* **99,** 1245–1250.
Johnson, M. L., Turner, B. W., and Ackers, G. K. (1984). *Proc. Natl. Acad. Sci. U.S.A.* **81,** 1093–1097.
Karplus, M. (1959). *J. Chem. Phys.* **30,** 11–18.
Karplus, M. (1963). *J. Am. Chem. Soc.* **85,** 2870–2871.
Karplus, M., Case, D. A., Gelin, B., Huynh, B. H., Lee, A. W.-M., and Sazbo, A. (1989). "Allosteric Enzymes," pp. 27–59. CRC Press, Boca Raton, Florida.
Kilmartin, J. J., Breen, J. J., Roberts, G. C. K., and Ho, C. (1973). *Proc. Natl. Acad. Sci. U.S.A.* **70,** 1246–1249.
Kilmartin, J. V., Hewitt, J. A., and Wootton, J. F. (1975). *J. Mol. Biol.* **93,** 203–218.
Koshland, D. E., Jr., Némethy, G., Filmer, D. (1966). *Biochemistry* **5,** 365–385.
Kurland, R. J., and McGarvey, B. R. (1970). *J. Magn. Reson.* **2,** 286–301.
La Mar, G. N., Budd, D. L., and Goff, H. (1977). *Biochem. Biophys. Res. Commun.* **77,** 104–110.
La Mar, G. N., Nagai, K., Jue, T., Budd, D. L., Gersonde, K., Sick, H., Kagimoto, T., Hayashi, A., and Taketa, F. (1980). *Biochem. Biophys. Res. Commun.* **96,** 1172–1177.

Lee, A. W.-M., and Karplus, M. (1983). *Proc. Natl. Acad. Sci. U.S.A.* **80,** 7055–7059.
Liddington, R., Derewenda, Z., Dodson, G., and Harris, D. (1988). *Nature (London)* **331,** 725–728.
Lindstrom, T. R., and Ho, C. (1972). *Proc. Natl. Acad. Sci. U.S.A.* **69,** 1707–1710.
Lindstrom, T. R., and Ho, C. (1973). *Biochemistry* **12,** 134–139.
Lindstrom, T. R., Olson, J. S., Nock, N. H., Gibson, Q. H., and Ho, C. (1971). *Biochem. Biophys. Res. Commun.* **45,** 22–26.
Lindstrom, T. R., Ho, C., and Pisciotta, A. V. (1972a). *Nature (London), New Biol.* **237,** 263–264.
Lindstrom, T. R., Noren, I. B. E., Charache, S., Lehman, H., and Ho, C. (1972b). *Biochemistry* **11,** 1677–1681.
Lindstrom, T. R., Baldassare, J. J., Bunn, H. F., and Ho, C. (1973). *Biochemistry* **12,** 4212–4217.
Luisi, B., and Shibayama, N. (1989). *J. Mol. Biol.* **206,** 723–736.
Luisi, B., Liddington, B., Fermi, G., and Shibayama, N. (1990). *J. Mol. Biol.* **214,** 7–14.
Madrid, M., Simplaceanu, V., Ho, N. T., and Ho, C. (1990). *J. Magn. Reson.* **88,** 42–59.
Maeda, T., and Ohnishi, S. (1971). *Biochemistry* **10,** 1177–1180.
Maeda, T., Imai, K., and Tyuma, I. (1972). *Biochemistry* **11,** 3685–3689.
Mansouri, A., and Winterhalter, K. H. (1973). *Biochemistry* **12,** 4946–4949.
Marden, M. C., Hazard, E. S., and Gibson, Q. H. (1986). *Biochemistry* **25,** 7591–7596.
Marquandt, D. W. (1963). *J. Soc. Ind. Appl. Math.* **11,** 431–436.
Mathews, A. J., Rohlfs, R., Olson, J. S., Tame, J., Renaud, J.-P., and Nagai, K. (1989). *J. Biol. Chem.* **264,** 16753–16583.
Mathews, A. J., Olson, J. S., Renaud, J.-P., Tame, J., and Nagai, K. (1991). *J. Biol. Chem.* **266,** 21631–21639.
Mayer, A., Ogawa, S., Shulman, R. G., and Gersonde, K. (1973). *J. Mol. Biol.* **81,** 187–197.
McConnell, H. M. (1956). *J. Chem. Phys.* **24,** 764–766.
McConnell, H. M. (1971). *Annu. Rev. Biochem.* **40,** 227–236.
McConnell, H. M., and Robertson, R. E. (1958). *J. Chem. Phys.* **29,** 1361–1365.
McConnell, H. M., Ogawa, S., and Horwitz, A. (1968). *Nature (London)* **220,** 787–788.
McDonald, C. C., and Phillips, P. W. (1967). *J. Am. Chem. Soc.* **89,** 6332–6341.
Mims, M. P., Olson, J. S., Russu, I. M., Miura, S., Cedel, T. E., and Ho, C. (1983a). *J. Biol. Chem.* **258,** 6125–6134.
Mims, M. P., Porras, A. G., Olson, J. S., Noble, R. W., and Peterson, J. A. (1983b). *J. Biol. Chem.* **258,** 14219–14232.
Minton, A. P. (1974). *Science* **184,** 577–579.
Minton, A. P., and Imai, K. (1974). *Proc. Natl. Acad. Sci. U.S.A.* **71,** 1418–1421.
Miura, S., and Ho, C. (1982). *Biochemistry* **21,** 6280–6287.
Miura, S., and Ho, C. (1984). *Biochemistry* **23,** 2492–2499.
Miura, S., Ikeda-Saito, M., Yonetani, T., and Ho, C. (1987). *Biochemistry* **26,** 2149–2155.
Moffat, K., Deatherage, J. F., and Seybert, D. W. (1979). *Science* **206,** 1035–1042.
Monod, J., Wyman, J., and Changeux, J.-P. (1965). *J. Mol. Biol.* **12,** 88–118.
Mozzarelli, A., Rivetti, C., Rossi, G. L., Henry, E. R., and Eaton, W. A. (1991). *Nature (London)* **351,** 416–419.
Nagai, K. (1977). *J. Mol. Biol.* **111,** 41–53.
Nagai, K., and Thørgersen, H. C. (1984). *Nature (London)* **309,** 810–812.
Nagai, K., La Mar, G. N., Jue, T., and Bunn, H. F. (1982). *Biochemistry* **21,** 842–847.

Nagai, K., Perutz, M. F., and Poyart, C. (1985). *Proc. Natl. Acad. Sci. U.S.A.* **82,** 7252–7255.
Nagai, K., Luisi, B., Shih, D., Miyazaki, G., Imai, K., Poyart, C., De Young, A., Kwiatkowski, L., Noble, R. W., Lin, S.-H., and Yu, N. T. (1987). *Nature (London)* **329,** 858–860.
Nagel, R. L., Gibson, Q. H., and Charache, S. (1967). *Biochemistry* **6,** 2395–2402.
Nasuda-Kouyama, A., Tachibana, H., and Wada, A. (1983). *J. Mol. Biol.* **164,** 451–476.
Noggle, J. H., and Schirmer, R. E. (1971). "The Nuclear Overhauser Effect: Chemical Applications." Academic Press, New York.
Ogata, R. T., and McConnell, H. M. (1971). *Cold Spring Harbor Symp. Quant. Biol.* **36,** 325–336.
Ogata, R. T., and McConnell, H. M. (1972a). *Proc. Natl. Acad. Sci. U.S.A.* **69,** 335–339.
Ogata, R. T., and McConnell, H. M. (1972b). *Biochemistry* **11,** 4792–4799.
Ogawa, S., and Shulman, R. G. (1971). *Biochem. Biophys. Res. Commun.* **42,** 9–15.
Ogawa, S., and Shulman, R. G. (1972). *J. Mol. Biol.* **70,** 315–336.
Ogawa, S., McConnell, H. M., and Horwitz, A. F. (1968). *Proc. Natl. Acad. Sci. U.S.A.* **61,** 401–405.
Olson, J. S. (1981). *In* "Methods in Enzymology" (E. Antonini, L. Rossi-Bernardi, and E. Chiancone, eds.) Vol. 76, pp. 631–651. Academic Press, New York.
Olson, J. S., and Gibson, Q. H. (1971). *J. Biol. Chem.* **246,** 5241–5253.
Olson, J. S., and Gibson, Q. H. (1972). *J. Biol. Chem.* **247,** 1713–1726.
Olson, J. S., and Gibson, Q. H. (1973a). *J. Biol. Chem.* **248,** 1616–1622.
Olson, J. S., and Gibson, Q. H. (1973b). *J. Biol. Chem.* **248,** 1623–1630.
Olson, J. S., Mims, M. P., and Reisberg, P. I. (1982). *In* "Hemoglobin and Oxygen Binding" (C. Ho, W. A. Eaton, J. P. Collman, Q. H. Gibson, J. S. Leigh, Jr., E. Margoliash, K. Moffat, and W. R. Scheidt, eds.), pp. 393–398. Elsevier/North-Holland, New York.
Olson, J. S., Mathews, A. J., Rohlfs, R. J., Springer, B. A., Egeberg, K. D., Sligar, S. G., Tame, J., Renaud, J.-P., and Nagai, K. (1988). *Nature (London)* **336,** 265–266.
Pauling, L. (1935), *Proc. Natl. Acad. Sci. U.S.A.* **21,** 186–191.
Pauling, L. (1936). *J. Chem. Phys.* **4,** 673–677.
Peller, L. (1982). *Nature (London)* **300,** 661–662.
Perkins, S. J. (1980). *J. Magn. Reson.* **38,** 297–312.
Perrella, M., Benazzi, L., Cremonesi, L., Veseley, S., Viggiano, V., and Rossi-Bernardi, L. (1983). *J. Biol. Chem.* **258,** 4511–4517.
Perrella, M., Sabbioneda, L., Samaga, M., and Rossi-Bernardi, L. (1986). *J. Biol. Chem.* **261,** 8391–8396.
Perrella, M., Benazzi, L., Shea, M. A., and Ackers, G. K. (1990a). *Biophys. Chem.* **35,** 97–103.
Perrella, M., Colosimo, A., Benazzi, L., Ripamonti, M., and Rossi-Bernardi, L. (1990b). *Biophys. Chem.* **37,** 211–223.
Perrella, M., Davids, N., and Rossi-Bernardi, L. (1992). *J. Biol. Chem.*, **267,** 8744–8751.
Perutz, M. F. (1968). *J. Cryst. Growth* **2,** 54–56.
Perutz, M. F. (1970). *Nature (London)* **228,** 726–739.
Perutz, M. F. (1976). *Br. Med. Bull.* **32,** 195–208.
Perutz, M. F. (1979). *Annu. Rev. Biochem.* **48,** 327–386.
Perutz, M. F. (1989). *Q. Rev. Biophys.* **22,** 139–236.
Perutz, M. F. (1990). *Annu. Rev. Physiol.* **52,** 1–25.

Perutz, M. F., and Ten Eyck, L. F. (1971). *Cold Spring Harbor Symp. Quant. Biol.* **36**, 295–309.
Perutz, M. F., Muirhead, H., Cox, J. M., and Goaman, L. C. G. (1968). *Nature (London)* **219**, 131–139.
Perutz, M. F., Pulsinelli, P. D., and Ranney, H. M. (1972). *Nature (London), New Biol.* **237**, 259–263.
Perutz, M. F., Ladner, J. E., Simon, S. R., and Ho, C. (1974). *Biochemistry* **13**, 2163–2173.
Perutz, M. F., Fermi, G., Luisi, B., Shaanan, B., and Liddington, R. C. (1987). *Acc. Chem. Res.* **20**, 309–321.
Pisciotta, A. V., Ebbe, S. N., and Hinz, J. E. (1959). *J. Lab. Clin. Med.* **54**, 73–87.
Plateau, P., Dumas, C., and Guéron, M. (1983). *J. Magn. Reson.* **54**, 46–53.
Pople, J. A. (1956). *J. Chem. Phys.* **24**, 1111.
Pople, J. A., Schneider, W. G., and Bernstein, H. J. (1959). "High-Resolution Nuclear Magnetic Resonance," Chapter 10. McGraw-Hill, New York.
Reisberg, P. I., and Olson, J. S. (1980a). *J. Biol. Chem.* **225**, 4144–4150.
Reisberg, P. I., and Olson, J. S. (1980b). *J. Biol. Chem.* **225**, 4151–4158.
Reisberg, P. I., and Olson, J. S. (1980c). *J. Biol. Chem.* **225**, 4159–4169.
Russu, I. M., and Ho, C. (1982). *Biophys. J.* **39**, 203–210.
Russu, I. M., Ho, N. T., and Ho, C. (1980). *Biochemistry* **19**, 1043–1052.
Russu, I. M., Ho, N. T., and Ho, C. (1982). *Biochemistry* **21**, 5031–5043.
Russu, I. M., Lin, A. K.-L. C., Ferro-Dosch, S., and Ho, C. (1984). *Biochim. Biophys. Acta* **785**, 123–131.
Russu, I. M., Ho, N. T., and Ho, C. (1987). *Biochim. Biophys. Acta* **914**, 40–48.
Russu, I. M., Wu, S.-S., Ho, N. T., Kellogg, G. W., and Ho, C. (1989). *Biochemistry* **28**, 5298–5306.
Russu, I. M., Wu, S.-S., Bupp, K. A., Ho, N. T., and Ho, C. (1990). *Biochemistry* **29**, 3785–3792.
Ryan, T. M., Behringer, R. R., Townes, T. M., Palmiter, R. D., and Brinster, R. L. (1989). *Proc. Natl. Acad. Sci. U.S.A.* **86**, 37–41.
Samaja, M., Rovida, E., Niggeler, M., Perrella, M., and Rossi-Bernardi, L. (1987). *J. Biol. Chem.* **262**, 4528–4533.
Saroff, H. A. (1972). *J. Phys. Chem.* **76**, 1597–1607.
Saroff, H. A., and Minton, A. P. (1972). *Science* **175**, 1253–1255.
Saroff, H. A., and Yap, W. T. (1972). *Biophysics* **11**, 957–971.
Schaeffer, C., Craescu, C. T., Mispelter, J., Garel, M.-C., Rosa, J., and Lhoste, J.-M. (1988). *Eur. J. Biochem.* **73**, 313–325.
Scheidt, W. R. (1977). *Acc. Chem. Res.* **10**, 339–345.
Shaanan, B. (1983). *J. Mol. Biol.* **171**, 31–59.
Sharma, V. S. (1989). *J. Biol. Chem.* **264**, 10582–10588.
Sharma, V. S., Schmidt, M. S., and Ranney, H. M. (1976). *J. Biol. Chem.* **251**, 4267–4272.
Sharma, V. S., Bandyopadhyay, D., Berjis, M., Rifkind, J., and Boss, G. R. (1991). *J. Biol. Chem.* **266**, 24492–24497.
Shibayama, N., Morimoto, H., and Miyazaki, G. (1986a). *J. Mol. Biol.* **192**, 323–329.
Shibayama, N., Morimoto, H., and Kitagawa, T. (1986b). *J. Mol. Biol.* **192**, 331–336.
Shibayama, N., Inubishi, T., Morimoto, H., and Yonetani, T. (1987). *Biochemistry* **26**, 2194–2201.
Shibayama, N., Imai, K., Hirata, H., Hiraiwa, H., Morimoto, H., and Saigo, S. (1991). *Biochemistry* **30**, 8158–8165.
Shih, D. T.-B., and Perutz, M. F. (1987). *J. Mol. Biol.* **195**, 419–422.
Shih, D. T.-B., Perutz, M. F., Gronenborn, A. M., and Clore, G. M. (1987). *J. Mol. Biol.* **195**, 453–455.

Shulman, R. G., Ogawa, S., Wüthrich, K., Yamone, T., Peisach, J., and Blumberg, W. E. (1969). *Science* **165,** 251–257.
Shulman, R. G., Glarum, S. H., and Karplus, M. (1971). *J. Mol. Biol.* **57,** 93–115.
Shulman, R. G., Ogawa, S., and Hopfield, J. J. (1972). *Arch. Biochem. Biophys.* **151,** 68–74.
Shulman, R. G., Hopfield, J. J., and Ogawa, S. (1975). *Q. Rev. Biophys.* **8,** 325–420.
Silva, M. M., Roger, P.H., and Arnone, A. (1992). *J. Biol. Chem.* **267,** 17248–17256.
Simolo, K., Stucky, G., Chen, S., Bailey, M., Scholes, C., and McLendon, G. (1985). *J. Am. Chem. Soc.* **107,** 2865–2872.
Slichter, C. P. (1990). "Principles of Magnetic Resonance," 3rd ed. Springer-Verlag, Berlin.
Smith, F. R., and Ackers, G. K. (1985). *Proc. Natl. Acad. Sci. U.S.A.* **82,** 5347–5351.
Smith, F. R., Gingrich, D., Hoffman, B. M., and Ackers, G. K. (1987). *Proc. Natl. Acad. Sci. U.S.A.* **84,** 7089–7093.
Smith, F. R., Lattman, E. E., and Carter, C. W., Jr. (1991). *Proteins: Struct., Funct., Genet.* **10,** 81–91.
Solomon, I. (1955). *Phys. Rev.* **99,** 559–565.
Speros, P. C., LiCata, V. J., Yonetani, T., and Ackers, G. K. (1991). *Biochemistry* **30,** 7254–7262.
Sternlicht, H., and Wilson, D. (1967). *Biochemistry* **6,** 2881–2892.
Stetzkowski, F., Cassoly, R., and Banerjee, R. (1979). *J. Biol. Chem.* **254,** 11351–11356.
Szabo, A. (1978). *Proc. Natl. Acad. Sci. U.S.A.* **75,** 2108–2111.
Szabo, A., and Karplus, M. (1972). *J. Mol. Biol.* **72,** 163–197.
Takahashi, S., Lin, A. K.-L. C., and Ho, C. (1980). *Biochemistry* **19,** 5196–5202.
Takano, T. (1977). *J. Mol. Biol.* **110,** 569–584.
Tame, J., Shih, D. T.-B., Pagnier, J., Fermi, G., and Nagai, K. (1991). *J. Mol. Biol.* **218,** 761–767.
Thomas, J. O., and Edelstein, S. J. (1972). *J. Biol. Chem.* **247,** 7870–7874.
Tjandra, N., Simplaceanu, V., Cottam, P. F., and Ho, C. (1992). *J. Biomol. NMR* **2,** 149–160.
Triesman, R., Orkin, S. H., and Maniatis, T. (1983). *Proc. Natl. Acad. Sci. U.S.A.* **80,** 7428–7432.
Tyuma, I., Imai, K., and Shimuzu, K. (1973). *Biochemistry* **12,** 1491–1498.
Udem, L., Ranney, H. M., Bunn, H. F., and Pisciotta, A. V. (1970). *J. Mol. Biol.* **48,** 489–498.
Valdes, R., Jr., and Ackers, G. K. (1977). *J. Biol. Chem.* **252,** 74–81.
Valdes, R., Jr., and Ackers, G. K. (1978a). *Proc. Natl. Acad. Sci. U.S.A.* **75,** 311–314.
Valdes, R., Jr., and Ackers, G. K. (1978b). *In* "Biochemical and Clinical Aspects of Hemoglobin Abnormalities" (W. S. Caughey, ed.)pp. 527–532. Academic Press, New York.
Viggiano, G., and Ho, C. (1979). *Proc. Natl. Acad. Sci. U.S.A.* **76,** 3673–3677.
Viggiano, G., Wiechelman, K. J., Chervenick, P. A., and Ho, C. (1978). *Biochemistry* **17,** 795–799.
Viggiano, G., Ho, N. T., and Ho, C. (1979). *Biochemistry* **18,** 5238–5247.
Wagenbach, M., O'Rourke, K., Vitez, L., Wieczorek, A., Hoffman, S., Durfee, S., Tedesco, J., and Stetler, G. (1991). *Bio/Technology* **9,** 57–61.
Walder, J. A., Zaugg, R. H., Walder, R. Y., Steel, J. M., and Klotz, I. M. (1979). *Biochemistry* **18,** 4265–4270.
Walder, J. A., Walder, R. Y., and Arnone, A. (1980). *J. Mol. Biol.* **141,** 195–216.
Waller, D. A., and Liddington, R. C. (1990). *Acta Crystallogr. Sect. B* **B46,** 409–418.
Ward, K. R., Wishner, B. C., Lattman, E. E., and Love, W. E. (1975). *J. Mol. Biol.* **98,** 161–171.
Weatherall, D. J., Clegg, J. B., Callender, S. T., Welk, R. M. G., Gale, R. E., Huehns,

E. R., Perutz, M. F., Viggiano, G., and Ho, C. (1977). *Br. J. Haematol.* **35,** 177–191.
Weber, G. (1982). *Nature (London)* **300,** 603–607.
Wiechelman, K. J., Charache, S., and Ho, C. (1974). *Biochemistry* **13,** 4772–4777.
Wüthrich, K. (1976). "NMR in Biological Research: Peptides and Proteins." Elsevier/North-Holland, Amsterdam, and American Elsevier, New York.
Wüthrich, K. (1986). "NMR of Proteins and Nucleic Acids." Wiley, New York.
Wüthrich, K., Shulman, R. G., and Peisach, J. (1968). *Proc. Natl. Acad. Sci. U.S.A.* **60,** 373–380.
Wyman, J. (1964). *Ad. Protein Chem.* **19,** 223–286.
Wyman, J., Bishop, G., Richey, B., Spokane, R., and Gill, S. (1982). *Biopolymers* **21,** 1735–1747.
Yao, C., Simplaceanu, V., Lin, A. K.-L. C., and Ho, C. (1986). *J. Magn. Reson.* **66,** 43–57.
Yonetani, T., Yamamoto, H., and Woodrow, G. V. (1974). *J. Biol. Chem.* **249,** 682–690.

NOTE ADDED IN PROOF. Recently, Arnone and co-workers (Silva *et al.*, 1992) reported that a new R-type structure for HbCO A crystals grown in low salt conditions (i.e., in 100 mM sodium cacodylate plus 75 mM chloride in the presence of ~16% polyethylene glycol 6000 at pH 5.8) has been obtained at 1.7 Å resolution. This structure is distinctly different from the classical R-structure of ligated hemoglobin reported by Perutz and co-workers. Arnone and co-workers have named this new quaternary structure for ligated Hb the "R2-state". This new X-ray crystallographic finding supports one of our main conclusions reached in this review that there are more than two quaternary structures of Hb A and that a two-state allosteric description of the cooperative oxygenation of hemoglobin cannot satisfactorily account for the properties of hemoglobin during the transition from the deoxy to the oxy state.

Viggiano *et al.* (1978) assigned +6.4-ppm resonance to the hydrogen bond between β98Val and β145Tyr. It has been pointed out by Ho and co-workers that this assignment is tentative and needs further verification. Very recently, Ishimori *et al.* (1992) reported their spectroscopic and biochemical studies on two recombinant hemoglobins, Hb(β37Trp→Phe) and Hb (β145Tyr→Phe). They found that the exchangeable proton resonance at +6.4 ppm is missing in deoxy-Hb(β37Trp→Phe), but is present in deoxy-Hb(β145Tyr→Phe). Based on their ^{1}H NMR results on deoxy-Hb(β145Tyr→Phe), they concluded that the +6.4-ppm resonance cannot originate from the hydrogen bond between β98Val and β145Tyr. Because this resonance is missing in deoxy-Hb(β37Trp→Phe), they concluded that the +6.4-ppm resonance is due to the intersubunit hydrogen bond between α94Asp and β37Trp. It should be mentioned that there are inconsistencies between their spectroscopic and biochemical results on these two recombinant Hbs. Ishimori *et al.* (1992) need to ascertain the nature of the proton resonances over the spectral region from +5 to +10 ppm from H$_2$O, i.e., which ones are due to the hfs proton resonances and which ones are due to the exchangeable proton resonances, or both. Without a knowledge of the nature of these resonances, they cannot make conclusions regarding the tertiary and quaternary structures of these two mutant Hbs from their ^{1}H NMR results and cannot make definitive resonance assignments. It should be noted that Fung and Ho (1975) first suggested that the +6.4-ppm resonance could arise from the hydrogen bond between α94Asp and β37Trp. Based on available information, the origin of the +6.4-ppm resonance needs additional verification.

THERMODYNAMICS OF STRUCTURAL STABILITY AND COOPERATIVE FOLDING BEHAVIOR IN PROTEINS

By KENNETH P. MURPHY and ERNESTO FREIRE

Department of Biology and Biocalorimetry Center, The Johns Hopkins University, Baltimore, Maryland 21218

I.	Introduction	313
II.	Folding/Unfolding Partition Function	314
III.	Estimation of Forces Required to Specify Partition Functions	316
	A. Results of Model Compound Studies	318
	B. Protein Denaturation Studies	327
	C. Origins of Convergence Temperatures	329
IV.	Calculation of Folding/Unfolding Thermodynamics from Protein Structure	335
V.	Thermodynamic Stability and Cooperative Interactions	340
	A. Hierarchical Approach to Partition Functions	340
	B. Thermodynamic Dissection of Cooperative Interactions	342
	C. Classification of Cooperative Forces and Mechanisms	344
VI.	Cooperativity of Two-Domain Proteins	347
	Yeast Phosphoglycerate Kinase	348
VII.	Cooperativity of Single-Domain Proteins	351
	Myoglobin	352
VIII.	Energetics of Molten Globule Intermediates	355
IX.	Concluding Remarks	358
	References	358

I. INTRODUCTION

The problem of predicting the structure of a protein from its amino acid sequence constitutes the classical "protein folding problem." A solution to this problem has eluded the efforts of many investigators ever since Anfinsen (1973) first demonstrated that proteins fold spontaneously into their functional native conformations. A related but simpler problem is that of predicting the stability and folding mechanism of a protein if its structure is known. The thermodynamic aspects of this problem constitute the subject of this review. This area of research has acquired significant practical importance because it provides the theoretical framework for the development of rational strategies for protein design and protein modification.

The energetics of the fundamental forces that determine the stability of proteins have been refined to the point to which quantitative calculations are now possible. These calculations are directed to (1) predict the overall stability of proteins based on structural information, (2) predict the cooperative folding/unfolding behavior of proteins, (3) predict the magnitude of the interactions between structural domains in proteins, and (4) predict the presence of specific partially folded intermediates (e.g., molten globules). Recent advances in this area will be discussed in this article.

II. Folding/Unfolding Partition Function

From a statistical thermodynamic standpoint, the description of the folding/unfolding equilibrium in proteins requires the specification of the system partition function, Q, defined as the sum of the statistical weights of all the possible states of the molecule (see Freire and Biltonen, 1978a):

$$Q = \sum_{i=0}^{n} e^{-\Delta G_i/RT} \tag{1}$$

where ΔG_i is the Gibbs free energy difference between the ith state and the reference state, R is the gas constant, and T is the absolute temperature. Following the standard convention used in protein thermodynamics, the reference state will be the native state throughout this article. Evaluation of the partition function requires the identification and enumeration of the relevant folding states (folded, unfolded, and partially folded states) of the protein and their Gibbs free energies. The population of any particular state, P_i, is then calculated with Eq. (2)

$$P_i = \frac{e^{-\Delta G_i/RT}}{Q} \tag{2}$$

and can be used to calculate observed system properties. Of particular importance is the average excess enthalpy function, $\langle \Delta H \rangle$, defined as

$$\langle \Delta H \rangle = \sum_{i=0}^{n} P_i \Delta H_i \tag{3}$$

because it allows calculation of the excess heat capacity function measured by differential scanning calorimetry:

$$\langle \Delta C_p \rangle = \frac{\partial}{\partial T} \langle \Delta H \rangle = \sum_{i=0}^{n} \Delta H_i \frac{\partial}{\partial T} P_i + \sum_{i=0}^{n} P_i \Delta C_{p,i} \quad (4)$$

Equations (1)–(4) provide the basic statistical thermodynamic framework necessary to deal with the protein folding problem. Several years ago, Freire and Biltonen (1978a) showed that scanning calorimetry data could be used to evaluate the protein folding/unfolding partition function experimentally by a double integration procedure:

$$\ln Q(T) = \int_{T_0}^{T} \frac{1}{RT^2} \left[\int_{T_0}^{T} \langle \Delta C_p \rangle \, dT \right] dT \quad (5)$$

where T_0 is a temperature at which the protein molecules are in the reference (folded) state. It was also shown that this experimentally obtained partition function could be used to assess the existence of equilibrium folding intermediates and to characterize them from a thermodynamic point of view (Freire and Biltonen, 1978a). Because the enthalpy and inverse temperature are conjugate thermodynamic variables, the calculation of the folding/unfolding partition function from calorimetric data requires no *a priori* assumptions regarding the number of states that become populated during the transition.

Calorimetric data provide a complete thermodynamic characterization as well as a direct experimental evaluation of the folding/unfolding partition function and the population of intermediate states. This approach has been used in numerous calorimetric applications during the last decade and will not be reviewed here (the reader is referred to Freire and Biltonen, 1978b; Privalov, 1982; Freire, 1989; Freire *et al.*, 1990, for reviews in this area).

It has been shown that the folding/unfolding partition function can also be approximated from structural information (Freire and Murphy, 1991; Freire *et al.*, 1991). Essentially, the strategy to calculate the folding/unfolding partition function starting from the molecular structure of the folded state involves (1) generation of the set of most probable partially folded states using the crystallographic structure as a template and (2) evaluation and assignment of the relative Gibbs free energies of those states. Once this task is accomplished, the evaluation of the partition function is straightforward.

III. Estimation of Forces Required to Specify Partition Functions

In order to specify the energetics of each state in the partition function the following information is necessary.

1. One must identify and enumerate the interactions that characterize each of the states relevant to the folding/unfolding partition function. This can be accomplished from the structure, if known, utilizing accessible polar and apolar surface area calculations and specific interactions as described below.

2. One must have accurate values for the magnitude of the fundamental forces that characterize the interactions in (1). These forces include primarily the hydrophobic effect, hydrogen bonding and electrostatic effects, and configurational entropy changes. Additionally, interactions with ligands, such as metal ions, urea, and protons, must be characterized for those cases in which these additional interactions are present.

It is important from the outset to define clearly what we mean in the use of terms such as "the hydrophobic effect" and "hydrogen bonding." In this review the hydrophobic effect is defined in terms of the thermodynamics associated with the transfer of an apolar surface from the protein interior (or the initial phase in the case of model compound studies) into water. As such, it includes not only the energetics associated with the restructuring of water around apolar groups, but also the change in the van der Waals interactions of the apolar groups on transfer. Note that this is a purely operational definition and is not restricted to any particular transfer process (Herzfeld, 1991), nor to any distinguishing feature (Dill, 1990b).

Similarly, the disruption of a hydrogen bond on unfolding of a protein involves both a change in the intrinsic hydrogen bond per se and in the van der Waals interactions experienced by the polar groups involved in hydrogen bonds. As has been noted (Dill, 1990a), it is very difficult to separate these two contributions in any experimental sense. Additionally, the energetics of salt bridges buried in the protein interior are not considered separately but are averaged into the polar contributions in our analysis.

To determine how the partition function varies with temperature it is necessary to estimate not only the free energy changes associated with each type of interaction, ΔG^0, but also the thermodynamic terms that specify the temperature dependence of the free energy change, namely, the enthalpy change, ΔH^0, the entropy change, ΔS^0, and the heat capacity change, ΔC_p^0. At any temperature, the Gibbs free energy of any particular interaction can be written in terms of the standard thermodynamic relationship:

$$\Delta G^0 = \Delta H^0(T_R) - T\Delta S^0(T_R) + \Delta C_p^0[(T - T_R) - T\ln(T/T_R)] \quad (6)$$

where T_R is any appropriate reference temperature.

The free energy change at any temperature is composed of several contributions and may be written as the sum of the apolar, polar, configurational, and other, predominantly ligand binding, terms:

$$\Delta G^0 = \Delta G_{ap} + \Delta G_{pol} - T\Delta S_{conf} + \Delta G_{other} \quad (7)$$

It is conceptually convenient to define the enthalpy changes with reference to the temperature at which the apolar contribution to the overall ΔH^0 is zero (Baldwin, 1986), so that the apolar enthalpy change ΔH_{ap} is simply

$$\Delta H_{ap} = \Delta C_{p,ap}(T - T_H^\ddagger) \quad (8)$$

and the polar enthalpy change ΔH_{pol} is

$$\Delta H_{pol} = \Delta H\dagger + \Delta C_{p,pol}(T - T_H^\ddagger) \quad (9)$$

where the subscripts ap and pol refer to the apolar and polar components respectively, $T_H^\ddagger$ is the temperature at which the ΔH_{ap} is zero, and $\Delta H\dagger$ is the polar enthalpy change at $T_H^\ddagger$.

Likewise, the entropy changes are defined with reference to the temperature at which ΔS_{ap} is zero:

$$\Delta S_{ap} = \Delta C_{p,ap} \ln(T/T_S^\ddagger) \quad (10)$$

and

$$\Delta S_{pol} = \Delta S\dagger + \Delta C_{p,pol} \ln(T/T_S^\ddagger) \quad (11)$$

where $T_S^\ddagger$ is the temperature at which the apolar entropy change is zero and the polar entropy change has the value of $\Delta S\dagger$.

Equations (6)–(11) are then combined to give the Gibbs free energy terms:

$$\Delta G_{ap} = \Delta C_{p,ap}[(T - T_H^\ddagger) - T \ln(T/T_S^\ddagger)] \qquad (12)$$

and

$$\Delta G_{pol} = \Delta H\dagger - T\Delta S\dagger + \Delta C_{p,pol}[(T - T_H^\ddagger) - T \ln(T/T_S^\ddagger)] \qquad (13)$$

As defined in Eqs. (6)–(13), the formulation is entirely general. What is required to evaluate structural stability of a protein are values for the fundamental parameters $\Delta C_{p,ap}$, $\Delta C_{p,pol}$, $\Delta H\dagger$, $\Delta S\dagger$, ΔS_{conf}, $T_H^\ddagger$, and $T_S^\ddagger$ with reference to some measurable structural property of the protein. In the next sections we will see that the values of these fundamental parameters can be estimated from studies on protein denaturation and on the dissolution of model compounds.

A. Results of Model Compound Studies

Owing to the complexity of protein structures, model compounds have often been employed in estimating thermodynamic parameters for protein unfolding. The utility of these parameters depends on the validity of the choice of a model system, i.e., how well the model system mimics the process of transferring different functional groups from the protein interior into the solvent. In principle, one would like to choose a reference state that is similar energetically to the protein interior.

There are several indications that a crystalline solid is the most appropriate state to model the protein interior (Chothia, 1984). The very fact that protein structures can be determined to high resolution by X-ray diffraction is indicative of the crystalline nature of the protein. Additionally, the packing density and volume properties of amino acid residues in proteins are characteristic of amino acid crystals (Richards, 1974, 1977). In spite of the apparent crystallinity of the protein interior, most model compound studies have investigated either the transfer of compounds from an organic liquid into water (see, for example, Nozaki and Tanford, 1971; Gill *et al.*, 1976; Fauchere and Pliska, 1983), or the association of solute molecules in aqueous solution (see, for example, Schellman, 1955; Klotz and Franzen, 1962; Susi *et al.*, 1964; Gill and Noll, 1972). Both these approaches tacitly assume a liquidlike protein interior.

The study of the detailed energetics of dissolution of amino acid compounds from the crystalline phase into water has received less attention, due largely to experimental difficulties associated with low solubility. Only recently the dissolution thermodynamics for a series of cyclic dipeptides, chosen to remove the effects of terminal charges, have been published (Murphy and Gill, 1989a,b, 1990) and provide preliminary estimates for transfer of amino acid groups from the solid phase into water.

1. Dissection of Specific Contributions Using Group Additivity Thermodynamics

One of the key observations obtained from numerous model compound studies, including solid, liquid, and gas dissolution, is that a thermodynamic value for the transfer of a given compound is given, to a good approximation, by the sum of the contributions of the constituent groups that make up that compound (Gill and Wadsö, 1976; Nichols *et al.*, 1976; Cabani *et al.*, 1981; Murphy and Gill, 1990). This property, known as *group additivity*, allows one to assign contributions to specific fundamental interactions such as hydrogen bonding or the hydrophobic effect.

The strategy used in this review to dissect specific fundamental interactions is to isolate first the hydrophobic effects and subsequently other interactions. Within this context the hydrophobic effect is defined as that contribution to the overall thermodynamics of a process that is proportional to the amount of apolar surface that becomes exposed to the solvent (Hermann, 1972; Gill and Wadsö, 1976; Livingstone *et al.*, 1991). The apolar surface area exposed to solvent can be computed using various algorithms (Lee and Richards, 1971; Hermann, 1972; Shrake and Rupley, 1973; Connolly, 1983) that yield the accessible surface area (ASA) in units of square angstroms ($Å^2$).

However, because various algorithms utilizing various sets of atomic and probe radii have been used in generating the data available in the literature, it is sometimes convenient to use a separate measure of apolar surface area, the apolar hydrogen. An apolar hydrogen is defined simply as a hydrogen atom bonded to carbon and will be designated here as aH. It has been shown that the number of apolar hydrogens is proportional to the ASA (Jorgensen *et al.*, 1985; Murphy and Gill, 1991). We have further determined the ratio of apolar hydrogens to ASA from various other ASA determinations: for the ASA values of Miller *et al.* (1987) (data kindly provided by C. Chothia) the ratio is 16.8 aH $Å^{-2}$; for the data of Livingstone *et al.* (1991), using their atomic radii set 1, the ratio is 14.2 aH $Å^{-2}$; for the data of Livingstone *et al.*, using their atomic radii set 2, the ratio is 15.1 aH $Å^{-2}$; and our own calculation, using the Lee and Richards algorithm (implementation by Scott R. Presnell, University of California,

San Francisco), provides a ratio of 14.7 aH Å^{-2}. We will use this latter value throughout this review.

With the hydrophobic effect defined as the contribution to the thermodynamics that is proportional to the exposure of apolar surface area, it is then possible for a set of homologous compounds to separate the hydrophobic contribution from all other effects by plotting any thermodynamic function (for instance, ΔH^0) versus the number of apolar hydrogens (or the apolar surface area) that become exposed to the solvent on transfer. To the extent that the other interactions make a constant contribution to the thermodynamics,

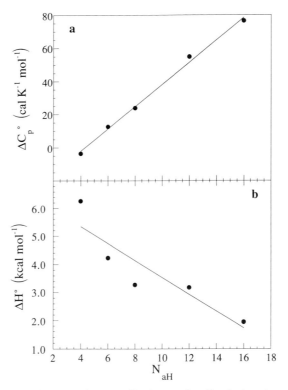

FIG. 1. Analysis of the apolar contribution to the dissolution thermodynamics of cyclic dipeptides into water. Each thermodynamic quantity is plotted against the number of apolar hydrogens (aH) (i.e., hydrogens bonded to carbon) (a) ΔC_p^0, (b) ΔH^0, (c) ΔS^0, and (d) ΔG^0. Lines are the linear regression fit of the data. As described in the text, the slope gives the hydrophobic contribution. Data are from Murphy and Gill (1990).

the slope will then give the hydrophobic contribution per aH or per unit area, and the intercept will be equal to the contribution of all other effects (hydrogen bonding, etc.). This approach has been used in the analysis of the model compound data described below and is illustrated for the case of the cyclic dipeptides in Fig. 1.

2. Heat Capacity Changes in Model Compounds

The measured heat capacity increment associated with the transfer of an apolar hydrogen from a crystalline amino acid solid into water is 6.7 ± 0.3 cal K^{-1} mol^{-1} or 0.45 ± 0.02 cal (mol Å^2)$^{-1}$ of apolar surface (Murphy and Gill, 1990, 1991). This same value is observed for the transfer of 1-alkanols into water (Hallén et al., 1986). The same value is also observed for the transfer of liquid hydrocarbons into water (Gill and Wadsö, 1976), and for the dissolution of alkane

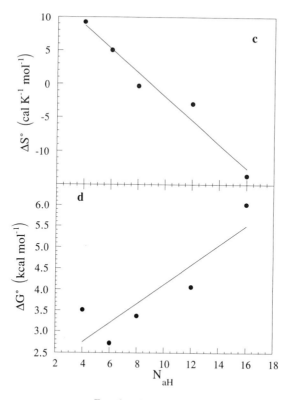

FIG. 1. (continued)

gases (Dec and Gill, 1985), as seen in Table I. In contrast to the positive contribution to ΔC_p^0 arising from the exposure of an apolar surface, the exposure of polar groups makes a negative contribution to ΔC_p^0 (Makhatadze and Privalov, 1990; Murphy and Gill, 1990, 1991). This value is -14.3 ± 1.4 cal K^{-1} mol^{-1} of hydrogen bonds in the cyclic dipeptides (Murphy and Gill, 1990).

Whereas the increase in heat capacity associated with the exposure of apolar groups to water has long been considered in protein unfolding studies (Sturtevant, 1977), the negative contribution from the exposure of polar surface area has, to our knowledge, only recently been considered in discussions of the ΔC_p^0 of denaturation (Murphy and Gill, 1990, 1991; Privalov and Makhatadze, 1990). In the past, most studies have used the approximation that the apolar heat capacity change is equal to the total ΔC_p^0 of unfolding (Baldwin, 1986; Spolar et al., 1989; Murphy et al., 1990; Livingstone et al., 1991). This has led to an underestimation of the apolar contribution to ΔC_p^0.

These model compound studies provide us with estimates of the first two fundamental parameters of $\Delta C_{p,ap} = 0.45$ cal K^{-1} (mol Å^2)$^{-1}$ of buried apolar surface area and $\Delta C_{p,pol} = -14.3$ cal K^{-1} mol^{-1} of hydrogen bonds. The average polar surface area buried per hydrogen bond in globular proteins is 54 ± 7 Å^2 (see below) so that $\Delta C_{p,pol} = -0.26$ cal K^{-1} (mol Å^2)$^{-1}$ of buried polar surface.

3. Enthalpic Effects

a. Estimation of Apolar Contributions. The hydrophobic effect has long been considered to be the predominant force in stabilizing protein structures (Kauzmann, 1959; Dill, 1990a). Most model compound studies of the hydrophobic effect have determined transfer free energies, ΔG_{tr}, in which the partitioning of an apolar solute between water and an organic solvent is determined. The primary difficulty with the interpretation of such results is that the value of ΔG_{tr} depends on the choice of concentration units (see Ben-Naim, 1978). Traditionally, the preferred standard state has been unit mole fraction (Tanford, 1980), but statistical arguments and excluded volume effects suggest that the entropy of mixing is not properly treated by this approach (Ben-Naim, 1978; DeYoung and Dill, 1990; Sharp et al., 1991a).

In contrast to ΔG^0 and ΔS^0, ΔH^0 and ΔC_p^0 are independent of the choice of concentration units and thus are easier to interpret. The hydrophobic contribution to ΔH^0 at 25°C from the solid dissolution data, given in Table I, is -300 cal (mol aH)$^{-1}$ or -20 cal (mol Å^2)$^{-1}$ of apolar surface area (Murphy and Gill, 1990). Very few direct

TABLE I
Contributions of Peptide and Apolar Hydrogen Groups to the Thermodynamics of Dissolution into Water[a]

Group	ΔC_p^0 (cal K^{-1} mol^{-1})	ΔH^0 (cal mol^{-1})	ΔS^0 (cal K^{-1} mol^{-1})	ΔG^0 (cal mol^{-1})
Peptide-CONH-(solid cyclic dipeptides)	-14.3 ± 1.4	3300 ± 400	7.9 ± 1.0	900 ± 400
Apolar hydrogen (solid cyclic dipeptides)	6.7 ± 0.3 [0.45 ± 0.02 cal K^{-1} (mol Å^2)$^{-1}$]	-300 ± 80 [-20 ± 5 cal (mol Å^2)$^{-1}$]	-1.8 ± 0.2 [-0.12 ± 0.01 cal K^{-1} (mol Å^2)$^{-1}$]	230 ± 70 [16 ± 5 cal (mol Å^2)$^{-1}$]
Apolar hydrogen (liquid hydrocarbons)	6.6 ± 0.5 [0.45 ± 0.04 cal K^{-1} (mol Å^2)$^{-1}$]	-78 ± 52 [-5 ± 4 cal (mol Å^2)$^{-1}$]	-1.5 ± 0.2 [-0.10 ± 0.01 cal K^{-1} (mol Å^2)$^{-1}$]	375 ± 17 [25 ± 1 cal (mol Å^2)$^{-1}$]
Apolar hydrogen (alkane gases)	6.6 ± 0.6 [0.45 ± 0.04 cal K^{-1} (mol Å^2)$^{-1}$]	-600 ± 90 [-41 ± 6 cal (mol Å^2)$^{-1}$]	—	—

[a] For solid cyclic dipeptides (Murphy and Gill, 1991), liquid hydrocarbons (Gill et al., 1976), and alkane gases (Dec and Gill, 1985) at 298.15 K. Confidence intervals are the standard errors as determined from the linear regression.

measurements are available on heats of dissolution of apolar liquids. The available data (Gill et al., 1975, 1976) are for a set of seven hydrocarbons, four of which are aromatic (benzene, toluene, ethylbenzene, and propylbenzene). As a combination of alkane and aromatic compounds, the liquid data do not constitute a homologous series and thus the group analysis is not strictly valid; however, these data have been widely used as a single set and it is worth analyzing them accordingly. The liquid data yield a total ΔH^0 of dissolution close to zero near 25°C (Gill et al., 1976; Gill and Wadsö, 1976), and the apolar contribution is accordingly small, -80 cal (mol aH)$^{-1}$ or -5 cal (mol Å^2)$^{-1}$. Finally, the best data on the alkane gases [methane, ethane, and propane (Dec and Gill, 1985)] yield a value of -600 cal (mol aH)$^{-1}$ or -40 cal (mol Å^2)$^{-1}$.

These values, in conjunction with the ΔC_p^0 values, allow determination of $T_H^{\ddagger}$ for each of the model compound systems as $T_H^{\ddagger} = 25 - \Delta H_{ap}/\Delta C_{p,ap}$. The value of $T_H^{\ddagger}$ is 71°C for the dipeptide solids, 36°C for the hydrocarbon liquids, and 117°C for the alkane gases. At these temperatures, the enthalpies of interactions of the apolar surfaces with water, including the enthalpy of water restructuring and the solute/solvent van der Waals interactions, are equal and opposite to the enthalpic interactions that the apolar surface experiences in the respective initial phases. As the apolar interaction with water is independent of the initial phase, greater apolar interactions in this phase result in a lower value for $T_H^{\ddagger}$.

It should be noted that $T_H^{\ddagger}$ for the crystalline solids is greater than that for the hydrocarbon liquids, indicating poorer apolar interactions in the initial phase. This initially seems contradictory, because ΔH^0 of fusion is positive and one would expect the apolar surfaces in the solid to have better interactions than in apolar liquids. This effect apparently arises because of the dominant role that hydrogen bonds play in determining the crystal lattice in the cyclic dipeptides studied. All these compounds show a nearly identical hydrogen bonding structure (Murphy, 1990), and it would appear that the crystal gives up good packing interactions between apolar surfaces in order to make good, but geometrically constraining, hydrogen bonds. This same conclusion was reached by Creighton (1991). Consequently, the tight packing observed in globular proteins (Richards, 1974, 1977), where hydrogen bond constraints must also be satisfied, does not necessarily imply better (or even equal) van der Waals interactions of apolar groups within the protein interior than with solvent water.

Because the apolar contribution to ΔH^0 of dissolution into water at 25°C is negative for the solids rather than near zero as seen in the liquids, the hydrophobic free energy change is only 230 cal (mol aH)$^{-1}$ [16 cal (mol Å^2)$^{-1}$] rather than the 375 cal (mol aH)$^{-1}$ [25 cal (mol Å^2)$^{-1}$] observed for the liquids. The volume correction suggested by Honig and co-workers (Sharp *et al.*, 1991a) would bring this value to about 800 cal (mol aH)$^{-1}$ [54 cal (mol Å^2)$^{-1}$] for the hydrophobic ΔG^0 at 25°C.

In summary, the model compound systems suggest that T_H^+ for the proteins should lie in the range from 30 to 120°C, depending on the packing interactions of the apolar groups. As mentioned above, the protein interior, due to its packing and the geometric constraints of satisfying hydrogen bonds, is expected to be more like the amino acid crystal than like an apolar liquid. In consequence, T_H^+ is expected to lie in the upper portion of the range given above.

b. Estimation of Polar Effects. At times, it has been considered that hydrogen bonding does not make a significant contribution to protein stability (Kauzmann, 1959; Creighton, 1984). This view has been based on the idea that hydrogen bonds that are formed in the native protein will be compensated by hydrogen bonds to water when the protein unfolds. Though some model compound studies of hydrogen-bonded dimer formation in solution appeared to support this conclusion (Klotz and Franzen, 1962), many others did not (Schellman, 1955; Susi *et al.*, 1964; Susi and Ard, 1969; Gill and Noll, 1972; Murphy and Gill, 1989a,b). Most of these studies estimate the enthalpy of breaking a hydrogen bond to be between 1.5 and 3 kcal mol^{-1} at 25°C. The recent studies on crystalline cyclic dipeptide dissolution provide an estimate for the hydrogen bond ΔH^0 of 3.3 kcal mol^{-1} at 25°C (Murphy and Gill, 1990, 1991). That this value is higher than estimates from other model compound studies probably reflects the increased hydrogen bond strength in the crystal as opposed to hydrogen bonds in solution; however, the peptide bond in these compounds is in the cis rather than the usual trans conformation found in proteins, which might also contribute to the higher value.

The importance of hydrogen bonding in protein stability is also supported by recent calorimetric experiments on the α-helix-to-coil transition. Scholtz *et al.* (1991) found a significant enthalpy change associated with the thermal unfolding of a 50-residue peptide composed primarily of alanine residues. In spite of difficulties in the quantitative evaluation of the data, an enthalpy change near 1.5 kcal mol^{-1} of hydrogen bonds was estimated at 50°C. Additionally, mutation studies of proteins (Alber *et al.*, 1987; Shirley *et al.*, 1991) also indicate the importance of hydrogen bonding to protein stability. In particular, Shirley *et al.* found that a hydrogen bond contributes, on average, 1.2 ± 0.6 kcal mol^{-1} to ΔG^0 of stabilization.

It is important to note, however, that the contribution of hydrogen bonding cannot be easily separated from van der Waals interactions between polar groups (Dill, 1990a). Furthermore, as hydrogen atoms do not have sufficient electron density to be positioned in protein crystal structures, it is difficult to quantify hydrogen bonds in proteins (Baker and Hubbard, 1984). Within this context, it is useful to consider the energetics of hydrogen bonding as an operational concept reflecting not only the intrinsic energy of a hydrogen bond per se, but the energetics of the overall process of disrupting the interaction between hydrogen bond pairs and the transfer to solvent of the constituent groups. Once exposed to the solvent, these hydrogen bond donor and acceptor groups will form hydrogen bonds with water.

Based on the range of enthalpy values for hydrogen bonding, the polar contribution to ΔH^0 of denaturation should be between 1.5 and 3.3 kcal mol^{-1} of hydrogen bonds at 25°C. This corresponds to values ranging between 27.5 and 60.4 cal (mol Å^2)$^{-1}$ of buried polar surface area. Estimation of the fundamental parameter $\Delta H^\dagger$ requires the additional specification of $T_H^\ddagger$ because $\Delta C_{p,pol}$ is not zero, as discussed above.

4. Entropic Effects

The apolar contribution to ΔS^0, ΔS_{ap}, is better characterized than ΔH_{ap}. The value of $T_S^\ddagger$ has been shown to be a universal temperature for all processes involving the transfer of an apolar surface into water and has a value of 112°C (Murphy *et al.*, 1990). At this temperature the ΔS^0 of transfer, $\Delta S^\dagger$, represents the mixing entropy of the process. The universal value of $T_S^\ddagger$ was determined using mole fraction concentration units, so that the liquid transfer $\Delta S^\dagger$ takes on a value of zero. The value of $T_S^\ddagger$ remains the same using the local standard state of Ben-Naim (i.e., molar concentration units) (Ben-Naim, 1978), but the value of $\Delta S^\dagger$ is increased by $R \ln(55.5)$, where R is the gas constant and 55.5 is the molarity of water.

This value of $T_S^\ddagger$ is the same as the entropic convergence temperature, T_S^*, observed in proteins (Baldwin, 1986; Murphy *et al.*, 1990). This is the temperature at which the denaturational entropies of globular proteins, normalized to the molecular weight or to the number of amino acid residues, take on nearly the same value when extrapolated under the assumption of constant ΔC_p^0 (Privalov and Khechinashvili, 1974). The significance and interpretation of this observation are discussed in more detail below.

Honig and co-workers (Sharp, 1991; Sharp *et al.*, 1991a,b) have proposed an excluded volume correction to the dissolution entropy changes for model compounds based on statistical mechanical arguments. Using their correction, $T_S^{\ddagger}$ is no longer a universal temperature for processes involving transfer of apolar surfaces into water, and there is no obvious connection between hydrophobic entropies and the convergence of denaturational entropies in proteins. It should be noted that in the absence of association–dissociation processes the thermodynamics of protein denaturation are free of concentration units, because denaturation is a phase change. Under these changes it is not clear how an excluded volume correction relates to this process.

An important class of model compound study that we have not discussed here is the determination of transfer free energies mentioned above. Though the study of transfer of amino acids and their analogs from water into various organic compounds provides a wealth of information about various interactions, the current data base includes only values ΔG^0 (usually relative to glycine) and not values for ΔH^0, ΔS^0, and ΔC_p^0, and thus is not suitable for the temperature-dependent information required within the context of this review. The thermodynamics from liquid hydrocarbon, crystalline cyclic dipeptide, and alkane gas dissolution are summarized in Table I.

B. Protein Denaturation Studies

Our primary source of information on the structural energetics of globular proteins comes from studies of protein denaturation. In particular, detailed thermodynamic information regarding protein structures has been obtained from high-sensitivity differential scanning calorimetry (DSC) (Privalov, 1979). Numerous globular proteins have been studied by DSC (Privalov and Gill, 1988), and these results provide a data set from which we can begin to understand the contributions of the various interactions that stabilize protein structures, and to which model compound results can be compared.

Convergence Temperatures and Hydration

One of the key observations resulting from calorimetric studies of the denaturation of globular proteins is that both ΔH^0 and ΔS^0 of denaturation, when normalized to the number of amino acid residues in the protein (or the molecular weight), converge to common values at specific temperatures when extrapolated under the assumption of constant ΔC_p^0 (Privalov and Khechinashvili, 1974; Privalov, 1979; Privalov and Gill, 1988). Although ΔC_p^0 values show some variation with temperature, this variation is small and the enthalpy and entropy values calculated for the temperature interval of interest (0–100°C) are not significantly affected by this variation (Privalov and Gill, 1988). The convergence values of ΔH^0 and ΔS^0 are designated ΔH^* and ΔS^* and the temperatures at which they converge are designated T_H^* and T_S^* (Murphy et al., 1990). Note that these are distinguished from the temperatures at which ΔH^0 and ΔS^0 are zero, designated T_h and T_s (Baldwin, 1986; Becktel and Schellman, 1987), and from the temperatures at which the apolar contribution to ΔH^0 and ΔS^0 are zero, designated $T_H^{\ddagger}$ and $T_S^{\ddagger}$. Any breakdown of the energetics should be consistent with the observed convergence behavior as well as the values of ΔH^* and ΔS^* observed for globular proteins.

It was originally supposed that T_S^* and T_H^* were equal and had a value near 110°C (Privalov and Khechinashvili, 1974; Privalov and Gill, 1988; Murphy et al., 1990). It would now appear however that ΔH^0 and ΔS^0 converge at separate temperatures (Murphy and Gill, 1990, 1991), as is observed in model compound dissolution studies of liquids, gases, and solids. For the globular proteins the best-fit value for T_H^* is 100.5°C (373.6 K), at which ΔH^*, per residue (res), is 1.35 ± 0.11 kcal (mol res)$^{-1}$ (Murphy and Gill, 1991). The best-fit value for T_S^* is 112°C (385.2 K), at which ΔS^* is 4.30 ± 0.12 cal K^{-1} (mol res)$^{-1}$, as previously reported (Privalov and Gill, 1988; Murphy et al., 1990).

The values of the convergence temperatures, as well as the convergence itself, reflect fundamental properties of the system (Murphy and Gill, 1990; Lee, 1991). As mentioned above, it has been shown that the value of 112°C for T_S^* is common to processes involving the transfer of apolar groups from solid, liquid, and gas phases, as well as from the protein interior into water using mole fraction or molar concentration units for the model compound data (Baldwin, 1986; Murphy et al., 1990). At this temperature, the ΔS^0 of transfer is the same for all solutes from a homologous series of compounds (e.g., normal alcohols), so that ΔS^* represents the size-independent entropy change associated with the process. Because in a homologous series of compounds only the hydrophobic surface area is varied, ΔS^* also represents the nonhydrophobic contribution to ΔS^0. For example, ΔS^* for the transfer of an apolar gas into water is -19 cal K^{-1} mol^{-1} (Murphy et al., 1990), very close to the value of -17 cal K^{-1} mol^{-1} estimated for the transfer of a mathematical point at T_S^* (Lee, 1991).

As discussed above, the thermodynamics of the hydrophobic effect are seen to be proportional to the apolar surface area exposed to the solvent. Based on the absence of a size dependence of ΔS^0 on transfer of apolar solutes into water at T_S^*, it is seen that the convergence temperature, T_S^*, and the temperature at which the apolar contribution to ΔS^0 is zero, $T_S^{\ddagger}$, are the same.

In contrast to the case of the entropy, however, the temperature $T_H^{\ddagger}$, at which the apolar enthalpy change, ΔH_{ap}, is zero, is not common to all transfer processes. For the case of the dissolution of homologous series of model compounds, it has been shown that T_H^* and $T_H^{\ddagger}$ are equivalent (Murphy and Gill, 1990, 1991; Lee, 1991); however, it is not clear initially whether the ΔH^0 convergence temperature observed in proteins necessarily coincides with $T_H^{\ddagger}$. The convergence temperature is near 100–120°C for a wide variety of gaseous compounds (Privalov and Gill, 1988; Murphy, 1990), about 100°C for protein denaturation (Murphy and Gill, 1991), near room temperature for hydrocarbon liquids (Gill et al., 1976; Gill and Wadsö, 1976), and about 71°C for the dissolution of solid cyclic dipeptides (Murphy and Gill, 1990).

C. Origins of Convergence Temperatures

Analysis of the dependence of the structural thermodynamics of globular proteins on apolar surface area provides an estimation of the role of various contributions to protein stability. However, as mentioned above, proteins also show convergence temperatures that can yield similar information, given certain assumptions.

Although the presence of a convergence temperature for both the specific enthalpy and the specific entropy of protein denaturation has been known for nearly two decades (Privalov and Khechinashvili, 1974), the possible origins of this behavior have only recently been discussed in detail (Murphy and Gill, 1990, 1991; Lee, 1991), although they were surmised earlier (Privalov and Khechinashvili, 1974). The initial observation by Baldwin (1986) that T_S^* for proteins occurred at the same temperature as T_s, the temperature at which ΔS is zero, for liquid hydrocarbon dissolution emphasized the importance of the hydrophobic effect in the convergence temperature. Privalov and Gill (1988) later noticed that T_H^* for proteins corresponds to the temperature at which ΔH^0 for apolar gas dissolution is zero, which is also the temperature at which ΔH^0 for liquid hydrocarbon dissolution is equal to ΔH^0 of vaporization. As these temperatures had similar values, near 110°C, Privalov and Gill (1988) defined "hydrophobic hydration" with reference to this temperature. Hydrophobic hydration was viewed as the process of transferring an apolar substance from a hypothetical condensed phase of noninteracting molecules into water (Privalov and Gill, 1988).

Subsequently it was observed that the presence of convergence temperatures is also easily seen in plots of ΔH^0 or ΔS^0 at 25°C versus ΔC_p^0. If a convergence temperature exists, then this plot will be linear. The slope is equal to $(298.15 - T_H^*)$ or $\ln(298.15/T_S^*)$ and the intercept is ΔH^* or ΔS^*. Using such plots it was shown that T_S^* for the transfer of apolar compounds from any phase, as well as for protein denaturation, was a universal temperature near 112°C (Murphy et al., 1990). This observation lent further support to the view that the convergence temperatures were associated with the hydrophobic effect. It must be noted also that these plots do not require the assumption that ΔC_p^0 be constant with temperature.

Recently it has been shown that convergence of thermodynamic quantities at some temperature will occur when there are two predominant interactions that independently contribute to the thermodynamics (i.e., group additivity), provided that one of these interactions is constant for the set of compounds being investigated (Murphy and Gill, 1990, 1991). For example, the 1-alkanols have varying amounts of apolar surface, but each compound has only one —OH group. Under these conditions it was demonstrated that the apolar contribution to ΔH^0 is zero at T_H^* (i.e., $T_H^* = T_H^\dagger$) and that the apolar contribution to ΔS^0 is zero at T_S^* (i.e., $T_S^* = T_S^\dagger$). Consequently, the values of ΔH^* and ΔS^* represent the nonapolar contribution at the appropriate temperature, T_H^* or T_S^*.

It would appear that these conditions are met for the set of globular proteins that have been thermodynamically studied in detail (Privalov, 1979; Privalov and Gill, 1988). In particular, it was noted that those proteins have an average of 0.73 hydrogen bonds per residue (Privalov and Gill, 1988). In order to characterize further the hydrogen bonding statistics and to determine the buried polar surface per hydrogen bond, this number was recalculated using the same criteria for all proteins. For this purpose we used the molecular graphics program Quanta (Polygen, Waltham, Massachusetts) and considered 11 globular proteins ranging in size from 76 to 261 residues and for which thermodynamic parameters are available. These proteins (and protein data bank file names) are ubiquitin (1UBQ), cytochrome c (5CYT), RNase T1 (1RNT), RNase A (7RSA), lysozyme (1LYM), myoglobin (4MBN), papain (9PAP), β-trypsin (1TLD), α-chymotrypsin (4CHA), carbonic anhydrase B (2CAB), and α-lactalbumin (1ALC). Polar hydrogen atoms were first added to the structure using standard geometries, following the recommendations of Baker and Hubbard (1984), and the cut-off angle about the hydrogen atom was set to 120°. A plot of the number of hydrogen bonds versus the number of residues yields a linear dependence with a slope equal to 0.72 ± 0.07, in good agreement with the previous estimate (Privalov and Gill, 1988). Additionally, a plot of the buried polar surface area versus the number of hydrogen bonds yields an average of 54 ± 7 Å^2 per hydrogen bond. As hydrogen bonds should be the primary contributor to the polar enthalpy change, ΔH_{pol} will be constant when normalized to the number of residues for this set of proteins. However, this analysis takes no account of the probable variability of the strength of hydrogen bonds, which depends on their specific geometry.

In contrast to the relatively constant number of hydrogen bonds per residue, a set of proteins must bury variable amounts of apolar surface area in order to show convergence (Murphy and Gill, 1991). At the temperature at which the apolar contribution to ΔH^0 is zero, no variation would be observed in ΔH^0 per residue and the constant polar contribution is all that should be observed. The breakdown into polar and apolar interactions can also be viewed in terms of buried surface area. Proteins bury an increasing amount of surface area per residue with increasing size, but the increase is due to increased burial of apolar surface, whereas the polar surface buried remains constant. This is illustrated in Fig. 2 for 12 globular proteins that show convergence of ΔH^0. These proteins bury a constant 39 ± 2 Å^2 of polar

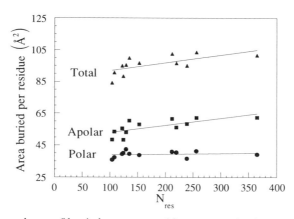

FIG. 2. Dependence of buried area per residue on protein size. The lines are the linear least-squares fits. The total area buried per residue increases with increasing number of residues as does the apolar surface area buried per residue. In contrast, the polar area buried per residue is independent of the size of the protein. The proteins plotted are listed in Table IV.

surface area per residue, whereas the apolar surface buried per residue increases with the size of the protein.

As noted above, the value of ΔH^* is 1.35 kcal (mol res)$^{-1}$ at 100.5°C for the globular proteins studied. This corresponds to 1.85 kcal mol^{-1} of hydrogen bonds. Using the value of ΔC_p per hydrogen bond from the cyclic dipeptide data, the value of ΔH^0 for hydrogen bonding at 25°C is 2.9 kcal mol^{-1}, in good agreement with the model compound estimates as discussed above.

The value of ΔH^* can also be compared to the helix unfolding ΔH^0 of Scholtz et al. (1991). The buried surface area, relative to the extended chain, was calculated for a 50-residue alanine α helix. An average of 19.5 Å^2 of polar surface is buried per residue and an average of 3.2 Å^2 of apolar surface is overexposed (i.e., is less accessible in the extended chain than in the helix). Using the fundamental parameters for the polar and apolar ΔC_p described above, a value of -6.5 cal K^{-1} (mol res)$^{-1}$ is estimated for ΔC_p for the helix denaturation. At 100°C the extrapolated value of ΔH^0 is about 1.0 kcal (mol res)$^{-1}$, again in reasonable agreement with the value of ΔH^* of 1.35 kcal (mol res)$^{-1}$. These results strongly support the assertion that the apolar contribution to ΔH^0 is close to zero at T_H^*.

The convergence of thermodynamic properties occurs because, within a set of proteins with nearly constant polar contribution per residue, the apolar contribution varies. If, for a single protein, the hydrophobic contribution could be modified without significantly affecting the hydrogen bonding, the same convergence behavior should be observed. This can, to some extent, be accomplished by the addition of organic cosolvents.

In the case of the addition of alcohols to the solvent, it might be expected that low alcohol concentrations would predominantly affect the hydrophobic effect, although higher alcohol concentrations will also have a pronounced effect on hydrogen bonding, because alcohols can form fewer hydrogen bonds than water. Velicelebi and Sturtevant (1979) studied the effect of a number of alcohols on the denaturation of lysozyme. Also, an extensive study of the effects of methanol on cytochrome c and ubiquitin stability has been carried out in our laboratory (L. Fu and E. Freire, unpublished results, 1992). These studies reveal that at low alcohol concentrations, ΔH^0 of denaturation increases with increasing alcohol while T_m and ΔC_p^0 decrease. This is in contrast to the normal situation in which a decrease in T_m coincides with a decrease in ΔH^0, due to the positive ΔC_p of unfolding. This result can be understood in terms of the model presented above. Because ΔH^0 is composed of two components, a positive hydrogen bonding component and a negative hydrophobic component, it is expected that low concentrations of alcohol, by decreasing the hydrophobic effect while not largely perturbing hydrogen bonding, result in an overall increase in ΔH^0.

If ΔH^0 at 25°C is plotted versus ΔC_p^0 at the four lowest concentrations of 1-propanol studied for lysozyme, or for cytochrome c from 0 to 10% methanol, a linear relationship is observed, indicating a convergence of ΔH^0 values. Linear least-squares analysis yields T_H^* and ΔH^* values in very good agreement with the values obtained from the analysis of globular proteins in aqueous solution.

It is important to note, however, that although group additivity with a constant component will always lead to convergence temperatures, the presence of convergence temperatures may not require this condition. Recently Lee (1991) proposed an alternative explanation of the protein convergence temperatures. In the analysis Lee showed that if the protein data are normalized to the total amount of buried surface area, the variable fraction of the total buried surface that is apolar will lead to a convergence temperature at which the apolar and polar ΔH^0 contributions are the same on a per square angstrom basis.

As has been discussed recently (Murphy et al., 1992), the formalism developed by Lee (1991) predicts not only the convergence behavior discussed by the author (i.e., when the apolar and polar contributions to the enthalpy are identical). For the case in which the buried polar area per residue is constant it also predicts convergence at the point at which the apolar contribution to the enthalpy is zero. In Fig. 3 we have plotted ΔH^0 versus ΔC_p^0 normalized either per residue (Fig. 3a) or per buried total surface area (Fig. 3b), in order to compare the results of the two approaches. It is clear that the linearity is better when the data are normalized to the number of residues than when they are normalized to the buried surface area. This is presumably due to variabilities in the surface area calculation. The slope of the line in Fig. 3a is -72.4, which corresponds to a convergence temperature, T_H^*, of 97.4°C for this set of proteins. If the above analysis is correct, then this temperature corresponds to $T_H^\dagger$. The value of ΔH^* is 1.32 kcal (mol res)$^{-1}$ or 33.6 cal (mol Å^2)$^{-1}$ of polar surface area.

The slope in Fig. 3b should correspond to the temperature at which the apolar and polar contributions to ΔH^0 are equal per square angstrom. This temperature can be calculated to be about 140°C (assuming that $T_H^* = T_H^\dagger$ as discussed above), as shown in the figure, and is about 40°C higher than the observed convergence temperature for the residue normalized analysis.

Thus, the protein data are consistent with a $T_H^\dagger$ close to T_H^* and

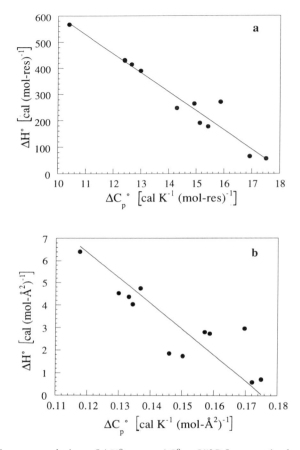

FIG. 3. Linear correlation of ΔH^0 versus ΔC_p^0 at 25°C for protein denaturation for the proteins listed in Table IV. (a) Normalized per number of residues; (b) normalized per total buried area. The line in (a) is the linear regression fit with a slope of -72.4 corresponding to a convergence temperature of 97.4°C. The line in (b) represents the line calculated from the parameters in Table II for convergence at the temperature at which the polar and apolar contributions to ΔH are equal per unit area. See text for details.

around 100°C. At this temperature, the apolar contribution to the enthalpy change, ΔH_{ap}, approaches zero. Therefore, the residual enthalpy change at this temperature, ΔH^*, arises from polar interactions and has a value of 35 cal (mol Å^2)$^{-1}$ of buried polar surface, or 1.85 kcal mol^{-1} of hydrogen bonds.

IV. Calculation of Folding/Unfolding Thermodynamics from Protein Structure

The data presented above allow for estimations of all the fundamental thermodynamic parameters necessary to calculate protein stability from Eqs. (6)–(13). The values for $\Delta C_{p,ap}$ and $\Delta C_{p,pol}$ are taken from the solid model compound data, along with accessible surface area calculations, and are 0.45 and -0.26 cal K^{-1} (mol Å^2)$^{-1}$ of buried apolar and polar area, respectively. Values of $T_H^\dagger$ and $T_S^\dagger$ are taken from the protein convergence temperatures and have values of 100 and 112°C, respectively. The value of $\Delta H^\dagger$ is equal to ΔH^* normalized to the average buried polar surface area (39.3 Å^2 res^{-1}) and is close to 34–35 cal (mol Å^2)$^{-1}$ of buried polar surface area. For monomeric globular proteins under standard conditions [i.e., low ionic strength, corrected for pH effects (Privalov and Gill, 1988)] it is assumed that ΔS^* contains primarily the configurational entropy change on unfolding. The value of ΔS_{conf} is thus 4.3 cal K^{-1} (mol res)$^{-1}$ (Murphy et al., 1990; Murphy and Gill, 1991). The additional entropic contribution of specific ionization or protonation effects or entropic effects arising from the presence of disulfide bridges must be considered explicitly (Freire and Murphy, 1991). The values of the fundamental thermodynamic parameters are summarized in Table II.

The thermodynamic parameters in Table II can be used to calculate the thermodynamic quantities necessary to define the partition function. Essentially, the strategy to calculate the folding/unfolding partition function from structural data involves the identification and enumeration of states and the assignment of Gibbs free energies to those states. Once this task is accomplished, the evaluation of the partition function is straightforward, as illustrated in Fig. 4.

TABLE II
Fundamental Thermodynamic Parameters for Estimating Protein Stability from Crystal Structure

Parameter	Estimated value
$\Delta C_{p,ap}$	0.45 ± 0.02 cal K^{-1} (mol Å^2)$^{-1}$
$\Delta C_{p,pol}$	-0.26 ± 0.03 cal K^{-1} (mol Å^2)$^{-1}$
$T_H^\dagger$	100 ± 6°C
$T_S^\dagger$	112 ± 1°C
$\Delta H^\dagger$	35 ± 3 cal (mol Å^2)$^{-1}$ of buried polar surface
$\Delta S^\dagger, \Delta S_{conf}$	4.3 ± 0.1 cal K^{-1} (mol res)$^{-1}$

INDEX	STATE	FREE ENERGY	ΔG	STAT. WEIGHT
0		G_0	0	1
1	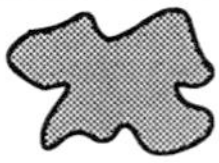	G_1	ΔG_1	$e^{-\Delta G_1/RT}$
2		G_2	ΔG_2	$e^{-\Delta G_2/RT}$
i	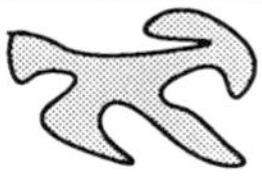	G_i	ΔG_i	$e^{-\Delta G_i/RT}$
n-1		G_{n-1}	ΔG_{n-1}	$e^{-\Delta G_{n-1}/RT}$
n	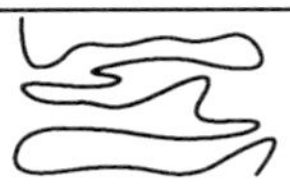	G_n	ΔG_n	$e^{-\Delta G_n/RT}$

FIG. 4. Schematic representation of the partition function [Eq. (1)] for protein folding/unfolding. Each state, from the native state ($i = 0$) to the unfolded state ($i = n$) and all intermediates ($i = 1$ to $n - 1$), is assigned a ΔG_i^0 relative to the native state from which the statistical weights are obtained. The partition function, Q, is simply the sum of the statistical weights of all the states. Other important parameters, including the population of each state [Eq. (2)], the excess enthalpy [Eq. (3)], and the excess heat capacity [Eq. (4)], are determined from the partition function as described in the text.

Equations (6)–(13) allow calculation of the free energy change at any temperature using the parameters in Table II, the number of residues, N_{res}, and the buried polar and apolar surface areas evaluated from the crystallographic structure using standard algorithms. The equations can be applied to the entire protein, a single domain, or to interfaces between structural elements.

The change in both polar and apolar buried surface area on denaturation can be estimated from the difference between the polar or apolar ASA of the folded protein and of the extended chain, respectively (Eisenberg and McLachlan, 1986; Ooi et al., 1987; Spolar et al., 1989). Although the denatured protein may not be a random coil, it has been argued that globular proteins behave experimentally as if fully solvated (Privalov et al., 1989; Creighton, 1991). Additionally, deuterium exchange studies on thermally denatured ribonuclease indicate no residual secondary structure (Robertson and Baldwin, 1991), lending further support to the practice of using the ASA of the extended chain in modeling the unfolded protein. The nonrandom coil nature of the denatured state is primarily reflected in the magnitude of the configurational entropy change associated with unfolding, which is somewhat smaller than the value observed for helix-to-random coil transitions (Freire and Murphy, 1991).

The structural parameters for 12 globular proteins are given in Table III. These parameters were determined from the composition of the proteins and from the X-ray crystal structures. The result of calculations of the overall folding/unfolding thermodynamics are illustrated in Table IV. As seen in Table IV, the predicted ΔC_p^0 and ΔH^0 values are on the average within 12 and 10% of the experimental values. The agreement is substantially better if only the subset of proteins that have been studied more systematically is used (e.g., myoglobin, cytochrome c, lysozyme, α-lactalbumin). The calculated entropy values are also within 10% of the experimental values. It should be noted that the calculated values do not include any additional specific contributions to the entropy change. Accurate predictions of the transition temperatures require the specification of these terms (Freire and Murphy, 1991; Freire et al., 1991).

TABLE III
Structural Parameters for Globular Proteins[a]

Protein	N_{res}	Buried apolar surface area (Å^2)	Buried polar surface area (Å^2)
Cytochrome c (5CYT)	103	5039	3726
Carbonic anhydrase B (2CAB)	256	15,950	10,590
Chymotrypsin (4CHA)	239	13,970	8762
α-Lactalbumin (1ALC)	122	6773	4814
Lysozyme (1LYM)	129	6844	5473
Myoglobin (4MBN)	153	8873	5927
Staphylococcus nuclease (1SNC)	135	8151	5344
Papain (9PAP)	212	13,070	8692
Parvalbumin (5CPV)	108	5770	4027
Pepsinogen (1PSG)	365	22,810	13,430
RNase A (7RSA)	124	6004	4963
Trypsin (1TLD)	220	12,370	8903

[a]Protein data bank file numbers are indicated in parentheses. Surface areas were calculated using the Lee and Richards algorithm (1971) as described in the text.

TABLE IV
Comparison of Calculated and Experimental Values at 60°C[a]

Protein	$\Delta C_{p,calc}$	$\Delta C_{p,exp}$	ΔH_{calc}	ΔH_{exp}	ΔS_{calc}	ΔS_{exp}	Ref.
Cytochrome c	1.3	1.6	78	78	257	208	b
Carbonic anhydrase B	4.4	3.8	195	182	467	530	c
Chymotrypsin	4.0	3.0	147	205	453	615	d
α-Lactalbumin	1.8	1.8	96	96	267	241	e
Lysozyme	1.7	1.6	124	112	318	310	f
Myoglobin	2.5	2.6	107	101	307	282	g
Staphylococcus nuclease	2.3	2.4	95	92	254	280	h
Staphylococcus nuclease	—	2.3	—	101	—	317	i
Papain	3.6	3.3	160	152	393	395	j
Parvalbumin	1.6	1.3	77	93	243	241	k
Pepsinogen	6.8	6.1	198	233	599	654	l
RNase A	1.4	1.4	118	111	332	325	m
Trypsin	3.3	2.9	180	186	481	544	n

[a]The predicted values were calculated as described in the text. Units are kcal mol^{-1} for ΔH, kcal K^{-1} mol^{-1} for ΔC_p, and cal K^{-1} mol^{-1} for ΔS; 60°C is the median temperature for denaturation of the proteins and consequently minimizes the extrapolation errors.
[b]Privalov and Khechinashvili (1974).
[c]Tatunashvili and Privalov (1986).
[d]Tischenko et al. (1974).
[e]Xie et al. (1973).
[f]Khechinashvili et al. (1973).
[g]Privalov et al. (1986).
[h]Calderon et al. (1985).
[i]Griko et al. (1988).
[j]Tiktopulo and Privalov (1978).
[k]Filimonov et al. (1978).
[l]Mateo and Privalov (1981).
[m]Privalov et al. (1973); M. Straume and E. Freire, unpublished results, 1991.
[n]Tischenko and Gorodnov (1979).

Figure 5 illustrates the contributions of the polar free energy change, the configuration entropy change, and the apolar free energy change to the overall free energy change of protein denaturation using myoglobin as an example. As discussed previously (Murphy and Gill, 1991), the primary stabilizing contribution is the polar enthalpy change whereas the main destabilizing component is the configurational entropy change. Because these two opposing forces are almost equal in magnitude, the overall balance is determined to a large extent by the hydrophobic effect.

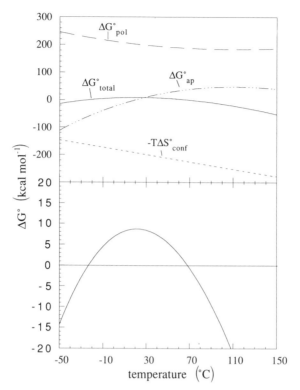

FIG. 5. Free energy of stabilization for myoglobin, calculated as described in the text for conditions of maximal stability (Privalov and Gill, 1988). The upper panel illustrates the contributions of the various interactions to the overall stability. Note that the primary stabilizing force is the polar ΔH resulting primarily from the interaction of hydrogen bonding groups. The main destabilizing force is the configurational entropy change, whereas the hydrophobic contribution is stabilizing above 20°C, but destabilizing below this temperature. The lower panel is an expanded view of the overall ΔG^0 showing the characteristic curvature with two temperatures at which denaturation is expected.

Figure 5 also illustrates other general features of globular protein stability. The positive overall ΔC_p results in the existence of two temperatures at which ΔG^0 is zero. The low temperature point defines the so-called cold denaturation of the protein and the high temperature point defines the heat denaturation. Additionally, the curvature in ΔG^0 implies the existence of a temperature of maximal stability. This temperature occurs at the point where ΔS^0 is equal to zero. The primary effect of changing environmental conditions is to move the curve up or down, thus moving the temperatures of cold and heat denaturation farther apart or closer together, but without large changes in the temperature of maximal stability. Also, it should be noted that the hydrophobic contribution to the free energy of stabilization is near zero at about 20°C; it contributes favorably to protein stability above this temperature and unfavorably at lower temperatures.

V. Thermodynamic Stability and Cooperative Interactions

It has been shown that under standard conditions most single-domain globular proteins exhibit a folding/unfolding behavior consistent with the two-state mechanism (Freire and Biltonen, 1978a; Privalov, 1979). From a statistical thermodynamic standpoint the implication is that the population of partially folded intermediate states is negligible and the partition function reduces to two terms:

$$Q = 1 + \sum_{i=1}^{n-1} e^{-\Delta G_i/RT} + e^{-\Delta G_n/RT} \tag{14}$$

$$Q \approx 1 + e^{-\Delta G_n/RT} \tag{15}$$

Equation (15) has been used implicitly or explicitly to analyze the folding/unfolding thermodynamics of single-domain small, globular proteins (Freire and Biltonen, 1978a; Privalov, 1979; Privalov and Gill, 1988; Freire, 1989). Under standard conditions, the population of the entire ensemble of partially folded states is negligible and the only states that ever become significantly populated are the native and the unfolded states. Cooperative interactions are responsible for this phenomenon. Under certain conditions, however (e.g., subzero temperatures, low pH, moderate concentrations of denaturants), partially folded states might become significantly populated, as will be discussed later.

A. Hierarchical Approach to Partition Functions

The formalism described above allows evaluation of the Gibbs free energy difference between specific structural states of the protein and the reference state, which is taken to be the native structure. The formalism requires access to the crystallographic structure of the protein under study so that, for each structurally defined state, the difference in hydrogen bonds, exposure of polar and apolar surfaces, ionized groups, etc., can be counted. Because the total number of possible structural states is astronomical, any attempt to evaluate the folding/unfolding partition function requires selection of a subset that is amenable to computation and that can be considered to have the highest probability of occurrence.

Fortunately, the number of states that ever become populated is relatively small, even under conditions that maximize the population of intermediates. It is apparent that the folding/unfolding partition function can be simplified so that it includes only those states that are relevant to the folding process. The approach that we have undertaken involves the use of the native conformation as a template to generate partially folded conformations, and to evaluate the Gibbs free energy of those conformations according to the rules described in Section III.

The key conceptual element is the introduction of hierarchical levels of cooperative folding units (Freire and Murphy, 1991). By definition, cooperative folding units are protein structural elements that exhibit cooperative folding behavior. At the most fundamental levels are those structural elements (e.g., α helices) that exhibit intrinsic cooperative behavior as a result of primarily local interactions, whereas at higher levels are those structures formed by two or more lower order cooperative folding units that exhibit cooperative behavior as a result of the interactions between their constituent lower order cooperative folding units. A protein is considered to be composed of one or more interacting cooperative folding units. Different structural states of the protein are formed by folding and unfolding the cooperative folding units in all possible combinations.

Cooperative folding units are defined from a functional rather than a purely structural point of view, even though in many cases a one-to-one correspondence can be found. Cooperative folding units can be as large as the entire protein (e.g., small globular proteins that exhibit two-state behavior effectively behave as cooperative folding units), entire protein domains, subdomains, and structural motifs, or as small as single α helices. The hierarchical level at which the description is made depends on the desired level of detail and is, in principle, only constrained by the availability of structural information.

According to the hierarchical approach, the number of states that need to be considered in the partition function is $2^{n_{cu}}$, where n_{cu} is the number of cooperative folding units. In order to develop a complete description of a system composed of n_{cu} interacting cooperative folding units it is necessary to evaluate the intrinsic energetics of each cooperative folding unit as well as the energetics of the interactions between cooperative folding units. All these energetic parameters determine the stability of the protein as well as the overall cooperativity of the folding/unfolding process.

It must be noted that stability and cooperativity are different concepts. Stability is dictated by the Gibbs free energy difference between the folded and unfolded states and is usually reflected in the transition temperature or the resistance of the native state to the effects of denaturing agents, whereas cooperativity is reflected in the population of intermediate states.

B. Thermodynamic Dissection of Cooperative Interactions

The same forces that stabilize the structure of proteins also give rise to well-defined cooperative folding/unfolding patterns, i.e., mechanisms that minimize the population of partially folded intermediates. These patterns will be illustrated for a hypothetical protein consisting of two cooperative folding units. The generalization to multiple cooperative folding units is straightforward as discussed by Freire and Murphy (1991). As shown in Fig. 6, the number of states associated with the folding/unfolding behavior of this hypothetical protein corresponds to the total number of combinations generated by folding and unfolding its constituent folding units. These states correspond to the folded structure, two partially folded intermediate states, and the unfolded state (Ramsay and Freire, 1990; Freire and Murphy, 1991). The partition function of such a molecule can be written as

$$Q = 1 + \Phi k_1 + \Phi_2 k_2 + \Phi_3 k_3 \tag{16}$$

where the $k_i = e^{-\Delta G_i/RT}$ terms represent the intrinsic statistical weights and the $\Phi_i = e^{-\Delta g_i/RT}$ terms represent the interaction parameters. In terms of the standard equilibrium constants for the two cooperative folding units, $k_1 = K_1$, $k_2 = K_2$, and $k_3 = K_1 K_2$. From a purely mathematical point of view, cooperativity is generated by the condition $\Phi_3 \neq \Phi_1 \Phi_2$. If $\Phi_3 = \Phi_1 \Phi_2$, then the partition function reduces to

$$Q = 1 + \Phi_1 K_1 + \Phi_2 K_2 + \Phi_1 \Phi_2 K_1 K_2 \tag{17a}$$

$$= (1 + \Phi_1 K_1)(1 + \Phi_2 K_2) \tag{17b}$$

which is the partition function for two independent domains.

STATE	FREE ENERGY	RELATIVE FREE ENERGY	STAT. WEIGHT
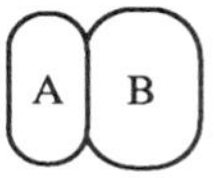 A B	$G_A + G_B + g_{A,B}$	0	1
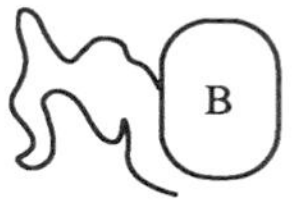 B	$G_{A'} + G_B + g_{A',B}$	$\Delta G_1 + \Delta g_1 = \Delta G_{A'} + \Delta g_{A',B}$	$\Phi_1 k_1$
 A	$G_A + G_{B'} + g_{A,B'}$	$\Delta G_2 + \Delta g_2 = \Delta G_{B'} + \Delta g_{A,B'}$	$\Phi_2 k_2$
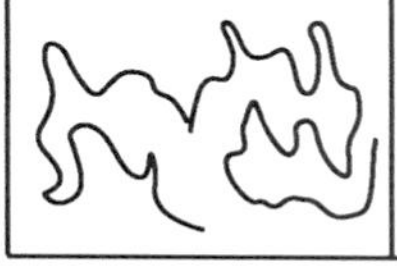	$G_{A'} + G_{B'} + g_{A',B'}$	$\Delta G_3 + \Delta g_3 = \Delta G_{A'} + \Delta G_{B'} + \Delta g_{A',B'}$	$\Phi_3 k_3$

FIG. 6. Representation of states for a hypothetical protein composed of two cooperative folding units. Free energy differences designated in uppercase letters represent the intrinsic stabilities of the cooperative units and have a contribution to the statistical weights designated by k. Free energy differences designated in lowercase letters represent the interaction between cooperative units and have a contribution to the statistical weights designated by Φ. [Reprinted from Freire and Murphy (1991).]

In general, the cooperative behavior of the system is determined by the properties of the Φ_i terms. For convenience, these terms can be expressed as the product of two terms ($\Phi_i = \Phi_i^* \Phi_i'$), where the Φ_i' terms contain the contributions from the protein region(s) undergoing unfolding and the Φ_i^* terms contain the contributions from the remaining regions of the protein (here termed complementary regions) (Freire and Murphy, 1991). This operational definition allows a mechanistic dissection of the different types of interactions that contribute to the cooperative folding/unfolding behavior of a protein. The partition function for the protein in Fig. 6 is then written as

$$Q = 1 + \Phi_1^* \Phi_1' k_1 + \Phi_2^* \Phi_2' k_2 + \Phi_3' k_3 \qquad (18)$$

or, in general,

$$Q = 1 + \sum_{i=1}^{n-1} \Phi_i^* \Phi_i' k_1 + \Phi_n' k_n \qquad (19)$$

It must be noted that, unlike the partially folded states, the totally unfolded state lacks a complementary region. The immediate consequence of this property is that partially folded states always expose to the solvent structural regions for which there is not a corresponding configurational entropy gain (Freire and Murphy, 1991).

C. Classification of Cooperative Forces and Mechanisms

As mentioned above, only recently has the magnitude of the individual forces that contribute to the stabilization of protein structure been evaluated with enough precision to permit reasonable predictions of protein stability from structural information. In the following paragraphs we summarize our current understanding of the influence of these forces on the cooperativity of the folding/unfolding transition. For clarity, we will use the example of a protein composed of two cooperative folding units.

1. Hydrophobic Interfaces

The interactions between domains or subdomains in proteins often involve the close apposition of hydrophobic surfaces. As illustrated in Fig. 7A, the unfolding of one of the domains elicits the exposure of apolar residues in the domain undergoing unfolding, as well as apolar residues from the rest of the protein. Though the solvent exposure of the unfolded domain is compensated by the favorable free energy of unfolding and especially its positive conformational entropy, the solvent exposure of the complementary regions of the protein is not compensated by a similar entropy change. Energetically, all partially folded states carry an extra free energy term arising from this uncompensated exposure of apolar groups. This uncompensated free energy constitutes the source of cooperativity for the hydrophobic interaction. The unfolded state exposes to the solvent what is unfolded, whereas partially folded states necessarily expose more than what is unfolded. As far as the exposure to the solvent is energetically unfavorable, partially folded states have a low probability of being populated.

Using the thermodynamic values discussed in Section IV it can be shown that ΔG_{ap} is positive at temperatures higher than 20°C, exhibits a maximum at 112°C, and is negative at temperatures lower than 20°C. This result allows one to predict that cooperative behavior based on purely hydrophobic interactions will be stronger at high temperatures and diminish at lower temperatures.

A) HYDROPHOBIC INTERFACE

Reference

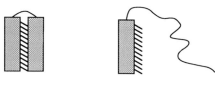
$\Delta G_2 + \Delta g_2' + \Delta g_2^*$

$\Delta G_1 + \Delta G_2 + \Delta g_1' + \Delta g_2'$

B) BONDED INTERFACE

Reference

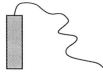
$\Delta G_2 + \Delta g'$

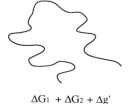
$\Delta G_1 + \Delta G_2 + \Delta g'$

C) LIGANDED INTERFACE

Reference

$\Delta G_2 + RT\ln(1+KX)$

$\Delta G_1 + \Delta G_2 + RT\ln(1+KX)$

FIG. 7. Schematic illustration of free energy assignments for three different types of interactions. The examples illustrate the situation existing for folded, unfolded, and one partially folded intermediate. For hydrophobic interfaces, partially folded intermediates always include an extra term Δg^* corresponding to the solvent exposure of protein regions that have not undergone unfolding. The unfolded state lacks this uncompensated exposure term. For a bonded or liganded interface, cooperative behavior is created when the unfolding of either domain results in the disruption of the bonded interface or the dissociation of the ligand molecule. [Reprinted from Freire *et al.* (1991)].

In general, the hydrophobic contribution to the cooperative behavior is dictated by the magnitude of the uncompensated solvent exposure of apolar surfaces. The magnitude of these terms is determined by the free energy change (ΔG_{ap}) associated with the exposure to the solvent of the apolar surface on the complementary region of the protein. Because the apolar surface area or the number of apolar hydrogens can be obtained from structural data, Eqs. (6)–(10) allow calculation of ΔG_{ap} for any structurally defined state of the protein.

2. Bonded Interfaces

The interaction between domains or subdomains often involves hydrogen bonds and their associated van der Waals contacts, salt bridges, and other noncovalent interactions. This situation is illustrated in Fig. 7B, where the change in Gibbs free energy associated with the unfolding of one or two domains is shown. It must be noted that the unfolding of only one of the domains is sufficient to break the bonds at the interface. This situation triggers the cooperative behavior. Contrary to the situation with the hydrophobic interface, the free energy of stabilization of hydrogen bonds increases at lower temperatures. For example, at very low or subzero temperatures, secondary structure elements like α helices have increased intrinsic stabilities, but the hydrophobic effect will not drive the protein into a compact structure (Freire and Murphy, 1991). Under those conditions it would be energetically favorable for a protein to adopt extended conformations characterized by different degrees of secondary structure. In this respect, Biringer and Fink (1982, 1988) have observed the presence of this type of intermediate during the refolding process of ribonuclease at subzero temperatures.

The above considerations suggest that the folding pathway of proteins might not be unique but dependent on temperature and other physical variables. At temperatures higher than 20°C, where the hydrophobic effect is strong and the intrinsic α-helix stabilities are poor, the hydrophobic collapse of the protein is expected to precede the formation of secondary structure elements. On the contrary, at low temperatures, where the hydrophobic effect is weak or nonexistent and the intrinsic α-helix stabilities are high, secondary structure elements might precede the formation of a compact structure. Kinetic measurements of the refolding of cytochrome c at 10°C using deuterium exchange techniques (Roder *et al.*, 1988) have revealed that at this temperature the first event is the formation of the N- and C-terminal helices.

3. Liganded Interfaces

As shown in Fig. 7C, the energetics of a liganded interface is mathematically very similar to that of a bonded interface except for the dependence on ligand concentration. In this case also, the ligand molecule dissociates after one of the domains unfolds, giving rise to the nonlinear dependence of the Gibbs free energy. It must be noted that a liganded interface contributes to the cooperative behavior if and only if the ligand dissociates on unfolding.

4. Other

Hydrophobic interactions, hydrogen bonds, and liganded interfaces are common to many protein systems and yield characteristic contributions to the free energy of interaction. They are not, however, exhaustive of all interactions defining interfaces. For example, in some cases, one of the domains at the interface may contain groups that become protonated on unfolding. In this case protonation occurs either on unfolding of the structural domain that contains the ionizable groups or on unfolding of the corresponding complementary region. The corresponding free energy must be included in the Φ terms of all the states that expose the ionizable group. All the appropriate contributions can be accounted for if the structure of the protein is known, as previously demonstrated for the case of myoglobin (Freire and Murphy, 1991).

Finally, the cooperative effects listed above are not exclusive of each other. In fact, it is perfectly feasible for two or more of them to exist simultaneously. In such cases, all the free energy contributions must be included in the appropriate Φ terms.

VI. Cooperativity of Two-Domain Proteins

Unlike most small globular proteins that show two-state folding/unfolding behavior (Privalov, 1979), the folding/unfolding transition of two-domain proteins has been shown to exhibit primarily three different types of behavior: (1) two-state behavior in which the only two states that become populated are those in which the two domains are both folded or unfolded—the heat denaturation of phosphoglycerate kinase (Griko *et al.*, 1988) belongs to this category; (2) independent folding/unfolding of the two domains such as in the case of pepsinogen (Mateo and Privalov, 1981); and (3) folding/unfolding of interacting domains, such as in the case of diphtheria toxin (Ramsay and Freire, 1990) or enzyme I from the phosphotransferase system (LiCalsi et al., 1991). In the three cases the domains behave as cooperative folding units, thus the observed behavior can be modeled and analyzed with Eqs. (16)–(18) for a protein composed of two interacting cooperative folding units.

Yeast Phosphoglycerate Kinase

Phosphoglycerate kinase is a protein composed of two structural domains of approximately equal molecular weight (N-terminal domain M_r 20,814; C-terminal domain M_r 23,735). The crystallographic structure has been determined (Watson et al., 1982) and can be used to calculate the hydrophobic surface exposed to water on total or partial unfolding as well as the number of hydrogen bonds and other bonds involved in the stabilization of the folded conformation.

The thermal unfolding of phosphoglycerate kinase has been studied by differential scanning calorimetry (Hu and Sturtevant, 1987; Griko et al., 1988; Brandts et al., 1989; Galisteo et al., 1991; Freire et al., 1991). In aqueous solution the transition is irreversible; however, in the presence of increasing concentrations of guanidinium hydrochloride (GuHCl), the transition becomes reversible and can be analyzed using thermodynamic methods (Griko et al., 1989; Freire et al., 1991).

At GuHCl concentrations higher than 0.5 M, the cold denaturation as well as the heat denaturation transitions are clearly visible in the calorimetric scans (Griko et al., 1989). Analysis of the calorimetric data indicates that the heat denaturation transition approaches a two-state transition whereas the cold denaturation is characterized by the presence of two peaks, indicating that the two domains undergo separate transitions. As shown in Fig. 8, at 0.7 M GuHCl, the heat denaturation transition is centered at ~40°C and characterized by an enthalpy change of 140 kcal mol^{-1} and a ΔC_p° of 7.3 cal K^{-1} mol^{-1}. The cold denaturation peaks are centered at about 7 and 20°C. The dramatic change in cooperative behavior for the heat and cold denaturation transitions within a single calorimetric scan occurs in a temperature interval of less than 20°C, indicating that the Φ terms in the partition function are strongly temperature dependent.

Analysis of the crystallographic structure indicates that on complete unfolding of the phosphoglycerate kinase (PGK) molecule a total of 1828 apolar hydrogens (814 from the N domain and 1014 from the C domain) become exposed to the solvent. Also, an average of 73% of the amino acid residues are hydrogen bonded (Chothia, 1976) and

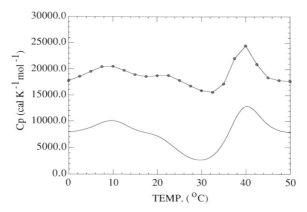

FIG. 8. Experimental (●—●) and theoretical heat capacity functions for the thermal folding/unfolding transition of phosphoglycerate kinase at pH 6.5 in the presence of 0.7 M GuHCl. The heat denaturation transition is characterized by a single peak, whereas the cold denaturation displays two peaks corresponding to the independent unfolding of the N and C domains. The experimental curve has been published before (Griko *et al.*, 1989). As discussed in the text, the theoretical curve does not represent the best fit to the experimental data, but rather the calculated curve using structural information in conjunction with thermodynamic information for elementary interactions. [Reprinted from Freire *et al.* (1991).]

contribute -14.3 cal K^{-1} (mol res)$^{-1}$ to the heat capacity change on unfolding (Murphy and Gill, 1990, 1991). These values predict a ΔC_p on unfolding of 7.78 kcal K^{-1} mol^{-1} of PGK (3.43 kcal K^{-1} mol^{-1} contributed by the N domain and 4.35 kcal K^{-1} mol^{-1} by the C domain). The experimental value obtained from the temperature dependence of the enthalpy change under conditions in which PGK undergoes reversible thermal unfolding is 7.5 kcal K^{-1} mol^{-1}. The enthalpy change extrapolated to the reference temperature T_H^* is 564 kcal mol^{-1}. This number is consistent with an average value of 1.36 kcal mol^{-1} per amino acid residue, very close to the average value estimated for ΔH^*.

Analysis of the crystallographic structure also reveals that the two domains interact primarily through hydrophobic and hydrogen bond interactions at the interface. The number of apolar hydrogens that become exposed on the C domain on unfolding of the N domain is 49.4, and the number of apolar hydrogens exposed on the N domain on unfolding of the C domain is 44.8. These values correspond to 726 and 659 Å^2 of apolar surface area, respectively. In addition, nine hydrogen bonds from the overlapping carboxy and amino terminals are broken following the unfolding of the N or C domain.

The structural calculations for the interdomain interaction predict a ΔH_{ap} of -24.8 kcal mol^{-1} at 25°C for the uncompensated exposure generated by the unfolding of the N domain, and -22.5 kcal mol^{-1} for the uncompensated exposure generated by the unfolding of the C domain. The corresponding ΔS_{ap} values at 25°C are -84.6 cal K^{-1} mol^{-1} and -76.7 cal K^{-1} mol^{-1}, and the ΔC_p values are 330.5 cal K^{-1} mol^{-1} and 299.7 cal K^{-1} mol^{-1}, respectively. The hydrogen bonds at the interface contribute a total enthalpy of 20.3 kcal mol^{-1}, an entropy change of 50.5 cal K^{-1} mol^{-1} and a ΔC_p^0 of -128.7 cal K^{-1} mol^{-1}. These thermodynamic values completely specify the Φ^* and Φ' terms in the partition function, allowing one to predict the cooperative behavior of the folding/unfolding transition and to compare the results with the experimental data (Freire and Murphy, 1991). When combined with the stability parameters, which in this case include the differential binding effects of GuHCl, it is possible to model the overall structural stability and cooperativity. It has been shown (see Fig. 8) that the above analysis predicts a heat capacity versus temperature transition curve that accurately reproduces the features observed experimentally with regard to the location of the peaks, the shapes, and the areas of both the heat and cold denaturation transitions (Freire et al., 1991).

Figure 9 shows the calculated overall free energy of stabilization of PGK at 0.7 M GuHCl as well as the cooperative Gibbs free energies ΔG_1^* and ΔG_2^* corresponding to the uncompensated exposure of the N and C domains (Freire et al., 1991). As is the case with other proteins (Privalov and Gill, 1988), the overall free energy of stabilization displays a characteristic curvature with two temperatures (the cold and heat denaturation temperatures) at which it is equal to zero. Below and above these temperatures, the protein is in the denatured state. As seen in Fig. 9, the cooperative free energy is large and positive for the heat denaturation transition and close to zero for the cold denaturation transition. As a result, at the temperature of the heat denaturation transition, the Φ^* terms are equal to 0.08 and 0.09 for the N and C domains, respectively, whereas at the temperature of the cold denaturation transition they are equal to 0.99 and 0.98. These values indicate that the cold denaturation is very close to two independent transitions whereas the heat denaturation approaches a two-state transition. Also, these Φ^* values result in a small population (<15%) of partially folded intermediates during the heat denaturation but a significant one (~60%) during the cold denaturation.

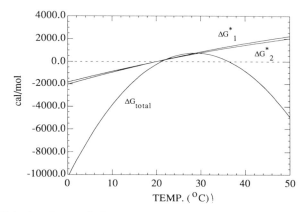

FIG. 9. Calculated overall free energy of stabilization (ΔG_{total}) for yeast phosphoglycerate kinase at pH 6.5 and 0.7 M GuHCl. This curve displays two zeros, corresponding to the temperatures of cold and heat denaturation. Also shown in the curve are the cooperative Gibbs free energies (ΔG^*) associated with the uncompensated exposure of apolar surfaces on unfolding of each of the domains. For both domains, ΔG^* is positive for the heat denaturation and close to zero for the cold denaturation. This behavior results in a cooperative heat denaturation and a noncooperative cold denaturation. [Reprinted from Freire et al. (1991).]

The only intermediate that becomes populated is the one in which the N domain is unfolded and the C domain is folded.

VII. Cooperativity of Single-Domain Proteins

One of the most striking observations regarding the folding/unfolding of single-domain globular proteins is the high degree of cooperativity. Cooperative interactions essentially reduce to zero the probability of hundreds or thousands of partially folded intermediates with a potential to become populated during the folding transition. From a statistical thermodynamic point of view, the most fundamental questions deal with the molecular origin of this cooperative behavior: How is the probability of intermediates reduced to negligible levels? Why do single-domain proteins undergo two-state transitions?

It has been shown that the hierarchical approach illustrated above for the case of a two-domain protein can be performed at a more fundamental level in order to account for the cooperative folding behavior of single-domain proteins (Freire and Murphy, 1991). This approach involves the use of the crystallographic structure of a protein as a template to identify low-order cooperative folding units (e.g., α helices) and to generate the set of most probable partially folded conformations necessary to define the partition function.

The folding/unfolding behavior of pure α-helical structural motifs can be defined in terms of intrinsic helix stability free energies as well as interaction free energies. In general, α helices in proteins are either formed or not formed at all (i.e., they exhibit two-state behavior), and as such they can be considered as first-order cooperative folding units. This property of α helices in most proteins greatly reduces the number of states that need to be considered because partially folded helical states need not be counted in the partition function. Within this context α helices can be used as building blocks to construct higher order cooperative folding units (e.g., a four-α-helix bundle). According to the criteria just described, this hypothetical four-α-helix bundle will give rise to 16 different configurations (2^4) defined in terms of all possible ways of folding and unfolding its constituent α helices.

An α-helix bundle may become a second-order cooperative folding unit if the interaction energy terms are such that the intermediate terms in the partition function become negligibly small [Eq. (14)] and the entire partition function reduces to a two-state partition function (i.e., a partition function of the form $1 + e^{-\Delta G/RT}$). If such is the case, the α-helix bundle will be either completely folded or unfolded. Higher order cooperative folding units can be constructed from lower order ones following the same rules. The most immediate application of this approach is to proteins exhibiting pure α-helical structural motifs.

Myoglobin

Myoglobin is a small globular protein of M_r 17,900, with a well-characterized structure (Kendrew *et al.*, 1960, 1961) and for which very precise thermodynamic data regarding its folding/unfolding equilibrium are available (Privalov and Khechinashvilli, 1974; Privalov, 1979; Privalov *et al.*, 1986). Myoglobin is an all α-helical protein made up of eight different right-handed α helices (A–H) that involve 121 out of a total of 153 amino acid residues. The heme group is located in a hydrophobic crevice between helices E and F. Myoglobin was the first protein to be analyzed with the hierarchical formalism (Freire and Murphy, 1991).

According to the hierarchical approximation to the partition function there are a total of 256 states (2^8) corresponding to all the possible states that can be generated by independently folding and unfolding the eight helices in the myoglobin molecule. These states were generated with the computer, using the crystallographic structure as a template (Freire and Murphy, 1991). The enthalpy, entropy, heat capacity, and Gibbs free energy of each partially folded state relative to that of the completely folded state were calculated from the following data: (1) the intrinsic energetics associated with the unfolding of the helix or set of helices that define that particular state of the protein; (2) the energetics of exposing the unfolded helix or helices to the solvent; and (3) the energetics of exposing the complementary regions of the protein to the solvent. In performing these calculations, it was assumed that the folded regions in partially folded states had the same structure as in the native state.

The thermal stability of metmyoglobin and apomyoglobin has been extensively studied under different solvent conditions (Privalov et al., 1986). In particular, it was shown that at low pH values both the heat and cold denaturation peaks are clearly visible in the calorimetric scans. Figure 10 shows the excess heat capacity function for apomyoglobin predicted by the hierarchical partition function and the thermodynamic parameters described above. In order to simulate the experimental curve obtained at pH 3.83 (Privalov et al., 1986), the protonation of five specific histidine residues on unfolding was

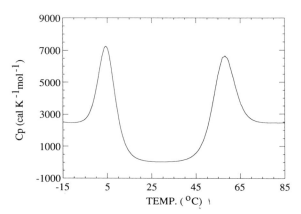

FIG. 10. Predicted excess heat capacity function versus temperature for myoglobin. The curve simulates the experimental curve obtained at pH 3.83 by Privalov et al. (1986). Under those conditions both the cold and heat denaturation curves can be studied experimentally. The predicted values are $T_{m,cold} = 4°C$; $T_{m,heat} = 58°C$; $\Delta H = 59$ kcal mol^{-1}; $\Delta C_p = 2.45$ kcal K^{-1} mol^{-1}. The experimental values are $T_{m,cold} = 3°C$; $T_{m,heat} = 57.5°C$; $\Delta H = 53$ kcal mol^{-1}; $\Delta C_p = 2.5$ kcal K^{-1} mol^{-1} (Privalov et al., 1986). [Reprinted from Freire and Murphy (1991).]

taken into consideration (Carver and Bradbury, 1984). This was done by including the protonation enthalpy (-6940 cal mol^{-1}) and entropy (4.2 cal K^{-1} mol^{-1}) changes associated with the exposure of these histidines to the solvent and by adjusting the intrinsic α-helix unfolding entropy change in order to match the experimental transition temperature of the heat denaturation peak. This adjustment is necessary to locate accurately the overall transition temperature because the existing entropy values from the protein data base are not precise enough to narrowly define transition temperatures. Nevertheless, it should be mentioned that the adjusted value at this pH was 4.0 cal K^{-1} mol^{-1} of amino acid, which is within the range 4.3 ± 0.3 cal K^{-1} mol^{-1} from the protein data base at that same temperature (Privalov and Gill, 1988).

As seen in Fig. 10, the model accurately predicts the presence, location, and area of the cold and heat denaturation peaks. Under these conditions, the hierarchical partition function predicts a heat denaturation peak centered at 58°C and a cold denaturation peak centered at 4°C. The enthalpy change for the heat denaturation peak is 59 kcal mol^{-1} and the ΔC_p is equal to 2.45 kcal K^{-1} mol^{-1}. The experimental values reported by Privalov *et al.* (1986) are 57.5 and 3°C for the heat and cold denaturation transition temperatures, 53 kcal mol^{-1} for the enthalpy change, and 2.5 kcal K^{-1} mol^{-1} for ΔC_p. Analysis of the theoretical curve indicates that it corresponds to a two-state transition, in agreement with the experimental data. The population of partially folded intermediates is never greater than 10^{-5} during the heat denaturation transition.

The high level of cooperativity of the folding/unfolding transition of myoglobin is a consequence of the almost negligible value of the Φ^* terms in the partition function. There appear to be two main contributions to the cooperative interactions. First is the unfavorable free energy resulting from the solvent exposure of complementary apolar areas located in partially folded states. Second is the reduced conformational entropy change associated with the unfolding of α helices in partially folded states; e.g., the number of degrees of freedom gained by the unfolding of a single α helix in an otherwise fully or partially folded protein is smaller than the number of degrees of freedom gained by that same helix when the entire protein unfolds. In other words, the conformational entropy change for helix unfolding is a function of the overall degree of unfolding and is expected to be minimal at the earliest stages of unfolding.

Whereas the hydrophobic effect by itself is the main force contributing to the cooperativity of the heat denaturation peak, the situation is not the same for the cold denaturation peak. As discussed above, the purely hydrophobic contributions to the free energy of stabilization become negative below 20°C. Thus hydrophobic interactions cannot generate the cooperativity observed in the cold denaturation transition. The reduced conformational entropy change associated with helix unfolding in partially folded states and tertiary H bonds account for the cooperativity of the cold denaturation by lowering to $\sim -35°C$ the temperature at which the cooperative free energy becomes negative. Even so, the population of partially folded intermediate states is about 1000 times larger for the cold denaturation transition than for the heat denaturation transition. This result is in agreement with the detection of partially folded apomyoglobin intermediates at low temperatures and low pH values (Hughson and Baldwin, 1989; Hughson et al., 1990).

VIII. Energetics of Molten Globule Intermediates

As discussed above, the folding of single-domain proteins is generally a highly cooperative process in which only the completely folded and unfolded states become significantly populated. Under certain circumstances, however, several proteins have been shown to exhibit a stable folding intermediate known as a "molten globule" (Kim and Baldwin, 1982, 1990; Kuwajima, 1989; Baum et al., 1989; Hughson et al., 1990; Christensen and Pain, 1991; Ewbank and Creighton, 1991). The molten globule state of apo-α-lactalbumin (MW 14,200) is the most extensively characterized and has been shown to be stable at low ionic strength, low pH, or in the presence of intermediate concentrations of denaturants (Kuwajima et al., 1976; Semisotnov et al., 1987; Baum et al., 1989; Jeng et al., 1990).

The molten globule intermediates found for different proteins appear to have some common features, primarily, a nativelike secondary structure, high degree of compactness, and a disrupted tertiary structure (Ohgushi and Wada, 1983; Ptitsyn, 1987; Kuwajima, 1989). The CD spectrum of the molten globule state of apo-α-lactalbumin is similar to that of the native protein in the far-UV region but is nearly absent in the near-UV region, consistent with an almost intact secondary structure but disrupted tertiary structure (Dolgikh et al., 1985; Kuwajima, 1989). The molten globule state has also been shown to exhibit a rotational correlation time, and viscosity close to that of the native state indicating that the molten globule is highly compact (Dolgikh et al., 1981). Ewbank and Creighton (1991) have recently shown that the molten globule state is compatible with a variety of disulfide bond pairings, suggesting the absence of many specific tertiary structure interactions.

Kuwajima (1989) recently characterized the molten globule "as a nonspecific assembly of secondary structure segments brought about by nonspecific hydrophobic interactions." Despite the existence of numerous structural studies of the molten globule state, a direct calorimetric characterization of its energetics has not been possible yet, due primarily to the fact that the molten globule does not undergo a cooperative unfolding transition. As discussed earlier, cooperativity and overall energetics are different concepts. Even though the absence of a cooperative unfolding transition makes direct measurements very difficult (e.g., differential scanning calorimetric curves are extremely broad), it does not imply that the molten and unfolded states are enthalpically similar. An inspection of the known structural features of the molten globule in conjunction with the fundamental thermodynamic parameters discussed in this article should provide an approximate idea of the energetic level of this state.

Let us consider the well-known molten globule state of apo-α-lactalbumin as an example. Most studies indicate that the molten globule state has an almost intact secondary structure. According to the crystallographic structure of baboon α-lactalbumin (Acharya *et al.*, 1989), there are about 69 secondary structure hydrogen bonds out of a total of 84. If one assumes that the secondary structure elements in the molten globule are intact but exposed to the solvent (as would be the case if the ΔC_p between the molten and unfolded states were negligible), then a ΔH^0 of about 150 kcal mol^{-1} at 25°C would be expected between the unfolded and molten globule states. This is certainly a ridiculous figure, leading to the absurd conclusion that the ΔH between the unfolded and the molten globule state is much larger than the one between the unfolded and the native state. At 25°C the experimental enthalpy difference between the unfolded and native state is only 30 kcal mol^{-1} (Pfeil and Sadowski, 1985; Xie *et al.*, 1991). Clearly, the molten globule must bury a significant fraction of its apolar surfaces from the solvent.

Even though it was reported earlier (Dolgikh *et al.*, 1985) that the molten globule state had a heat capacity similar to that of the unfolded state, implying that the molten globule was fully exposed to water, this view appears to be at odds with thermodynamic considerations and new experimental data. Also, if the molten globule is fully hydrated and its tertiary structure interactions are largely disrupted, why does it assume a compact conformation? What are the forces that oppose the configurational entropy and its tendency to drive the molten globule into multiple extended conformations?

Recently, Xie et al. (1991) concluded from the analysis of the temperature–GuHCl heat capacity surface of apo-α-lactalbumin that the ΔC_p for the molten globule is only 326 cal K^{-1} mol^{-1} compared to a ΔC_p of 1825 cal K^{-1} mol^{-1} for the unfolded state. These results are in agreement with the results of Chen et al. (1989) for T4 lysozyme. These authors concluded that the ΔC_p between the molten globule transition state and the native state of T4 lysozyme is 500 cal K^{-1} mol^{-1} compared to the ΔC_p of 2200 cal K^{-1} mol^{-1} between the unfolded and native state. These results suggest that the molten globule maintains a large fraction of the hydrophobic contacts existing in the native state and that water does not significantly penetrate into the interior of the molten globule. If this is so, hydrophobic interactions will be primarily responsible for maintaining the molten globule in a compact state. Also, the presence of significant hydrophobic interactions is consistent with a small ΔH at 25°C.

It is important to note that the experimental apparent enthalpy values for the molten globule contain additional contributions. Because molten globule states are experimentally obtained either in the presence of GuHCl or at low pH values, the experimentally measured values contain contributions arising from differential GuHCl binding effects or the protonation of several ionizable groups. These effects are exothermic and contribute to lower the apparent enthalpy change (Xie et al., 1991).

The absence of a significant number of tertiary structure hydrogen bonds coupled to the presence of significant hydrophobic interactions is consistent with the known structural features of the molten globule state. Because structures stabilized by the apposition of purely hydrophobic surfaces lack the rigid geometric constraints created by hydrogen bonds at specific locations in the molecule, they can be expected to have a larger conformational entropy. This gain in conformational entropy could compensate the loss of tertiary structure hydrogen bonds, resulting in the stabilization of the molten globule state. This larger entropy is also in agreement with the recent observation that the molten globule is consistent with a variety of disulfide bond pairings (Ewbank and Creighton, 1991) and with the observation that the partially folded intermediate of apomyoglobin found at low pH lacks the specific tertiary structure contacts existing in the native state (Hughson and Baldwin, 1989; Hughson et al., 1990).

IX. Concluding Remarks

We have attempted to show that quantitative predictions of the stability and cooperative folding behavior of proteins are within reach. Experimental thermodynamic values for the fundamental forces that determine protein stability have been refined to a point in which accurate calculations of the free energy of stabilization are possible, provided that the protein structure is known. Estimation of the folding/unfolding partition function using a hierarchical approximation and subsequent prediction of cooperative behavior is also increasingly feasible, as demonstrated for myoglobin and yeast phosphoglycerate kinase. It is likely that future work will extend this capability to proteins exhibiting other than pure α-helical structural motifs.

Acknowledgments

We would like to thank Dr. Vinod Bhakuni for his assistance in the hydrogen bond and α-lactalbumin calculations, Prof. Cyrus Chothia for making ASA data available, and Prof. Peter Privalov for many helpful discussions.

Supported by National Institutes of Health Grants RR04328, NS24520, and GM37911. K.P.M. was supported by a postdoctoral fellowship from the NSF (DIR-8721059).

References

Acharya, K. R., Stuart, D. F., Walker, W. P. C., Lewis, M., and Phillips, D. C. (1989). *J. Mol. Biol.* **208**, 99–127.
Alber, T., Dao-pin, S., Wilson, K., Wozniak, J. A., Cook, S. P., and Matthews, B. W. (1987). *Nature (London)* **330**, 41–46.
Anfinsen, C. B. (1973). *Science* **181**, 223–230.
Baker, E. N., and Hubbard, R. E. (1984). *Prog. Biophys. Mol. Biol.* **44**, 97–179.
Baldwin, R. L. (1986). *Proc. Natl. Acad. Sci. U.S.A.* **83**, 8069–8072.
Baum, J., Dobson, C. M., Evans, P. A., and Hanley, C. (1989). *Biochemistry* **28**, 7–13.
Becktel, W. J., and Schellman, J. A. (1987). *Biopolymers* **26**, 1859–1877.
Ben-Naim, A. (1978). *J. Phys. Chem.* **82**, 792–803.
Biringer, R. G., and Fink, A. L. (1982). *J. Mol. Biol.* **160**, 87–116.
Biringer, R. G., and Fink, A. L. (1988). *Biochemistry* **27**, 315–325.
Brandts, J. F., Hu, C. Q., Lin, L.-N., and Mas, M. T. (1989). *Biochemistry* **28**, 8588–8596.
Cabani, S., Gianni, P., Mollica, V., and Lepori, L. (1981). *J. Solution Chem.* **10**, 563–595.
Calderon, R. O., Stolowich, N. J., Gerlt, J. A., and Sturtevant, J. M. (1985). *Biochemistry* **24**, 6044–6049.
Carver, J. A., and Bradbury, J. H. (1984). *Biochemistry* **23**, 4890–4905.
Chen, B. L., Baase, N. A., and Schellman, J. A. (1989). *Biochemistry* **28**, 691–699.
Chothia, C. (1976). *J. Mol. Biol.* **105**, 1–14.
Chothia, C. (1984). *Annu. Rev. Biochem.* **53**, 537–572.

Christensen, H., and Pain, R. H. (1991). *Eur. Biophys. J.* **19,** 221–229.
Connolly, M. L. (1983). *J. Appl. Crystallogr.* **16,** 548–558.
Creighton, T. E. (1984). Proteins: Structure and Molecular Properties." Freeman, New York.
Creighton, T. E. (1991). *Curr. Opin. Struct. Biol.* **1,** 5–16.
Dec, S. F., and Gill, S. J. (1985). *J. Solution Chem.* **14,** 827–836.
DeYoung, L. R., and Dill, K. A. (1990). *J. Phys. Chem.* **94,** 801–809.
Dill, K. A. (1990a). *Biochemistry* **29,** 7133–7155.
Dill, K. A. (1990b). *Science* **250,** 297.
Dolgikh, D. A., Gilmanshin, R. I., Brazhnikov, E. V., Bychkova, V. E., Semisotnov, G. V., Venyaminov, S. Yu., and Ptitsyn, O. B. (1981). *FEBS Lett.* **136,** 311–315.
Dolgikh, D. A., Abaturov, L. V., Bolotina, I. A., Brazhnikov, E. V., Bychkova, V. E., Bushuev, V. N., Gilmanshin, R. I., Levedev, Yu. O., Semisotnov, G. V., Tiktopulo, E. I., and Ptitsyn, O. B. (1985). *Eur. Biophys. J.* **13,** 109–121.
Eisenberg, D., and McLachlan, A. D. (1986). *Nature (London)* **319,** 199–203.
Ewbank, J. J., and Creighton, T. (1991). *Nature (London)* **350,** 518–520.
Fauchere, J. L., and Pliska, V. (1983). *Eur. J. Med. Chem.* **18,** 369–375.
Filimonov, V. V., Pfeil, W., Tsalkova, T. N., and Privalov, P. L. (1978). *Biophys. Biochem.* **8,** 117–122.
Freire, E. (1989). *Comments Mol. Cell. Biophys.* **6,** 123–140.
Freire, E., and Biltonen, R. L. (1978a). *Biopolymers* **17,** 481–494.
Freire, E., and Biltonen, R. L. (1978b). *CRC Crit. Rev. Biochem.* **5,** 85–124.
Freire, E., and Murphy, K. P. (1991). *J. Mol. Biol.* **222,** 687–698.
Freire, E., van Osdol, W. W., Mayorga, O. L., and Sanchez-Ruiz, J. M. (1990). *Annu. Rev. Biophys. Biophys. Chem.* **19,** 159–188.
Freire, E., Murphy, K. P., Sanchez-Ruiz, J.M., Galisteo, M. L., and Privalov, P. L. (1992). *Biochemistry* **31,** 250–256.
Galisteo, M. L., Mateo, P. L., and Sanchez-Ruiz, J. M. (1991). *Biochemistry* **30,** 2061–2066.
Gill, S. J., and Noll, L. (1972). *J. Phys. Chem.* **76,** 3065–3068.
Gill, S. J., and Wadsö, I. (1976). *Proc. Natl. Acad. Sci. U.S.A.* **73,** 2955–2958.
Gill, S. J., Nichols, N. F., and Wadsö, I. (1975). *J. Chem. Thermodyn.* **7,** 175.
Gill, S. J., Nichols, N. F., and Wadsö, I. (1976). *J. Chem. Thermodyn.* **8,** 445–452.
Griko, Y. V., Privalov, P. L., Sturtevant, J. M., and Venyaminov, S. Y. (1988). *Proc. Natl. Acad. Sci. U.S.A.* **85,** 3343–3347.
Griko, Y. V., Venyaminov, S. Y., and Privalov, P. L. (1989). *FEBS Lett.* **244,** 276–278.
Hallén, D., Nilsson, S.-O., Rothschild, W., and Wadsö, I. (1986). *J. Chem. Thermodyn.* **18,** 429–442.
Hermann, R. B. (1972). *J. Phys. Chem.* **76,** 2754–2759.
Herzfeld, J. (1991). *Science* **253,** 88.
Hu, C. Q., and Sturtevant, J. M. (1987). *Biochemistry* **26,** 178–182.
Hughson, F. M., and Baldwin, R. L. (1989). *Biochemistry* **28,** 4415–4422.
Hughson, F. M., Wright, P. E., and Baldwin, R. L. (1990). *Science* **249,** 1544–1548.
Jeng, M.-F., Englander, S. W., Elove, G. A., Wand, A. J., and Roder, H. (1990). *Biochemistry* **29,** 10433–10437.
Jorgensen, W. L., Gao, J., and Ravimohan, C. (1985). *J. Phys. Chem.* **89,** 3470–3473.
Kauzmann, W. (1959). *Adv. Protein Chem.* **14,** 1–63.
Kendrew, J. C., Dickerson, R. E., Strandberg, B. E., Hart, R. G., Davies, D. R., Phillips, D. C., and Shore, V. C. (1960). *Nature (London)* **185,** 422–427.

Kendrew, J. C., Watson, H. C., Standberg, B. E., Dickerson, R. E., Phillips, D. C., and Shore, V. C. (1961). *Nature (London)* **190**, 666–670.
Khechinashvili, N. N., Privalov, P. L., and Tiktopolu, E. I. (1973). *FEBS Lett.* **30**, 57–60.
Kim, P. S., and Baldwin, R. L. (1982). *Annu. Rev. Biochem.* **51**, 459–489.
Kim, P. S., and Baldwin, R. L. (1990). *Annu. Rev. Biochem.* **59**, 631–660.
Klotz, I. M., and Franzen, J. S. (1962). *J. Am. Chem. Soc.* **84**, 3461–3466.
Kuwajima, K. (1989). *Proteins: Struct., Funct., Genet.* **6**, 87–103.
Kuwajima, K., Nitta, K., Yoneyama, M., and Sugai, S. (1976). *J. Mol. Biol.* **106**, 359–373.
Kuwajima, K., Mitani, M., and Sugai, S. (1989). *J. Mol. Biol.* **206**, 547–561.
Lee, B., and Richards, F. M. (1971). *J. Mol. Biol.* **55**, 379–400.
Lee, B. (1991). *Proc. Natl. Acad. Sci. U.S.A.* **88**, 5154–5158.
LiCalsi, C., Crocenzi, T. S., Freire, E., and Roseman, S. (1991). *J. Biol. Chem.* **266**, 19519–19527.
Livingstone, J. R., Spolar, R. S., and Record, M. T., Jr. (1991). *Biochemistry* **30**, 4237–4244.
Makhatadze, G. I., and Privalov, P. L. (1990). *J. Mol. Biol.* **213**, 375–384.
Mateo, P. L., and Privalov, P. L. (1981). *FEBS Lett.* **123**, 189–192.
Miller, S., Lesk, A. M., Janin, J., and Chothia, C. (1987). *Nature (London)* **328**, 834–836.
Murphy, K. P. (1990). Ph.D. Thesis, University of Colorado, Boulder.
Murphy, K. P., and Gill, S. J. (1989a). *Thermochim. Acta* **139**, 279–290.
Murphy, K. P., and Gill, S. J. (1989b). *J. Chem. Thermodyn.* **21**, 903–913.
Murphy, K. P., and Gill, S. J. (1990). *Thermochim. Acta* **172**, 11–20.
Murphy, K. P., and Gill, S. J. (1991). *J. Mol. Biol.* **222**, 699–709.
Murphy, K. P., Privalov, P. L., and Gill, S. J. (1990). *Science* **247**, 559–561.
Murphy, K. P., Bhakuni, V., Xie, D., and Freire, E. (1992). *J. Mol. Biol.* (in press).
Nichols, N., Sköld, R., Spink, C., and Wadsö, I. (1976). *J. Chem. Thermodyn.* **8**, 1081–1093.
Nozaki, Y., and Tanford, C. (1971). *J. Biol. Chem.* **246**, 2211–2217.
Ohgushi, M., and Wada, A. (1983). *FEBS Lett.* **164**, 21–24.
Ooi, T., Oobatake, M., Némethy, G., and Scheraga, H. A. (1987). *Proc. Natl. Acad. Sci. U.S.A.* **84**, 3086–3090.
Pfeil, W., and Sadowski, M. L. (1985). *Stud. Biophys.* **109**, 163–170.
Privalov, P. L. (1979). *Adv. Protein Chem.* **33**, 167–239.
Privalov, P. L. (1982). *Adv. Protein Chem.* **35**, 1–104.
Privalov, P. L., and Gill, S. J. (1988). *Adv. Protein Chem.* **39**, 191–234.
Privalov, P. L., and Khechinashvili, N. N. (1974). *J. Mol. Biol.* **86**, 665–684.
Privalov, P. L., and Makhatadze, G. I. (1990). *J. Mol. Biol.* **213**, 385–391.
Privalov, P. L., Tiktopulo, E. I., and Khechinashvili, N. N. (1973). *Int. J. Pept. Protein Res.* **5**, 229–237.
Privalov, P. L., Griko, Y. V., Venyaminov, S. Y., and Kutyshenko, V. P. (1986). *J. Mol. Biol.* **190**, 487–498.
Privalov, P. L., Tiktopulo, E. I., Yenyaminov, S. Y., Griko, Y. V., Makhatadze, G. I., and Khechinashvili, N. N. (1989). *J. Mol. Biol.* **205**, 727–750.
Ptitsyn, O. B. (1987). *J. Protein Chem.* **6**, 272–293.
Ramsay, G., and Freire, E. (1990). *Biochemistry* **29**, 8677–8683.
Richards, F. M. (1974). *J. Mol. Biol.* **82**, 1–14.

Richards, F. M. (1977). *Annu. Rev. Biophys. Bioeng.* **6,** 151–176.
Robertson, A. D., and Baldwin, R. L. (1991). *Biochemistry* **30,** 9907–9914.
Roder, H., Elove, G. A., and Englander, S. W. (1988). *Nature (London)* **335,** 700–704.
Schellman, J. A. (1955). *C. R. Trav. Lab. Carlsberg, Ser. Chim.* **29,** 223–229.
Scholtz, J. M., Marqusee, S., Baldwin, R. L., York, E. J., Stewart, J. M., Santoro, M., and Bolen, D. W. (1991). *Proc. Natl. Acad. Sci. U.S.A.* **88,** 2854–2858.
Semisotnov, G. V., Rodionova, N. A., Kutyshenko, V. P., Ebert, B., Blanck, J., and Ptitsyn, O. B. (1987). *FEBS Lett.* **224,** 9–13.
Sharp, K. A. (1991). *Curr. Opin. Struct. Biol.* **1,** 171–174.
Sharp, K. A., Nicholls, A., Fine, R. F., and Honig, B. (1991a). *Biochemistry* **30,** 9686–9697.
Sharp, K. A., Nicholls, A., Fine, R. F., and Honig, B. (1991b). *Science* **252,** 106–109.
Shirley, B. A., Stanssens, P., Hahn, U., and Pace, C. N. (1992). *Biochemistry* **31,** 725–732.
Shrake, A., and Rupley, J. A. (1973). *J. Mol. Biol.* **79,** 351–372.
Spolar, R. S., Ha, J.-H., and Record, M. T., Jr. (1989). *Proc. Natl. Acad. Sci. U.S.A.* **86,** 8382–8385.
Sturtevant, J. M. (1977). *Proc. Natl. Acad. Sci. U.S.A.* **74,** 2236–2240.
Susi, H., and Ard, J. S. (1969). *J. Phys. Chem.* **73,** 2440–2441.
Susi, H., Timasheff, S. N., and Ard, J. S. (1964). *J. Biol. Chem.* **239,** 3051–3054.
Tanford, C. (1980). "The Hydrophobic Effect." Wiley, New York.
Tatunashvili, L. V., and Privalov, P. L. (1986). *Biofizika* **31,** 578–581.
Tiktopulo, E. I., and Privalov, P. L. (1978). *FEBS Lett.* **91,** 57–58.
Tischenko, V. M., and Gorodnov, B. G. (1979). *Biofizika* **24,** 334–335.
Tischenko, V. M., Tiktopulo, E. I., and Privalov, P. L. (1974). *Biofizika* **19,** 400–404.
Velicelebi, G., and Sturtevant, J. M. (1979). *Biochemistry* **18,** 1180–1186.
Watson, H. C., Walker, N. P. C., Shaw, P. J., Bryant, P. L., Fothergill, L. A., Perkins, R. E., Conroy, S. C., Dobson, M. J., Tuite, M. F., Kingsman, A. J., and Kingsman, S. M. (1982). *EMBO J.* **1,** 1635–1640.
Xie, D., Bhakuni, V., and Freire, E. (1991). *Biochemistry* **30,** 10673–10678.

AUTHOR INDEX

Numbers in italics refer to the pages on which the complete references are listed.

A

Abaturov, L. V., 355, *359*
Abelson, J. N., 22, *61*
Abragam, A., 179, *303*
Acharya, K. R., 86, *149*, 356, *358*
Ackers, G. K., 166, 209, 233, 281, 282, 283, 285, 288, 295, 298, 299, 302, *303*, *304*, *305*, *307*, *310*, *311*, *312*
Adachi, K., 199, 202, 242, *304*
Adams, J. A., 20, 23, *61*
Akhmanova, A. S., 51, 53, *61*
Alber, T., 325, *358*
Albertsson, P.-Å., 8, *56*
Alden, R. A., 20, 21, *61*
Alexander, J., 76, 77, 137, *143*
Allais, C., 91, *146*
Allan-Wojtas, P., 106, *146*
Allerhand, A., 184, *304*
Aloni, H., 5, 17, 18, 19, 37, *61*
Altekar, W., 14, 42, *56*, *59*
Amioni, G., 253, 295, *304*
Anderson, C. F., 2, *61*
Anderson, L., 295, *304*
Andrews, A. L., 88, *143*
Andrews, A. T., 74, *143*
Andrews, J., 20, 23, *56*
Anfinsen, C. B., 313, *358*
Antonini, E., 154, 217, 218, 229, 236, 253, 295, *304*
Aoki, T., 74, 98, 131, 133, 136, *143*
Aoyagi, S., 87, *145*
Ard, J. S., 318, 325, *361*
Arima, S., 96, *149*
Arnone, A., 164, 262, 264, 286, *304*, *312*
Asakura, T., 199, 202, 242, *304*
Atkinson, D., 88, *143*

Aubert, J. P., 90, 91, *147*
Avron, M., 4, *57*

B

Baase, N. A., 357, *358*
Bahdyopadhyay, D., 217, 220, *311*
Bailey, M., 284, *311*
Bailey, R. T., 103, 127, 128, 130, 131, *146*
Baker, E. N., 326, 330, *358*
Balch, W. E., 43, *58*
Baldacci, G., 41, 49, 55, *57*
Baldassare, J. J., 199, 221, 243, 244, 245, 295, 301, *304*, *305*, *306*, *308*
Baldwin, J. M., 25, 30, 55, 59, 158, 202, 207, 213, 218, 219, 244, 266, 288, 291, 292, 298, *304*
Baldwin, R. L., 317, 322, 325, 326, 328, 329, 332, 337, 355, *358*, *359*, *360*, *361*
Bandyopadhyay, D., 291, *304*
Banerjee, R., 253, *304*
Baratova, L. A., 51, *60*
Bartels, H., 28, *61*
Baum, J., 355, *358*
Baumeister, W., 30, *57*
Baumrucker, C. R., 83, *148*
Baumrucker, C. W., 82, 83, *143*
Bax, A., 185, *304*
Bayley, S. T., 14, *57*
Beattie, C. W., 68, 69, 77, 78, *149*
Becker, E. D., 176, *304*
Beckmann, E., 25, 30, 55, *59*
Becktel, W. J., 328, *358*
Beddell, C. R., 20, *61*
Beeby, R., 118, *143*
Behringer, R. R., 165, *310*
Bekhof, J.-J., 84, *144*

Ben-Amotz, A., 4, *57*
Ben-Naim, A., 322, 326, *358*
Benazzi, L., 218, 281, 282, 283, 298, *309*, *310*
Benerjee, R., 234, *311*
Benesch, R. E., 164, 209, *304*
Benesch, R., 164, 209, *304*
Benkovic, S. J., 20, 22, 23, 56, *57*, *58*, *61*, *62*
Bennett, W., 27, 28, *62*
Benovic, J. L., 164, *306*
Berg, E., 29, *61*
Berjis, M., 217, 220, 291, *304*, *311*
Bernal, J. D., 161, *304*
Bernstein, H. J., 176, 254, *310*
Berry, G. P., 106, *143*, *144*, *151*
Betlach, M. C., 52, *57*
Betlach, M., 44, 45, *57*, *60*
Betts, F., 107, 128, 129, *147*
Beverley, S. M., 21, *57*
Bhakuni, V., 356, 357, *361*
Biltonen, R. L., 314, 315, 340, *359*
Bingham, E. W., 80, *143*
Birdsall, B., 20, 23, *56*
Biringer, R. G., 346, *358*
Bishop, G., 218, *312*
Blake, C. C. F., 76, *143*
Blakemore, R., 43, *58*
Blakley, R. L., 20, *57*
Blanck, A., 49, 52, 53, 55, *57*, *59*
Blanck, J., 355, *361*
Blaseio, U., 45, 49, *60*
Blaurock, A., 29, *57*
Blecher, O., 48, 56, *57*
Bloembergen, N., 177, *304*
Bloomfield, V. A., 103, 112, 113, 114, 119, 120, 131, *143*, *144*, *147*, *150*
Blough, N. V., 281, 286, *304*
Blouquit, Y., 200, *305*
Blumberg, W. E., 253, 260, *311*
Blumenthal, N. C., 107, 128, 129, *147*
Boehm, G., 9, 10, 15, 21, 50, *62*
Bogomolni, R., 30, *61*
Bolen, D. W., 325, 332, *361*
Bolin, J. J., 20, 21, *57*, *58*
Bolotina, I. A., 355, *359*
Bonen, L., 43, *58*
Bonete, M. J., 10, *57*
Bonsing, J., 68, 69, 75, 77, 78, *143*, *149*
Booy, F. P., 106, 107, *150*

Borochov, N., 5, 8, 17, 37, 38, 39, 40, 41, *62*
Boss, G. R., 217, 220, *311*
Boswell, R. T., 101, 113, *150*
Botet, R., 141, *143*
Both, P., 70, 82, 104, 105, 113, 114, 131, *149*, *150*
Boulton, A. P., 75, 80, *143*, *147*
Bovey, F. A., 179, *307*
Boyer, H. W., 45, *57*, *60*
Bradbury, J. H., 354, *358*
Brandts, J. F., 348, *358*
Braunitzer, G., 238, *304*
Brazhnikov, E. V., 355, *359*
Breen, J. J., 199, 221, 295, *306*, *308*
Bringe, N. A., 116, *143*
Brink, J., 106, 107, *150*
Brinkhuis, J. A., 70, 82, 91, 97, 103, 112, 122, *148*, *150*
Brinkhuis, J., 139, *148*
Brinster, R. L., 165, *310*
Brooker, B. E., 111, *145*
Brooks, C. L., 80, *143*
Brown, D. H., 31, *61*
Brown, J. W., 43, *57*
Brown, T. L., 75, 76, 88, *146*
Brulé, G., 125, *148*
Brunner, J. R., 66, *150*
Brunori, M., 154, 217, 218, 229, 236, 253, 295, *304*
Bryant, P. L., 348, *361*
Brzozowski, A., 286, 290, 291, 300, *304*
Bubstov, P. M., 51, *60*
Buchheim, W., 106, 107, 110, 113, 114, 115, 117, 119, 128, 129, *143*, *144*, *147*, *149*
Budd, D. L., 206, *308*
Büldt, G., 31, *60*
Bumeister, W., 29, 30, *59*
Bunick, G. J., 17, 18, 37, *62*
Bunn, H. F., 154, 218, 243, 253, 261, 277, 295, *304*, *308*, *309*, *311*
Burchard, W., 95, 96, 97, 101, 102, 103, *150*
Busch, M. R., 155, 165, 166, 173, 199, *304*, *306*
Bushuev, V. E., 355, *359*
Bushway, A. A., 82, *147*
Butler, J. E., 66, 67, 68, 69, *144*
Bychkova, V. E., 355, *359*

Byler, D. M., 86, *143*
Bystroff, C., 20, *57*

C

Cabani, S., 319, *358*
Cadenas, E., 10, *57*
Calderon, R. O., 338, *358*
Caley, D. S., 2, *61*
Callender, S. T., 242, *312*
Calmettes, P., 37, *57*
Camaco, M. L., 10, *57*
Caplan, S. R., 14, *57*
Caputo, A., 295, *304*
Cardingley, J. S., 21, *57*
Carrington, A., 176, *304*
Carroll, R. J., 106, 118, *143*, *148*
Carter, C. W., Jr., 165, 280, 288, 289, 299, 302, *311*
Carver, J. A., 354, *358*
Casassa, E. F., 31, *57*
Case, D. A., 295, *308*
Caspar, D. L. D., 97, *143*
Cassoly, R., 217, 234, 253, 273, 291, *304*, *305*, *311*
Cebula, D. J., 109, 110, *149*
Cedel, T. E., 230, 231, 232, 234, 235, *308*
Cendrin, F., 5, 8, 17, 37, 38, 39, 40, 41, 56, *57*, *62*
Ceska, T. A., 25, 30, 55, *59*
Chada, K., 165, *305*
Champness, J. N., 20, *61*
Chang, S. H., 49, 52, *58*
Changeux, J.-P., 155, 158, 295, 296, 300, *309*
Chanutin, A., 164, *305*
Chao, B. H., 55, *58*
Chaplin, B., 117, 118, *143*
Chaplin, L. C., 125, 126, *143*
Charache, S., 191, 192, 194, 199, 206, 207, 210, 211, 219, 230, 238, 239, 240, 243, 244, 245, 295, 301, *304*, *305*, *306*, *308*, *309*, *312*
Charlebois, R. L., 46, 48, *57*
Chen, B. L., 357, *358*
Chen, J.-T., 22, *57*
Chen, K. N., 43, *58*
Chen, S., 284, *311*

Chervenick, P. A., 194, 199, 202, 242, *312*
Cheryan, M., 117, 118, *147*
Chevalet, C., 75, *145*
Chiba, H., 70, 77, 81, 82, 93, 98, 114, 116, *148*, *149*, *151*
Chothia, C., 91, *147*, 158, 213, 244, 266, 288, 291, 292, 298, *304*, 318, 319, 348, *358*, *360*
Chou, P. Y., 90, 91, *143*
Christensen, H., 355, *358*
Christian, J. H. B., 14, *57*
Chudgar, A., 112, 113, *144*
Chung, A. E., 165, *306*
Clegg, J. B., 242, *312*
Clegg, R. A., 84, *150*
Clift, S. M., 75, 76, 88, *146*
Cline, S. W., 5, 48, 50, *57*
Clore, G. M., 155, 185, *305*, *311*
Cohen, A., 46, 50, 51, 54, *59*
Cohen, S., 29, 30, *59*
Cole, F. X., 216, *305*
Collman, J. P., 154, 218, *307*
Colosimo, A., 218, 281, 283, *310*
Colvin, J. R., 106, 124, *149*
Connely, P. R., 215, 218, *305*
Connolly, M. L., 319, *358*
Conover, R. K., 50, *57*
Conroy, S. C., 348, *361*
Constantini, F., 165, *305*
Cook, S. P., 325, *358*
Cookson, D. J., 92, *143*
Cottam, P. F., 185, *311*
Cox, J. M., 238, *310*
Coyrell, C. D., 295, *305*
Cozart, P. E., 165, *307*
Craescu, C. T., 199, 200, *305*, *310*
Craig, R. K., 75, 80, *143*, *147*
Craik, C. S., 76, *143*
Creamer, L. K., 86, 89, 92, 93, 106, 111, 114, 113, *143*, *144*, *148*, *150*, *151*
Creighton, T. E., 324, 325, 337, *358*
Creighton, T., 355, 357, *359*
Cremonesi, L., 281, *309*
Criss, W. E., 80, *147*
Crocenzi, T. S., 348, *360*
Crutchfield, G., 134, *145*
Cummings, S., 40, *58*
Curnish, R. R., 164, *305*

D

Dalens, M., 75, *145*
Dalgleish, D. G., 84, 93, 94, 98, 99, 100, 102, 114, 115, 116, 118, 119, 121, 122, 131, 133, 137, 138, 139, 142, *144*, *145*, *146*, *148*
Dalton, B. P., 6, *59*
Dalvit, C., 199, 200, 207, 208, 209, 210, 211, 212, 213, 219, 230, *305*
Dalziel, K., 190, *305*
D'Ambrosio, C., 284, 285, *307*
Damerval, T., 49, *58*
Daniels, C. J., 43, 48, *57*, *60*
Dann, J. G., 20, *61*
Danon, A., 14, *57*
Danson, M. J., 6, 12, 13, *57*
Dao-pin, S., 325, *358*
D'Arcy, A., 20, *60*
Darling, D. F., 141, 142, *144*
Das Sarma, S., 44, 45, 48, 49, 51, *57*, *58*, *61*
Daugherty, M. A., 166, 281, 282, 285, 288, 295, 299, 302, *304*, *305*
Dave, J. R., 69, 76, 137, *150*
Davids, N., 281, 284, *310*
Davidson, C. M., 121, 142, *146*
Davies, D. R., 87, *144*, 352, *359*
Davies, D. T., 71, 73, 74, 103, 104, 114, 119, 124, 125, 128, 129, 131, 132, *144*, *146*, *150*
Davis, D. G., 199, 237, 238, 239, 240, 243, 244, 245, 295, 301, *305*
de Bernard, B., 92, *143*
de Bruyn, P. L., 103, 127, 130, 131, *146*
de Recondo, A.-M., 41, 49, 55, *57*
De Young, A., 165, 216, *309*
Deatherage, J. F., 214, 218, *309*
Dec, S. F., 322, 323, 324, *358*
Dejmek, P., 125, *150*
Delius, H., 45, *58*
Dencher, N. A., 31, *60*
Dennis, P. P., 10, 43, 49, 51, 52, 53, *58*, *60*, *61*
Derewenda, Z., 213, 214, 218, 219, 235, 286, 288, 290, 291, 300, *304*, *305*, *308*
Deutsch, R., 273, *307*

Dewan, R. K., 112, 113, 114, 131, *144*, *147*, *149*
DeYoung, L. R., 322, *358*
Di Cera, E., 215, 218, *305*
Dickerson, R. E., 154, 155, 156, 158, 159, 160, 161, 163, 191, 292, *305*, 352, *359*
Dickson, I. R., 92, 93, *144*
Dill, K. A., 316, 322, 326, *358*, *359*
Dobson, C. M., 355, *358*
Dobson, M. J., 348, *361*
Dodson, E., 286, 290, 291, 300, *304*
Dodson, G., 213, 214, 218, 219, 235, 286, 288, 290, 291, 300, *304*, *305*, *308*
Doi, T., 55, *59*
Dolgikh, D. A., 355, *359*
Donkai, N., 96, 97, *146*
Donnelly, W. J., 114, 115, 117, 119, *144*, *147*
Doolittle, W. F., 45, 46, 48, 50, 51, 54, *57*, *59*, *61*
Dorland, L., 70, *150*
Downing, K., 25, 30, 55, *59*
Downing, W. L., 51, 52, 53, *61*
Doyle, M. L., 166, 215, 218, 283, 285, 288, 295, 299, 302, *304*, *305*
Drakenberg, T., 125, *150*
Duke, M. V., 49, *61*
Dumas, C., 186, *310*
Dundas, I. E. D., 7, *58*
Dunn, R. J., 49, 52, 55, *58*
Durfee, S. L., 165, *307*
Durfee, S., 165, *312*
Dyall-Smith, M. L., 5, 48, *59*
Dyer, T. A., 43, *58*
Dylewski, D. P., 74, 82, *146*

E

Eaton, W. A., 154, 218, 273, 290, 295, 300, *305*, *307*, *309*
Ebbe, S. N., 253, *310*
Ebert, B., 355, *361*
Ebert, K., 44, 45, *58*
Ebner, K. E., 66, *150*
Echard, G., 75, *145*
Edalji, R., 164, *304*
Edelstein, S. J., 218, 295, *305*, *306*, *311*
Edsall, J. T., 154, 295, *305*

Edwards, C., 3, *58*
Egeberg, K. D., 165, 216, 217, *309*
Eigel, W. N., 66, 67, 68, 69, 82, *144, 147*
Eisenberg, D., 336, *359*
Eisenberg, E., 51, *60*
Eisenberg, H., 5, 8, 9, 11, 15, 17, 18, 19, 31, 32, 34, 35, 36, 37, 38, 39, 40, 41, 57, 58, 60, 61, 62
Ekstrand, B., 111, *144*
Eliopoulos, E., 20, *61*
Ellenberger, T. E., 21, *57*
Elove, G. A., 346, 355, *359, 360*
Emsley, P., 213, 214, 218, 219, 288, 291, *305*
Enderby, J. E., 40, *58*
Engelman, D. M., 30, 55, *60*
Englander, S. W., 346, 355, *359, 360*
Englert, C., 52, *58*
Enoki, Y., 253, *305*
Ernsrøm, C. A., 66, 67, 68, 69, *144*
Ernst, M. H., 140, *150*
Evans, M. T. A., 88, *143*
Evans, P. A., 355, *358*
Evard, R., 115, 116, *149*
Ewbank, J. J., 355, 357, *359*

F

Fahey, R. C., 11, 14, *61*
Faloona, F. A., 4, *60*
Farrar, T. C., 2, *61*
Farrell, H. M., Jr., 66, 67, 68, 69, 83, 85, 98, 101, 106, 113, 118, *143, 144, 150*
Fasman, G. D., 90, 91, *143*
Fauchere, J. L., 318, *359*
Fauquant, J., 125, *148*
Feeney, J., 20, 23, *56*
Fermi, G., 165, 202, 213, 216, 235, 285, 286, 287, 288, 289, 291, *308, 305, 310, 311*
Ferrando, E., 49, 52, 55, *57*
Ferrige, A. G., 190, *305*
Ferro-Dosch, S., 199, *310*
Fersht, A. R., 24, *61*
Fiat, A.-M., 70, *150*
Fierke, C. A., 20, 22, 23, 56, 57, *58*
Filimonov, V. V., 338, *359*
Filman, D. J., 20, 21, 57, *58*

Filmer, D., 156, 295, 300, 301, *308*
Fine, R. F., 322, 325, 327, *361*
Finer, E. G., 88, *143*
Fink, A. L., 346, *358*
Fiol, C. J., 80, *144*
Fitzpatrick, M. M., 114, *149*
Fletterick, R. J., 86, *149*
Fletterick, R., 76, *143*
Fomme, R., 213, 235, *305*
Ford, T. F., 114, *144*
Forget, B. G., 154, 218, 243, *304*
Foster, L. M., 20, *59*
Foster, P. G., 20, *59*
Foster, R. C., 75, *144*
Fothergill, L. A., 348, *361*
Fournet, G., 32, *58*
Fox, G. E., 3, 43, *58, 62*
Fox, P. F., 131, 136, *147, 149*
Franeshci, F., 28, *61*
Frank, R., 41, *58*
Franke, W. W., 82, *147*
Franzen, J. S., 318, 325, *360*
Freer, S. T., 20, 21, *61*
Freire, E., 314, 315, 335, 337, 340, 341, 342, 343, 345, 346, 348, 347, 349, 350, 351, 353, 356, 357, *359, 360, 361*
Friedman, J., 45, 57, *60*
Frolow, F., 39, 40, *58*
Fung, L. W.-M., 175, 181, 182, 186, 194, 199, 200, 201, 202, 242, 246, 253, 254, 255, 256, 257, 258, 259, 261, 263, 267, 277, 295, 297, 301, *304, 305, 306, 307*
Fushitani, K., 165, *307*

G

Gaind, D. K., 114, *149*
Gale, R. E., 242, *312*
Galisteo, M. L., 315, 337, 345, 348, 349, 350, 351, *359*
Gao, J., 319, *359*
Garel, M.-C., 200, *310*
Garnier, J., 88, *144*
Garrett, R. A., 49, 50, 51, 53, *60*
Gaskell, M., 165, *306*
Gautron, J.-P., 78, *144*
Gaye, P., 74, 78, 79, 80, *144, 147*
Geddes, A. J., 20, *61*

Geerts, J.-P., 84, *144*
Geis, I., 154, 155, 156, 158, 159, 160, 161, 163, 191, 292, *305*
Gekko, K., 131, *148*
Gelin, B. R., 295, *306*
Gelin, B., 295, *308*
Gellin, J., 75, *145*
Gerl, L., 49, 53, *58*
Gerlt, J. A., 338, *358*
German, B., 154, 155, *306*
Gersonde, K., 206, 260, 261, *308*
Ghahraman, P., 45, *60*
Gianni, P., 319, *358*
Gibbons, N. E., 12, *58*
Gibso, J., 43, *58*
Gibson, Q. H., 154, 215, 216, 218, 235, 236, 253, 273, 291, 295, *305*, *306*, *307*, *308*, *309*
Gilbert, W., 76, *145*
Gill, S. J., 215, 218, 290, *305*, *306*, 318, 319, 320, 321, 322, 323, 324, 325, 326, 327, 328, 329, 330, 331, 335, 338, 339, 340, 349, 350, 354, *358*, *359*, *360*
Gill, S., 218, *312*
Gilmanshin, R. I., 355, *359*
Gilmore, R., 74, *150*
Gingrich, D., 166, 281, 282, 283, 285, 288, 298, 302, *305*, *311*
Ginzburg, B. Z., 14, *58*
Ginzburg, M., 14, *58*
Glarum, S. H., 178, *311*
Glasoe, P. K., 187, *306*
Glatter, O., 32, *58*
Glimcher, M. J., 86, *148*
Globel, G., 74, *150*
Glöckner, J., 89, *148*
Glonek, T., 92, *147*
Goaman, L. C. G., 238, *310*
Goebel, W., 44, 45, *58*
Goff, H., 206, *308*
Goldmann, S., 51, *60*
Goldsmith, E. J., 86, *149*
Gollois, M., 75, *145*
Gordon, W. G., 101, 113, *150*
Gorkom, M., 127, *150*
Gorodetsky, S. I., 76, 77, 137, *143*
Gorodnov, B. G., 338, *361*
Grabowski, M., 286, 290, 291, 300, *304*
Graham, E. R. B., 86, 88, 89, *145*

Graves, D. R., 165, *306*
Gray, R. D., 215, *306*
Green, J. P., 88, *143*
Green, M. L., 116, 117, 118, 134, *143*, *145*
Griffin, M. C. A., 119, 120, 122, 131, 132, *145*
Griffin, W. G., 119, *145*
Griko, Y. V., 315, 337, 338, 347, 348, 349, 352, 353, 354, *359*, *360*
Groebe, D. R., 165, *306*
Gronenborn, A. M., 155, 185, *305*, *311*
Gropp, F., 3, 49, 51, 52, 53, *58*, *60*, *62*
Grosclaude, F., 66, 75, *145*, *148*
Grosveld, F., 165, *306*
Guéron, M., 179, 186, *306*, *310*
Guinet, F., 41, 49, 55, *57*, *58*
Guinier, A., 32, *58*
Gupta, R. K., 164, *306*
Gupta, R., 43, *58*
Gurel, O., 30, 55, *60*
Gutowsky, H. G., 184, *304*

H

Ha, J.-H., 322, 336, *361*
Haber, J. E., 253, 295, *306*
Hackett, N. R., 48, 55, *58*
Haga, M., 87, *145*
Hahn, U., 325, *361*
Haigh, C. W., 179, *306*
Haik, Y., 5, 8, 17, 37, 38, 39, 40, 41, *58*, *62*
Haire, R. N., 290, *306*
Hall, D. O., 13, 23, 31, 55, *59*
Hallén, D., 321, *359*
Halvorson, H. O., 6, *59*
Hamlin, R. C., 20, 21, *57*
Hanke, C., 45, *58*
Hanley, C., 355, *358*
Hanscombe, O., 165, *306*
Hansen, C., 20, 21, *61*
Hansen, H. A. S., 28, *61*
Harel, M., 39, 40, *58*
Harries, J. A., 131, *145*
Harries, J. E., 103, 127, 128, 130, 131, *145*, *146*
Harris, D., 213, 214, 218, 219, 235, 286, 288, 290, 291, 300, *305*, *308*
Hart, R. G., 352, *359*

Hartman, R., 12, 14, *58*
Harwalkar, V. R., 66, 67, 68, 69, *144*
Hase, T., 13, 23, 31, 55, *59*
Hasnain, S. S., 103, 127, 130, 131, *145, 146*
Hatfield, G. W., 7, *61*
Hayashi, A., 191, 206, 261, 295, *306, 308*
Hazard, E. S., 295, *308*
Háze, G., 78, *144*
Hecht, K., 16, 17, 18, *59*
Hegemann, P., 49, *59*
Heid, H. W., 75, 80, *147*
Heidner, E. J., 218, *306*
Helmers, N. H., 86, *146*
Henderson, R., 25, 30, 55, *59*
Henry, E. R., 273, 290, 295, 300, *305, 307, 309*
Henschen, A., 76, 78, 137, *146*
Herdman, G. J., 40, *58*
Heremans, K., 131, *148*
Hermann, G., 52, *61*
Hermann, R. B., 319, *359*
Herzfeld, J., 295, *306*, 316, *359*
Hespell, R. B., 43, *58*
Heth, A. A., 114, 116, 117, *145*
Hewitt, J. A., 275, 276, 278, *308*
Hill, R. J., 136, *145*
Hill, T. L., 123, *145*
Hill, V. A., 116, *145*
Hille, R., 217, 218, *306*
Hilschmann, N., 238, *304*
Hilse, K., 238, *304*
Hinz, A., 105, 107, *149*
Hinz, J. E., 253, *310*
Hiragi, Y., 96, 97, *146*
Hiraiwa, H., 268, 270, 297, *311*
Hirata, H., 268, 270, 297, *311*
Hirsch-Twizer, S., 51, *60*
Ho, C., 93, *145*, 154, 155, 164, 165, 166, 172, 173, 175, 181, 182, 183, 185, 186, 187, 190, 191, 192, 193, 194, 199, 200, 201, 202, 203, 204, 205, 206, 207, 208, 209, 210, 211, 212, 213, 215, 216, 218, 219, 220, 221, 222, 224, 225, 226, 227, 228, 230, 231, 232, 234, 235, 236, 237, 238, 239, 240, 241, 242, 243, 244, 245, 246, 247, 248, 249, 250, 252, 253, 254, 255, 256, 257, 258, 259, 261, 262, 263, 264, 265, 266, 267, 268, 269, 270, 271, 272, 273, 274, 277, 278, 279, 283, 286, 290, 291, 295, 296, 297, 298, 301, 302, *304, 305, 306, 307, 308, 309, 310, 311, 312*
Ho, N. T., 155, 164, 173, 181, 182, 183, 187, 190, 192, 193, 199, 201, 202, 203, 228, 245, 246, 247, 252, 286, 296, 301, *304, 308, 310, 312*
Hobbs, A. A., 78, 79, *145*
Hobbs, D. G., 116, *145*
Hochmann, J., 186, *307*
Hochstein, L. I., 6, 12, *59*
Hoffman, B. M., 166, 281, 282, 283, 285, 286, 288, 298, 302, *304, 305, 307, 311*
Hoffman, S. J., 165, *307*
Hoffman, S., 165, *312*
Hoffmann-Porsorske, E., 81, *147*
Hofman, R. L., 46, *57*
Hofrichter, J., 273, 295, *305, 307*
Holmes, M. L., 5, 48, *59*
Holmes, P. K., 6, *59*
Holt, C., 64, 84, 85, 87, 88, 89, 91, 92, 103, 104, 111, 112, 114, 119, 121, 122, 123, 124, 125, 126, 127, 128, 129, 130, 131, 132, 136, 142, *144, 145, 146*
Holter, H., 137, 139, *146*
Honig, B., 322, 325, 327, *361*
Hopfield, J. J., 175, 220, 253, 260, 295, *307, 311*
Horisberger, M., 118, *146*
Horne, D. S., 99, 100, 101, 113, 114, 115, 116, 119, 120, 121, 136, 142, *144, 146*
Horne, M., 49, 52, *58, 60*
Horowitz, P. M., 86, *148*
Horwitz, A. F., 295, *309*
Horwitz, A., 253, *308*
Howell, E. E., 20, 22, 23, *59, 61*
Howell, L. G., 7, *60*
Hu, C. Q., 348, *358, 359*
Huang, T.-H., 220, *307*
Hubbard, J. S., 6, *59*
Hubbard, R. E., 326, 330, *358*
Huedepohl, H., 52, *59*
Huedepohl, U., 51, 52, *61*
Huehns, E. R., 242, *312*

Huestis, W. H., 215, 216, 295, *307*
Hughson, F. M., 355, *359*
Hukins, D. W. L., 103, 127, 128, 130, 131, *145*, *146*
Humphrey, R. S., 89, 90, *146*
Hurley, J. H., 86, *146*
Husband, A. D., 84, *150*
Huynh, B. H., 295, *308*
Hyslop, D. B., 139, 142, *146*

I

Ifft, J. B., 39, *58*
Ikeda-Saito, M., 267, 268, 269, 270, 271, 272, 273, 283, 284, 285, 295, 297, *307*, *309*
Imade, T., 92, 93, *146*
Imai, K., 164, 165, 216, 244, 245, 247, 248, 253, 267, 268, 270, 273, 283, 295, 297, *306*, *307*, *308*, *309*, *311*
Imamura, T., 131, 133, 136, *143*
Inubushi, T., 284, 285, 288, *307*, *311*
Irlam, J. C., 131, *145*
Ishimori, K., 165, *307*
Israelachvili, J., 95, 100, 135, *146*
Iwata, S., 119, *147*

J

Jacrot, B., 34, *62*
Jaenicke, R., 9, 10, 15, 16, 17, 18, 21, 50, *59*, *62*, 79, *146*
Janin, J., 319, *360*
Jarasch, E.-D., 75, 80, *147*
Javor, B. J., 12, 14, *59*
Javor, B., 14, *60*
Jay, E., 55, *59*
Jeng, M.-F., 355, *359*
Jenness, R., 66, 67, 68, 69, 84, 113, 119, 120, *144*, *145*, *150*
Jensen, R. G., 83, *148*
Jesson, J. P., 177, 178, *307*
Jimenez-Flores, R., 77, *146*
Johns, R. B., 81, *147*
Johnson, C. E., Jr., 179, *307*
Johnson, J. A., 166, 281, 282, 285, 288, 295, 299, *305*
Johnson, K. A., 20, 23, *58*
Johnson, L. N., 86, *149*
Johnson, M. E., 175, 181, 182, 186, 215, 218, 219, 220, 221, 228, 242, 246, 263, 290, 296, *306*, *307*
Johnson, M. L., 295, *307*
Jollès, J., 91, *146*
Jollès, P., 70, 76, 78, 90, 91, 137, *146*, *147*, *150*
Jolley, K. W., 89, 90, *146*
Jones, J. G., 49, *58*
Jones, R. T., 199, 243, 244, 245, 295, 301, *304*, *305*
Jones, W. K., 75, 76, 88, *146*
Jontell, M., 92, *143*
Jorgensen, W. L., 319, *359*
Josephson, R. V., 66, 113, 114, *147*, *150*
Jue, T., 206, 277, *308*, *309*
Juez, G., 12, 43, *61*
Jullien, R., 141, *143*

K

Kaeda, J., 165, *306*
Kagimoto, T., 206, *308*
Kagrmanova, V. K., 51, *60*
Kajino, T., 131, *148*
Kajiwara, K., 96, 97, *146*
Kako, Y., 74, 98, 131, 133, *143*
Kalab, M., 106, *146*
Kam, Z., 37, *61*
Kamekura, M., 12, 43, *59*, *61*
Kaminogawa, S., 92, 93, 101, *146*, *148*
Kang, Y. C., 77, *146*
Kapelinskaya, T. V., 76, 77, 137, *143*
Kargamanova, V. K., 51, 53, *61*
Karnik, S. S., 55, *59*
Karplus, M., 178, 184, 295, *307*, *308*, *311*
Karpus, M., 295, *306*
Katagawa, T., 165, *307*
Kates, M., 12, 43, *59*, *61*
Kaufman, B. T., 20, 21, *61*
Kauzmann, W., 322, 325, *359*
Kearney, R. D., 107, 128, 129, *147*
Keegstra, W., 106, 107, *150*
Keenan, T. W., 74, 82, 83, *143*, *146*, *147*
Kegeles, G., 96, *147*, *149*
Kellerhals, H., 186, *307*
Kellogg, G. W., 164, 199, *310*
Kelly, R. B., 74, 78, *147*

Kendrew, J. C., 352, *359*
Kerling, K. E. T., 90, 91, *148*
Kerney, R. D., 114, *147*
Kerscher, L., 10, 11, 13, 23, 31, 55, *59*
Kessel, M., 10, 29, 30, *59*
Khechinashvili, N. N., 326, 327, 328, 329, 337, 338, 352, *359*, *360*
Khorana, H. G., 30, 44, 45, 49, 51, 52, 55, *57*, *58*, *59*, *60*, *61*
Kikuchi, G., 261, *306*
Kilmartin, J. J., 199, *308*
Kilmartin, J. V., 199, 202, 242, 275, 276, 278, *304*, *308*
Kim, P. S., 322, 355, *359*
Kim, Y. K., 73, *150*
Kimur, K., 55, *58*
King, T. P., 7, *59*
Kingsman, A. J., 348, *361*
Kingsman, S. M., 348, *361*
Kinsella, J. E., 116, *143*
Kirchmeier, O., 134, *147*
Kirk, K. J., 83, *150*
Kitagawa, T., 288, *311*
Klenk, H.-P., 3, 51, *60*, *62*
Klink, F., 10, *59*
Klotz, I. M., 262, *312*, 318, 325, *360*
Klug, A., 97, *143*
Knoop, A.-M., 106, 107, 123, 129, *147*
Knoop, E., 106, 107, 123, 129, *147*
Knoops, J., 104, 105, 123, *149*
Kobayashi, T., 49, *61*
Kodma, S., 66, 136, *147*
Koenig, H., 3, *59*
Koepke, A. K. E., 51, 53, *61*
Koga, Y., 125, *151*
Koike, K., 13, *59*
Koike, M., 13, *59*
Kolka, C., 2, *61*
Kollias, G., 165, *306*
Koops, J., 104, 105, 123, *149*
Koshland, D. E., Jr., 86, *146*, 156, 253, 295, 300, 301, *306*, *308*
Kratky, O., 32, *58*
Kraut, J., 20, 21, 22, 23, *57*, *58*, *59*, *61*
Kripphal, G., 14, *60*
Krishman, G., 42, *59*
Kudo, S., 119, 136, *147*
Kuhn, N. J., 82, 84, *147*, *148*
Kumosinski, T. F., 98, 101, *144*
Kurland, R. J., 177, *308*

Kurplus, M., 295, *308*
Kushner, D. J., 2, 14, 43, *59*
Kutyshenko, V. P., 315, 338, 352, 353, 354, 355, *360*, *361*
Kuwajima, K., 355, 356, *360*
Kuwata, T., 74, *143*
Kwiatkowski, L., 165, 216, *309*

L

La Mar, G. N., 206, 277, *308*, *309*
LaCour, T., 42, *60*
Lacy, E., 165, *305*
Ladner, J. E., 241, 242, 243, 246, *310*
Ladner, R. C., 218, *306*
Lam, C.-H. J., 199, 224, 225, 226, 228, 246, 247, 248, 252, 286, 291, 295, 296, 301, *307*
Lam, W. L., 46, 48, 50, 51, 54, *57*, *59*
Laman, V. R., 237, *305*
Landsman, S. G., 114, *144*
Landt, M., 80, *143*
Langworthy, T. A., 3, *59*
Lanyi, J. K., 14, *59*
Larsen, H., 2, 12, *59*, *60*
Larsson-Raznikiewicz, M., 111, *144*
Lattman, E. E., 165, 280, 286, 288, 289, 299, 302, *311*, *312*
Law, A. J. R., 71, 73, 74, 103, 104, 114, 115, 119, 124, 125, 127, 128, 129, 130, 131, 132, 133, *144*, *146*
Leberman, R., 41, *58*
Lechner, J., 29, 49, 52, *60*
Lee, A. W.-M., 295, *306*, *308*
Lee, B. K., 329, 332, *360*
Lee, B., 95, 100, 135, *148*, 319, 337, *360*
Lee, S. L., 92, *147*
Leffers, H., 49, 50, 51, 53, *60*
Lehman, H., 192, 194, 206, 207, 210, 211, 219, 230, *308*
Leicht, W., 7, 8, 9, 10, 11, 15, 39, *58*, *60*
Leidermana, L. J., 80, *147*
Leigh, J. S., Jr., 154, 218, *307*
Lennon, D. P. W., 81, *147*
Lenstra, J. A., 90, *147*
Leong, D. M., 52, 57
Leong, S. L., 114, 131, *147*
Léonil, J., 64, *147*

Lepori, L., 319, *358*
Leser, U., 6, *60*
Lesk, A. M., 319, *360*
Leskiw, B., 9, 10, 15, 21, 50, *62*
Levedev, Yu. O., 355, *359*
Levine, B. A., 92, *143*
Levitt, M., 91, *147*
Lewis, B. J., 43, *58*
Lewis, M., 356, *358*
Lhoste, J.-M., 200, *310*
LiCalsi, C., 348, *360*
LiCata, V. J., 166, 281, 282, 283, 285, 288, 295, 299, *305*, *311*
Liddington, B., 287, 288, 289, *308*
Liddington, R. C., 235, 285, 286, 290, 291, 300, *310*, *312*
Liddington, R., 218, 235, 286, 290, 291, 300, *304*, *308*
Liebhaber, S. A., 165, *306*
Lim, V. I., 90, *147*
Lin, A. K.-L. C., 186, 227, 228, *312*
Lin, A. K.-L., 199, 204, 205, 206, *310*, *311*
Lin, L.-N., 348, *358*
Lin, S. H. C., 112, 113, 114, 131, *147*
Lin, S.-H., 165, 216, *309*
Linde, A., 92, *143*
Linderstrøm-Lang, K., 66, 136, *147*
Lindon, J. C., 190
Lindstrom, T. R., 190, 192, 194, 199, 206, 207, 210, 211, 213, 215, 216, 219, 221, 228, 230, 235, 236, 243, 244, 245, 253, 254, 255, 256, 257, 258, 259, 261, 267, 295, 296, 297, 301, *305*, *306*, *308*
Linzell, J. L., 84, *147*
Livingstone, J. R., 319, 322, *360*
Locker, D. L., 165, *307*
Long, F. A., 187, *306*
Lottspeich, F., 49, 50, 51, 52, 53, 55, 57, 59, *60*
Loucheaux-Lefebvre, M. H., 76, 78, 90, 91, 137, *146*, *147*
Louie, A., 51, 52, 53, *61*
Love, W. E., 286, *312*
Luehrsen, K. R., 43, *58*
Luh, F.-Y., 165, *306*
Luisi, B., 165, 216, 285, 286, 287, 288, 289, 291, *308*, *309*, *310*
Luzzati, V., 32, 34, 35, *60*

Luzzatto, L., 165, *306*
Lyster, R. L. J., 129, 131, 132, *145*, *147*

M

Ma, D.-P., 49, *61*
Mace, J. E., 155, 173, 199, *304*
Mackinlay, A. G., 67, 68, 69, 75, 76, 77, 78, 137, *143*, *149*
Mada, M., 119, *147*
Madon, J., 6, *60*
Madrid, M., 182, 183, 187, *308*
Madsen, N. B., 86, *149*
Maeda, T., 273, 295, *308*
Magram, J., 165, *305*
Magrum, L., 43, *58*
Majumdar, A., 49, 52, *58*
Makhatadze, G. I., 322, 337, *360*
Maki, R., 77, *148*
Makinlay, A. G., 88, 89, 90, 92, *149*
Makowski, I., 28, *61*
Malcolm, G. N., 86, 88, 89, *145*
Mallion, R. B., 179, *306*
Mandelbrot, B. B., 141, *147*
Maniatis, T., 165, *311*
Maniloff, J., 43, *58*
Mankin, A. S., 51, 53, *60*, *61*
Mann, S., 129, *147*
Mansouri, A., 215, *308*
Marden, M. C., 295, *308*
Margoliash, E., 154, 218, *307*
Marquandt, D. W., 224, *308*
Marquesee, S., 325, 332, *361*
Marti, T., 55, *60*
Martin, S. R., 120, *145*
Mas, M. T., 348, *358*
Mateo, P. L., 338, 347, 348, *359*, *360*
Matheson, A. T., 51, 52, 53, *61*
Mathews, A. J., 165, 214, 216, 217, 290, *308*, *309*
Mathews, D. A., 20, *58*
Mathews, D. W., 165, 214, 216, *308*
Matsubara, H., 13, 23, 31, 55, *59*
Matthews, B. W., 325, *358*
Matthews, D. A., 20, 21, *57*, *61*
Maubois, J. L., 64, *147*
Mavarech, M., 21, 39, 40, 49, *58*, *61*
May, B. P., 10, 49, 51, 52, 53, *60*
Mayer, A., 260, 261, *308*
Mayhew, S. G., 7, *60*

Mayorga, O. L., 315, *359*
McCarthy, B. J., 44, *60*
McConnell, H. M., 178, 215, 253, 295, *308, 309*
McCoy, J. M., 49, 52, 55, *58*
McDonald, C. C., 179, *308*
McGann, T. C. A., 107, 114, 115, 117, 119, 125, 128, 129, 131, 132, *144, 147, 148*
McGarvey, B. R., 177, *308*
McGourty, J. L., 286, *304*
McIntosh, T. J., 135, *147*
McKenzie, H. A., 86, 88, 89, *145*
McKinlay, A. G., 103, 133, *149*
McLachlan, A. D., 176, *304*, 336, *359*
McLendon, G., 284, *311*
McNeil, G. P., 114, 115, 117, 119, *144*
McQuattie, A., 6, 13, *57*
Mead, R. J., Jr., 103, *143*
Mead, R., 112, 113, *144*
Meakin, P., 141, *150*
Meggio, F., 81, *147*
Mehaia, M. A., 117, 118, *147*
Mellander, O., 66, *147*
Mengele, R., 29, *61*
Mercier, J.-C., 66, 74, 78, 79, 80, 81, *144, 147, 148*
Mescher, M. F., 29, *60*
Mevarech, M., 7, 8, 9, 10, 11, 13, 15, 17, 18, 21, 23, 31, 39, 46, 50, 51, 55, *58, 59, 60, 61, 62*
Meyer, H. E., 81, *147*
Meyers, S., 28, *61*
Miller, A. B., 6, *59*
Miller, S., 319, *360*
Mims, M. P., 229, 230, 231, 232, 234, 235, *308, 309*
Minton, A. P., 199, 200, 253, 254, 255, 256, 257, 258, 259, 261, 267, 295, 297, 301, *305, 306, 309, 310*
Mitani, M., 355, *360*
Miura, S., 230, 231, 232, 234, 235, 262, 263, 264, 265, 266, 267, 268, 269, 270, 271, 272, 273, 274, 277, 278, 279, 295, 297, 298, 302, *308, 309*
Miyazaki, G., 165, 216, 287, *307, 309, 311*
Mock, N. H., 199, 243, 244, 245, 295, 301, *305*
Mock, N. L., 237, *305*

Moffat, K., 154, 214, 218, *307, 309*
Mogi, T., 55, *60*
Moir, P. D., 101, *146*
Mollica, V., 319, *358*
Monod, J., 155, 158, 295, 296, 300, *309*
Moore, A., 75, 80, *147*
Moore, R. L., 44, *60*
Morgan, G., 83, *150*
Morimoto, H., 165, 268, 270, 287, 288, 297, *307, 311*
Morishima, I., 165, *307*
Mornt, S. V., 116, *145*
Morr, C. V., 66, 112, 113, 114, 131, *144, 147, 150*
Morrié, D. J., 82, *147*
Morton, R. A., 14, *57*
Mossing, M. C., 2, *61*
Motokama, Y., 261, 295, *306*
Mozzarelli, A., 290, 300, *309*
Muir, D. D., 114, 136, *145*
Muirhead, H., 238, *310*
Mullis, K. B., 4, *60*
Munord, R. E., 113, *148*
Murphy, K. P., 315, 319, 320, 321, 322, 324, 325, 326, 328, 329, 330, 331, 335, 337, 338, 341, 342, 343, 345, 346, 347, 348, 349, 350, 351, 353, *359, 360*
Myers, D., 166, 285, 288, 295, 299, 302, *304*

N

Nagai, K., 165, 206, 213, 214, 216, 217, 218, 219, 277, 288, 290, 291, *305, 308, 309, 311*
Nagao, M., 77, *148*
Nagel, R. L., 199, 295, *306, 309*
Nagura, M., 96, 97, *146*
Nakai, S., 90, *148*
Nakhasi, H. L., 69, 76, 137, *150*
Nassal, M., 55, *59*
Nasuda-Kouyama, A., 215, *309*
Navaratnam, N., 84, *148*
Naylor, A. M., 22, 23, *57*
Neilson, G. W., 40, *58*
Nelson, L. S., Jr., 103, 127, 128, 130, 131, *146*
Némethy, G., 156, 295, 300, 301, *308*, 336, *360*

Neumann, E., 9, 11, 15, 17, 18, *60*
Neville, M. C., 83, 84, *148*
Newton, C. H., 51, 52, 53, *61*
Newton, G. L., 14, *60*
Niazi, G., 290, *306*
Nicholls, A., 322, 325, 327, *361*
Nichols, N. F., 318, 323, 324, 329, *359*
Nichols, N., 319, *360*
Nieuwlandt, D. T., 48, *60*
Niggeler, M., 291, *310*
Niki, R., 95, 96, 97, 101, 102, 103, *146, 149, 150*
Nilsson, S.-O., 321, *359*
Nissenbaum, A., 2, *60*
Nitta, K., 355, *360*
Noble, R. W., *150*, 165, 216, 229, 295, *304, 308, 309*
Nock, N. H., 216, 235, 236, *308*
Noggle, J. H., 185, *309*
Noguchi, H., 92, 93, *146*
Noll, L., 318, 325, *359*
Noren, I. B. E., 192, 194, 206, 207, 210, 211, 219, 230, *308*
North, A. C. T., 20, *61*
Nozaki, Y., 318, *360*
Nutting, G. C., 106, *143*
Nyborg, J., 42, *60*

O

Oatley, S. J., 20, 23, 57, *59*
Obata, T., 131, *148*
O'Brien, J. R. P., 190, *305*
O'Brien, T. L., 83, *148*
Odagiri, S., 92, 93, *148*
Oefner, C., 20, *60*
Oesterhelt, D., 10, 11, 12, 13, 14, 23, 29, 31, 46, 49, 53, 55, 57, 58, 59, 60, 61
Ogata, R. T., 215, 253, 295, *309*
Ogawa, S., 175, 220, 253, 260, 261, 264, 266, 295, *308, 309, 311*
Ogden, R. C., 22, *61*
Ogg, D., 20, *61*
Ohgushi, M., 355, *360*
Ohmia, K., 131, *148*
Ohnishi, S., 295, *308*
Oikawa, A., 49, *61*
Oka, T., 75, 80, 88, *147, 151*
Olson, J. S., 165, 214, 216, 217, 218, 229, 230, 231, 232, 234, 235, 236, 290, *306, 308, 309, 310*
Ono, T., 90, 92, 93, 131, *148*
Oobatake, M., 336, *360*
Ooi, T., 336, *360*
Oosthuizen, J. C., 119, *149*
Oren, A., 14, *60*
Orkin, S. H., 165, *311*
O'Rourke, K., 165, *312*
Osguthorpe, D. J., 86, *144*
Osterhelt, D., 49, 52, 55, *57*
Ostler, G., 20, 23, *56*
Overbeek, J. T. G., 141, *148*
Owen, A. J., 74, *143*

P

Pace, C. N., 325, *361*
Pagnier, J., 165, 216, *311*
Pain, R. H., 355, *358*
Palm, P., 3, 51, 52, 58, *60, 62*
Palmer, G., 217, 218, *306*
Palmiter, R. D., 165, *310*
Papadopoulos, G., 31, *60*
Parker, S. B., 129, *147*
Parker, T. G., 93, 94, 98, 99, 102, 116, 119, *144, 146, 148*
Parry, D. A. D., 86, 89, 92, *144*
Parry, R. M., Jr., 118, *148*
Parsegian, V. A., 95, 100, 135, *148*
Pascall, J. C., 80, *143*
Pashley, R., 95, 100, 135, *146*
Paterson, E., 99, *144*
Paterson, R. A., 190, *305*
Patton, S., 83, *148*
Pauling, L., 179, 295, *309*
Paulke, C., 28, *61*
Payens, T. A. J., 85, 88, 96, 97, 103, 131, 134, 137, 138, 139, 141, *148*
Peaker, M., 84, *147*
Peisach, J., 237, 239, 253, 260, *311, 312*
Peller, L., 295, *309*
Perich, J. W., 81, *147*
Perkins, D. J., 92, 93, *144*
Perkins, R. E., 348, *361*
Perkins, S. J., 179, *309*
Perrella, M., 218, 281, 282, 283, 284, 291, 298, *309, 310*
Pertuz, M. F., 199, 154, 155, 156, 158, 164, 165, 202, 213, 218, 235, 238,

241, 242, 243, 244, 246, 253, 254, 266, 273, 275, 285, 286, 287, 288, 291, 292, 295, 296, 298, *304*, *305*, *306*, *309*, *310*, *311*, *312*
Perutz, M., 213, 214, 218, 219, 288, 291, *305*
Petering, D. H., 283, *307*
Peterson, J. A., 229, *308*
Pfeifer, F., 43, 44, 45, 49, 52, *57*, *58*, *60*
Pfeil, W., 338, 356, *359*, *360*
Phillips, D. C., 352, 356, *358*, *359*
Phillips, M. C., 88, *143*
Phillips, P. W., 179, *308*
Phipps, B. M., 30, *57*
Phipps-Todd, B. E., 106, *146*
Piat, G., 175, 242, 246, *306*
Pierce, K. N., 113, *148*
Pierre, A., 125, *148*
Pinna, L. A., 81, *147*
Pisciotta, A. V., 253, 254, 255, 256, 257, 258, 259, 261, 267, 295, 297, 301, *306*, *308*, *310*, *311*
Plateau, P., 186, *310*
Pliska, V., 318, *359*
Poll, J. K., 104, *149*
Pond, J. L., 3, *59*
Pope, J. M., 88, 89, 90, 92, *149*
Pople, J. A., 176, 179, 254, *310*
Popot, J. L., 30, 55, *60*
Porras, A. G., 229, *308*
Postner, A. S., 107, 128, 129, *147*
Poyart, C., 165, 216, *309*
Prentice, J. H., 111, *145*
Price, J. C., 119, 120, 131, 132, *145*
Privalov, P. L., 315, 322, 326, 327, 328, 329, 330, 331, 335, 337, 338, 339, 340, 345, 347, 348, 349, 350, 351, 352, 353, 354, *359*, *360*, *361*
Provencher, S. W., 89, *148*
Ptitsyn, O. B., 355, *359*, *360*, *361*
Puehler, G., 3, 51, *60*, *62*
Pulaski, P., 98, 101, *144*
Pulsinelli, P. D., 253, 254, *310*
Pundak, S., 5, 9, 10, 15, 17, 18, 19, 36, 37, 39, *58*, *60*, *61*
Pyne, G. T., 114, 125, 126, 131, 132, *147*, *148*

R

Raap, J., 90, 91, *148*
Radice, G., 165, *305*
Raftery, M. A., 215, 216, 295, *307*
Rajagopalan, R., 14, *56*
RajBhandary, U. L., 44, 45, 49, 51, 52, *57*, *58*, *61*
Ramalingham, V., 86, *146*
Ramirez, C., 51, 52, 53, *61*
Ramsay, G., 342, 348, *360*
Ramsdell, G. A., 114, *144*
Ranney, H. M., 218, 253, 254, 261, *310*, *311*
Rao, K. K., 13, 23, 31, 55, *59*
Rapael, K., 165, *305*
Rau, D. C., 95, 100, 135, *148*
Ravimohan, C., 319, *359*
Record, M. T., Jr., 2, *61*, 319, 322, 336, *360*, *361*
Redfield, A. G., 220, *307*
Reerink, H., 141, *148*
Reeve, J. W., 43, *57*
Reich, M. H., 37, *61*
Reisberg, P. I., 234, 235, *309*, *310*
Reiter, W.-D., 3, 51, 52, *59*, *61*, *62*
Renaud, J.-P., 165, 213, 214, 216, 217, 218, 219, 288, 290, 291, *305*, *308*, *309*
Renugopalakrishnan, V., 86, *148*
Ribadeau Dumas, B., 66, 70, *148*, *151*
Richards, F. M., 318, 319, 324, 337, *360*
Richardson, B. C., 113, *148*
Richardson, T., 77, 86, 89, 92, *144*, *146*, *147*
Richey, B., 2, *61*, 218, 290, *306*, *312*
Rifkind, J., 217, 220, *311*
Rimerman, R. A., 7, *61*
Ripamonti, M., 218, 281, 283, *310*
Rivetti, C., 290, 300, *309*
Roach, P. J., 80, *144*
Roberts, G. C. K., 20, 23, *56*, 122, *145*, 199, *308*
Robertson, A. D., 337, *360*
Robertson, R. E., 178, *308*
Robertson, R. N., 88, *143*
Robson, B., 86, *144*
Roder, H., 346, 355, *359*, *360*
Rodionova, N. A., 355, *361*

Rodriguez-Valera, F., 3, 12, 43, *61*
Roehrich, J. M., 165, *307*
Roeske, R. W., 80, *144*
Rogers, P., 286, *304*
Rohlfs, R. J., 165, 216, 217, *309*
Rohlfs, R., 165, 214, 216, *308*
Rollema, H. S., 91, 97, 103, 122, *148*
Rosa, J., 200, *310*
Rose, D., 106, 114, 124, *149*
Rose, M. R., 45, *61*
Rose, Z. B., 164, *306*
Roseman, S., 348, *360*
Rosen, J. M., 75, 76, 78, 79, 88, *145, 146*
Rosenberg, A., 290, *306*
Rosenshine, I., 9, 10, 15, 21, 46, 49, 48, 50, *61, 62*
Rossi, G. L., 290, 300, *309*
Rossi-Bernardi, L., 218, 281, 283, 284, 291, *309, 310*
Rossi-Fanelli, A., 295, *304*
Roth, E. F., Jr., 199, *306*
Rothschild, W., 321, *359*
Rovida, E., 291, *310*
Rudloff, V., 238, *304*
Rupley, J. A., 319, *361*
Russel, A. J., 24, *61*
Russu, I. M., 155, 164, 172, 175, 181, 192, 193, 199, 201, 202, 203, 230, 231, 232, 234, 235, 243, 252, *306, 308, 310*
Rutter, W., 76, *143*
Ryan, J. J., 126, *148*
Ryan, T. M., 165, *310*

S

Sabbioneda, L., 218, 281, *310*
Sachs, L., 14, *58*
Sadowski, M. L., 356, *360*
Saigo, S., 268, 270, 297, *311*
Saito, Z., 114, *149*
Salin, M. L., 49, *61*
Salmon, P. S., 40, *58*
Samaga, M., 218, 281, *310*
Samaja, M., 291, *310*
Sanchez-Ruiz, J. M., 315, 337, 345, 348, 349, 350, 351, *359*
Sander, L. M., 141, *150*
Santoro, M., 325, 332, *361*

Sapienza, C., 45, *61*
Saroff, H. A., 295, *310*
Sasaki, M., 82, *147*
Sasaki, R., 70, 77, 81, 82, 93, 98, 114, 116, *148, 149, 151*
Sato, Y., 92, 93, *146*
Sawyer, L., 64, 85, 87, 88, 89, 91, 127, 128, *145, 146*
Sazbo, A., 295, *308*
Schachman, H. K., 32, 33, *61*
Schaeffer, C., 200, *310*
Schaier, R. W., 127, *150*
Schalkwyk, L. C., 46, 50, *57*
Schegk, E. S., 49, 52, 55, *57*
Scheidt, W. R., 154, 218, 281, *307, 310*
Schellman, J. A., 318, 325, 328, 357, *358, 360*
Scherag, H. A., 336, *360*
Scherjon, J. W., 84, *144*
Scherphof, G., 29, *57*
Schirmer, R. E., 185, *309*
Schleich, T., 17, *62*
Schmidt, D. G., 65, 85, 95, 104, 105, 106, 107, 108, 110, 113, 123, *143, 149, 150*
Schmidt, M. S., 218, *311*
Schneider, W. G., 176, 254, *310*
Scholes, C., 284, *311*
Scholtz, J. M., 325, 332, *361*
Scott-Blair, G. W., 119, *149*
Sculley, T. B., 103, 133, *149*
Selker, F., 83, 84, *148*
Semisotnov, G. V., 355, *359, 361*
Semple, K., 83, 84, *148*
Seybert, D. W., 214, 218, *309*
Sgaramella, V., 55, *59*
Shaanan, B., 213, 214, 218, 219, 235, 285, 286, 288, 291, *305, 310, 311*
Shah, F., 68, 69, 77, 78, *149*
Shaham, M., 31, *61*
Shand, R. F., 52, *57*
Sharma, V. S., 217, 218, 220, 291, *304, 311*
Sharp, K. A., 322, 325, 327, *361*
Shaw, P. J., 348, *361*
Shea, M. A., 166, 281, 282, 285, 288, 295, 298, 299, *305, 310*
Shears, S. B., 83, *150*

Shibayama, N., 268, 270, 287, 288, 289, 297, *308*, *311*
Shih, D. T.-B., 155, 165, 216, *307*, *311*
Shih, D., 165, 216, *309*
Shimizu, M., 101, *146*
Shimmin, L. C., 51, 52, 53, *61*
Shimuzu, K., 245, 247, 248, *311*
Shimzu, S., 131, *148*
Shin, M., 191, *306*
Shirley, B. A., 325, *361*
Shoham, M., 39, 40, *58*
Shore, V. C., 352, *359*
Shrake, A., 319, *361*
Shulman, R. G., 175, 178, 220, 237, 239, 253, 260, 261, 264, 266, 295, *305*, *308*, *309*, *311*, *312*
Sick, H., 206, *308*
Sickinger, H.-D., 12, 14, *58*
Sidhu, K., 112, 114, *149*
Silverman, M. P., 14, *59*
Simolo, K., 284, *311*
Simon, S. A., 135, *147*
Simon, S. R., 241, 242, 243, 246, *310*
Simplaceanu, V., 182, 183, 185, 186, 187, 227, 228, *308*, *312*, *311*
Simsek, M., 44, 45, 49, 52, *58*, *61*
Singh, H., 131, 136, *149*
Skarzynski, T., 286, 290, 291, 300, *304*
Skipper, N., 40, *58*
Sköld, R., 319, *360*
Slangen, C. J., 70, 71, 82, 91, *150*
Slattery, C. W., 93, 115, 116, *149*, *150*
Sleigh, R. W., 88, 89, 92, 103, 133, *149*
Slichter, C. P., 176, *311*
Sligar, S. G., 165, 216, 217, *309*
Smith, F. R., 166, 280, 281, 282, 285, 288, 289, 295, 298, 299, 302, *303*, *311*
Snoeren, T. H. M., 95, 96, *149*
Solomon, I., 180, *311*
Sommer, J. H., 273, *307*
Sood, S. M., 112, 114, *149*
Soppa, J., 46, *61*
Speros, P. C., 281, 283, *305*, *311*
Spink, C., 319, *360*
Spiridinova, V. A., 51, 53, *61*
Spokane, R., 218, *312*
Spolar, R. S., 319, 322, 336, *360*, *361*
Sprang, S. R., 86, *149*
Springer, B. A., 165, 216, 217, *309*

Stackebrandt, E., 43, *58*
Stahl, D. A., 43, *58*
Staiert, P. A., 84, *148*
Stammers, D. K., 20, *61*
Stanley, E., 295, *306*
Stanssens, P., 325, *361*
Stanton, E. K., 114, *149*
Steel, J. M., 262, *312*
Sternlicht, H., 179, *311*
Stetler, G., 165, *312*
Stetter, K. O., 3, *59*
Stettler, G. L., 165, *307*
Stetzkowski, F., 234, *311*
Stevenson, K. J., 6, 13, *57*
Stewart, A. F., 67, 68, 69, 76, 77, 78, 137, *143*, *149*
Stewart, J. M., 325, 332, *361*
Stoeckenius, W., 29, 30, *57*, *61*
Stolowich, N. J., 338, *358*
Stothart, P. H., 109, 110, *149*
Strandberg, B. E., 352, *359*
Strobel, I., 29, *61*
Strobel, M. S., 22, *61*
Strominger, J. L., 29, *60*
Stroud, R. M., 86, *146*
Stuart, D. F., 356, *358*
Stuart, D. L., 86, *149*
Stucky, G., 284, *311*
Sturtevant, J. M., 322, 332, 338, 347, 348, *358*, *359*, *361*
Sugai, S., 355, *360*
Sullivan, R. A., 114, *149*
Sumper, M., 12, 29, 49, 52, 53, *58*, *60*, *61*
Sundquist, A. R., 11, 14, *61*
Susi, H., 86, *143*, 318, 325, *361*
Sussman, J. L., 31, 39, 40, *58*, *61*
Suzuki, H., 131, 136, *143*
Suzuki, T., 191, 295, *306*
Swaisgood, H. E., 66, 85, 90, 114, 116, 117, *145*, *149*, *150*
Sweetsur, A. W. M., 136, *145*
Szabo, A., 218, 295, *311*

T

Tachibana, H., 215, *309*
Tai, M., 96, *149*
Taira, K., 22, *57*
Takahashi, S., 199, 204, 205, 206, 224,

225, 226, 228, 246, 247, 248, 252, 286, 291, 295, 296, 301, *307*, *311*
Takahata, K., 116, *151*
Takano, T., 231, *311*
Takao, M., 49, *61*
Takase, K., 96, *149*
Taketa, F., 206, *308*
Takeuchi, M., 70, 81, 82, 114, *149*, *151*
Tam, M. F., 165, *306*
Tam, P., 51, *60*
Tame, J., 165, 214, 216, 217, 290, *307*, *308*, *309*, *311*
Tandeau de Marsac, N., 49, *58*
Tanford, C., 322, *361*
Tanner, R., 43, *58*
Tanord, C., 318, *360*
Tardieu, A., 32, 34, 35, *60*
Tatunashvili, L. V., 338, *361*
Taylor, J. E., 84, *150*
Taylor, J. F., 295, *304*
Taylor, M. D., 74, *143*
Tchelet, R., 46, *61*
Tedesco, J. L., 165, *307*
Tedesco, J., 165, *312*
Ten Eyck, L. F., 273, 275, 298, *310*
ter Horst, M. G., 127, *150*
Termonia, Y., 141, *150*
Tetrina, N. L., 51, *60*
Thillet, J., 20, 23, *61*
Thomas, J. O., 218, *311*
Thompson, M. D., 69, 76, 137, *150*
Thompson, M. P., 83, 85, 98, 101, 106, 113, *143*, *144*, *150*
Thørgersen, H. C., 165, *309*
Thorness, P. E., 86, *146*
Thurn, A., 95, 96, 97, 101, 102, 103, *150*
Tiktopulo, E. I., 337, 338, 355, *359*, *360*, *361*
Tillit, J., 41, 49, 55, *57*
Timasheff, S. N., 318, *361*
Tischenko, V. M., 338, *361*
Tisel, W. A., 290, *306*
Tittor, J., 55, *60*
Tjandra, N., 185, *311*
Tkach, M., 76, 77, 137, *143*
Tomita, S., 253, *305*
Torreblanca, M., 12, 43, *61*
Townes, T. M., 165, *310*
Toyooka, K., 98, *143*

Triesman, R., 165, *311*
Tsalkova, T. N., 338, *359*
Tsao, T. Y. M., 165, *306*
Tsouluhas, D., 46, 50, 51, 54, *59*
Tsuda, E., 70, 81, *149*
Tsugo, T., 125, *151*
Tu, C. P., 22, *57*
Tuite, M. F., 348, *361*
Turner, B. W., 295, *305*, *307*
Turner, G. J., 166, 281, 282, 285, 288, 295, 299, *305*
Tyuma, I., 245, 247, 248, 273, *308*, *311*

U

Udem, L., 253, 261, *311*
Unwin, N., 30, *59*
Urakawa, H., 96, 97, *146*

V

Valdes, R., Jr., 209, 233, *312*
Valley, D., 286, 290, 291, 300, *304*
van Assendelft, G. B., 165, *306*
van Breemen, J. F. L., 106, 107, *150*
van Bruggen, E. F. J., 106, 107, *150*
van der Spek, C. A., 97, 103, 105, 107, 112, *149*, *150*
van der Spek, C., 70, 82, *150*
van Dongen, P. G. J., 140, *150*
van Halbeek, H., 70, *150*
van Hooydonk, A. C. M., 141, 142, *144*, *150*
van Kemenade, M. J. J. M., 85, 92, 103, 127, 128, 130, 131, *145*, *146*, *150*
van Markwijk, B. W., 113, 114, 131, *150*
van Markwijk, B., 95, 96, *149*
van Montort, R., 95, 96, *149*
van Osdol, W. W., 315, *359*
van Riel, J. A. M., 70, 71, 82, *150*
van Rooijen, P. J., 91, *150*
Varvill, K., 86, *149*
Vauthey, M., 118, *146*
Veis, A., 92, *147*
Velicelebi, G., 332, *361*
Ventosa, A., 12, 43, *61*
Venyaminov, S. Y., 315, 338, 347, 348, 349, 352, 353, 354, *359*, *360*
Venyaminov, S. Yu., 355, *359*

AUTHOR INDEX

Veseley, S., 281, *309*
Vidal, M., 165, *306*
Viggiano, G., 190, 194, 199, 202, 220, 221, 222, 224, 225, 226, 228, 242, 245, 246, 247, 248, 249, 250, 252, 283, 286, 291, 295, 296, 301, *307*, *312*
Vigianno, S., 281, *309*
Villafranca, J. E., 22, 23, *59*, *61*
Virk, S. S., 83, 84, *148*, *150*
Visser, S., 70, 71, 82, 90, 91, 127, *148*, *150*
Vitez, L., 165, *312*
Vliegenthart, J. F. G., 70, *150*
Voet, D. H., 22, *61*
Vogelsang-Wenke, H., 49, *59*
Volz, K. W., 20, 21, *61*
von Boehlen, K., 28, *61*
von der Haar, F., 7, *62*
von Hippel, H. P., 17, *62*
von Hippel, P. H., 66, *150*
von Smoluchowski, M., 138, 139, *150*
Vonlanthen, M., 118, *146*
Vreeman, H. J., 70, 71, 82, 85, 88, 90, 91, 96, 97, 103, 112, 113, 114, 122, 131, *148*, *150*

W

Wachtel, E., 36, 37, 40, *58*, *62*
Wada, A., 215, 355, *309*, *360*
Wada, Y., 165, *307*
Wadsö, I., 318, 319, 321, 323, 324, 329, *359*, *360*
Wagenbach, M., 165, *312*
Wagner, C. R., 22, *62*
Wahlgren, N. M., 125, *150*
Wakabayashi, S., 13, 23, 31, 55, *59*
Wake, R. G., 136, *145*
Walder, J. A., 262, 264, *312*
Walder, R. Y., 262, 264, *312*
Walker, N. P. C., 348, *361*
Walker, W. P. C., 356, *358*
Waller, D. A., 235, 290, 291, 300, *312*
Walstra, P., 65, 107, 113, 119, 120, 134, 136, 142, *150*
Walter, P., 74, *150*
Waltho, J. A., 14, *57*
Wand, A. J., 355, *359*
Wang, A., 80, *144*

Ward, K. R., 286, *312*
Ward, S., 84, *148*
Warren, M. S., 23, *59*
Watari, H., 295, *306*
Watson, H. C., 348, 352, *359*, *361*
Watters, C. D., 83, 84, *148*, *150*
Waugh, D. F., 64, 66, 93, 115, *145*, *150*
Weatherall, D. J., 242, *312*
Weber, G., 295, *312*
Wei, G. J., 113, 119, 120, *150*
Weinstein, S., 27, 28, *62*
Welk, R. M. G., 242, *312*
Welsch, U., 106, 107, *143*
Werber, M. M., 7, 8, 9, 10, 11, 13, 23, 31, 55, *59*, *60*, *62*
Werczberger, R., 21, 46, 49, *60*, *61*
West, D. W., 84, *150*
Westerbeek, D., 104, 123, *149*
White, A., 82, 84, *147*
White, J. C. D., 119, 124, *144*, *150*
Whitney, R. McL., 66, 67, 68, 69, *144*, *150*
Wiechelman, K. J., 175, 191, 194, 199, 202, 242, 246, *306*, *312*
Wiechen, A., 106, 107, 123, 129, *147*
Wieczorek, A., 165, *312*
Wiersma, A. K., 138, 139, *148*
Wildhaber, I., 29, 30, *57*, *59*
Wiley, J. S., 199, 202, 242, *304*
Wilkansky *Volcani*, B., 2, *62*
Williams, R. J. P., 92, 129, *143*, *147*
Willis, I. M., 67, 68, 69, 77, 78, *149*
Wilson, D., 179, *311*
Wilson, K., 325, *358*
Winkler, F. K., 20, *60*
Winterhalter, K. H., 215, *308*
Winterhalter, K., 253, 295, *304*
Wishner, B. C., 286, *312*
Witten, T. A., Jr., 141, *150*
Wittmann, H. G., 26, 27, 28, *62*
Woese, C. R., 3, 43, *58*, *62*
Wolfe, R. S., 43, *58*
Wooding, F. B. P., 83, *150*
Woodrow, G. V., 283, *312*
Wootton, J. F., 275, 276, 278, *308*
Wozniak, J. A., 325, *358*
Wright, P. E., 200, *305*, 355, *359*
Wu, S.-S., 164, 199, *310*
Wüthrich, K., 178, 184, 185, 237, 239, 253, 260, *311*, *312*

Wyman, J., 155, 158, 218, 234, 253, 295, 296, 300, *304, 309, 312*
Wyman, J., Jr., 154, 155, *306*

X

Xie, D., 356, 357, *361*

Y

Yada, R., 90, *148*
Yaguchi, M., 73, *150*
Yahagi, M., 92, *148*
Yamada, N., 74, 133, *143*
Yamamoto, H., 283, *307, 312*
Yamashita, S., 114, *144, 151*
Yamauchi, K., 87, 92, 93, 101, 125, *145, 146, 148, 151*
Yamone, T., 253, 260, *311*
Yao, C., 186, 227, 228, *312*
Yap, W. T., 295, *310*
Yasui, A., 49, *61*
Yee, J., 51, 52, 53, *61*
Yenyaminov, S. Y., 337, *360*
Yerle, M., 75, *145*
Yokabson, E., 51, *60*
Yonath, A., 26, 27, 28, 39, 40, *58, 61, 62*, 125, *151*
Yonetani, T., 267, 268, 269, 270, 271, 272, 273, 281, 283, 284, 285, 288, 295, 297, *307, 309, 311, 312*
Yoneyama, M., 355, *360*
York, E. J., 325, 332, *361*
Yoshikawa, M., 70, 81, 82, 93, 98, 114, 116, *149, 151*
Yoshimura, M., 75, 88, *151*
Yu, N. T., 165, 216, *309*
Yu-Lee, L.-Y., 75, 76, 88, *146*
Yutani, K., 90, *148*

Z

Zablen, L. B., 43, *58*
Zaccai, G., 5, 8, 17, 18, 30, 31, 34, 36, 37, 38, 39, 40, 41, 49, 55, *57, 60, 62*
Zaugg, R. H., 262, *312*
Zayzsev-Bashan, A., 28, *61*
Zemlin, F., 25, 30, 55, *59*
Zevaco, C., 70, *151*
Zillig, W., 3, 6, 49, 51, 52, 53, *58, 59, 60, 61, 62*
Zulak, I. M., 82, *147*
Zusman, T., 9, 10, 15, 21, 24, 49, 50, *62*, *61*

SUBJECT INDEX

A

Accessible surface area, 319
N-Acetylgalactosaminyltransferase, mammary gland, 81
Adair constants, for Hb A oxygenation with IHP, 248
Affinity chromatography, halophilic enzymes, 11
Aliphatic proton resonances, deoxy-Hb A and oxy-Hb A, 174
Alkyl isocyanides, binding to Hb A
  equilibrium binding parameters, 234–235
  ^{1}H NMR, 229–234
Alpha helices
  α_{S1}-casein, 86
  α_{S2}-casein, 87–88
  κ-caseins, 90–91
Amino acid composition, micellar calcium phosphate, 128
Amino acid sequences
  dihydrofolate reductase (*H. volcanii*), 21–22
  halobacteria, analysis levels, 53
Ammonium sulfate-mediated chromatography
  dihydrofolate reductase (*H. volcanii*), 10
  ferredoxin (*H. marismortui*), 10
  halophilic enzymes, 7–10
Apo-α-lactalbumin, molten globule state, 355–357
Apolar contributions, to thermodynamics of dissolution in water, 322–325
Apolar enthalpy change, 317
Apolar entropy change, 317
Apolar hydrogen, 319
  dissolution thermodynamics, 322

Archaebacteria
  distinguishing features, 3
  halophilic, *see* Halobacteria
Aromatic proton resonances, deoxy-Hb A and oxy-Hb A, 173
Aspartyl residues, and Ca^{2+} binding to caseins, 92
Average excess enthalpy function, 314
Azide, binding to met-Hb, ^{1}H NMR, 236–240
Azidomethemoglobin, ^{1}H NMR
  azide binding to met-Hb, 236
  low-spin ferric hfs proton resonances, 170

B

Bacillus stearothermophilus, ribosomal subunits, 26–28
Bacteriorhodopsin
  in halobacteria, 14, 55
  mutations, 44
  in purple membrane (*H. halobium*), 30–31
Bis-γ-glutamylcysteine reductase
  affinity chromatography, 11
  of halobacteria, 14
Bohr effect
  acid, 155
  alkaline, 154–155
Bonded interfaces, in cooperativity of protein folding/unfolding, 346
Bop$^-$ phenotype (*H. halobium*), 44–45

C

Calcium ion
  accumulation during casein micelle formation, 83–84

and association behavior of caseins, 97–103
binding to caseins, 91–94
Calcium phosphate
  and casein function, 85
  casein interactions, 103–105
  micellar
    amino acid composition, 128
    chemical characterization, 123–129
    physical characterization, 129–131
Carbon dioxide, *in vivo* fixation by halobacteria, 14
Carbon monoxide
  binding to Hb A, ^{1}H NMR, 218–229
  as ligand for hemoglobin, 207–211
Carbon monoxyhemoglobin, ^{1}H NMR
  α and β chains, 207–209
  ligand binding, comparison with HbO_2 A, 218–219
  sample preparation, 187–191
  spectral artifacts, 200
Casein kinases, mammary gland, 80
Casein micelles
  appearance of, 105–110
  and calcium phosphate
    chemical characterization, 123–129
    physical characterization, 129–131
  caseinate calcium and calcium phosphate—calcium fractions, 124–125
  κ-casein location in, 115–119
  chymosin effects, 137
  clusters, fractal dimension of, 140–141
  coagulation, 134–135
  dissociation, 131–133
  formation, 82–85
  glyco-κ-casein distribution, 117–118
  hairy micelle model, 64, 122, 136
  and particle aggregation theory, 140–141
  property variations, 113–115
  renneting reactions, 137–143
  size and size distribution, 111–112
  size variations, 113–115
  stability, 133–143
  steric stabilization, 135–136
  submicelles, 64–65, 107–110
  substructure of, 105–110
  surface area, 115
  surface hot spots, 141
  surface structure, 119–123
  voluminosity, 112–113
  whole, Ca^{2+} binding, 125
Caseins
  association behavior, 94–103
    with calcium, 97–103
    without calcium, 95–97
  binding to thioester-derivatized glass beads, 116–117
  calcium ion binding, 91–94
  and calcium phosphate growth, 85
  calcium phosphate interactions, 103–105
  definition of, 66
  functional domain hypothesis, 75–76
  gene family divergence, 78
  genomic structure, 75–78
  glycosylation, 81–82
  in milk, 71–74
  mRNAs, 75–78
  nomenclature for, 65–66
  phosphorus in, 80–81
  phosphorylation, 80
  secondary processing, 78–85
  secondary structures, 85–86
  signal peptides, 79
  small-angle neutron scattering, 109–110
  synthesis and secretion, 74–75
β-Caseins
  association in presence of Ca^{2+} ions, 102
  Ca^{2+} binding, 92–94
  characterization, 68–69
  in milk, 73
  primary structure and phosphorylation site for β-CN A^2-5P, 69
  secondary structures, 88–90
γ-Caseins
  in milk, 71
  nomenclature for, 68
κ-Caseins
  association in presence of Ca^{2+} ions, 102–103
  characterization, 69–71
  chymosin effects, 69
  κ-CN A-1P
    phosphorylation site, 69

primary structure, 69
coat—core model of, 115
gene ancestry, 78
glycosylation, 81–82
location in micelles, 115–119
in milk, 73
primary structure, 69, 90
α_{S1}-Caseins
  association in presence of Ca^{2+} ions, 97–101
  Ca^{2+} binding, 92–94
  characterization, 66–68
  α_{S1}CN B-8P, primary structure and phosphorylation siter, 67
  in milk, 73
  secondary structures, 86–87
α_{S2}-Caseins
  association in presence of Ca^{2+} ions, 101–102
  characterization, 68
  α_{S2}CN A-11P, primary structure and phosphorylation site, 68
  in milk, 73
  secondary structures, 87–88
Casomorphins, 64
Catalytic properties, dihydrofolate reductase $H.$ $volcanii$, salt concentration effects, 20–24
Cell envelope, halobacteria, 29–30
Chemical shift, proton, standard in ^{1}H NMR, 187
Chromatography, in hydroxylapatite gels, of halophilic enzymes, 11–12
Chromosome mapping, $H.$ $volcanii$, 46
Chymosin, 65
  effects on
    κ-casein, 69
    casein micelles, 69
Circular dichroism
  apo-α-lactalbumin molten globule state, 355–356
  α_{S1}-casein, 86
  α_{S2}-casein, 87–88
  κ-casein, 90–91
  β-casein backbone structure, 88–89
  far-UV, malate dehydrogenase ($H.$ $marismortui$), 18
Coagulation
  casein micelles, 134–135
  milk, ethanol effects, 119–121

Complementary DNA, coding sequences for bovine caseins, 77–78
Complementary regions, in proteins, 343–344
Consensus promoter structures, 51–52
Consensus sequences, signal peptides of Ca^{2+}-sensitive caseins, 79
Contact (Fermi) shifts, 177–178
Contrast variation experiments, in X-ray and neutron scattering, 34, 38
Convergence temperatures
  alternative explanation of, 333
  entropic, 326–327
  of globular proteins, 327–328
  origins of, 329–334
  plots of, 330
Cooperative folding units, 341
  interaction thermodynamics, 342–344
  stability vs. cooperativity, 342
Cosmids
  complementation of mutations in $H.$ $volcanii$ with, 50
  $H.$ $volcanii$ chromosome, 46
COSY technique, 184, 209
Cryoelectron microscopy, casein micelles, 106–107
Cyanide binding, to met-Hb, ^{1}H NMR, 236–240
Cyanomethemoglobin, ^{1}H NMR
  cyanide binding to met-Hb, 236
  low-spin ferric hfs proton resonances, 170
Cyclic dipeptides, dissolution thermodynamics, 319–320

D

Denaturation, see Protein denaturation
Deoxyhemoglobin, ^{1}H NMR
  aliphatic proton resonances, 174
  aromatic proton resonances, 173
  conformational differences with ligated forms, 161
  ferrous hfs proton resonances in D_2O, 171, 204
  hfs and exchangeable proton resonances in H_2O, 171
  hfs exchangeable, $N_\delta H$, of proximal histidyl residues, 168

hfs proton resonances in H_2O, 205
ring-current shifted and hfs proton resonances, 176
sample preparation, 187–191
spectra summary, 300- and 500 MHz, 167
subunit motion to oxy form, 156–158
Differential scanning calorimetry, for thermodynamics of protein structures, 327
Diffusion experiments, with halobacteria, 32–33
Dihydrofolate reductase, *H. volcanii*
catalytic properties, 23–26
crystal structure, 20
isolation of gene coding for, 49–50
primary structure, 21–22
purification by ammonium sulfate-mediated chromatography, 10
site-specific mutagenesis, 22
Dihydrolipoamide dehydrogenase
affinity chromatography, 11
purification, 6
2,3-Diphosphoglycerate, and O_2 binding to Hb, 164–165
Dispersion energy, and casein coagulation, 134–135
Dissolution thermodynamics, for cyclic peptides, 319–320
Disulfide bond pairings, in molten globule state, 356
DLVO theory, and casein micelles, 134
DNA, introduction in halobacterial spheroplasts, 47–48
DNA-dependent RNA polymerase, from halobacteria, 3, 6
2,3-DPG, *see* 2,3-Diphosphoglycerate

E

Electron microscopy
casein micelles, 106–107
H. halobium purple membrane, 30–31
halobacterial surface layers, 29–30
ribosomal subunits of halophilic proteins, 26–28
Elongation factor Tu (*H. marismortui*), 41–42
Enthalpy change, 317
apolar, 317

for myoglobin, 338
polar, 317
Entropic convergence temperature, 326–327
Entropy change
apolar, 317
configurational, for myoglobin, 338
polar, 317
Equilibrium constants, isonitrile binding with Hb A, 234–235
Equilibrium O_2 binding, asymmetrical cyanomet valency hybrid Hbs, 267–268
one cyanomet heme, 268–270
two cyanomet hemes, 270–273
Ethyl methane sulfonate, point mutations induced by, 46–47

F

Fast liquid chromatography, for caseins, 74
Fermi shifts, 177–178
Ferredoxin
H. marismortui, 10, 31
purification by ammonium sulfate-mediated chromatography (*H. marismortui*), 10
Flory–Stockmayer theory, 140
Fluorescence spectra, malate dehydrogenase (*H. marismortui*), 18
Fractal dimension, of rennetting reactions, 140–141
Free energies, Hb ligation intermediates, 281–283
cooperative, 281–283
Freeze-fracturing, casein micelles, 106
Functional domain hypothesis, for caseins, 75–76

G

Galactosyltransferase
and calcium accumulation, 84
mammary gland, 81
Gel-permeation chromatography, halophilic enzymes, 11–12
Gibbs free energy, interactions in partition function, 317–318

Glass beads, covalent linkage of caseins to, 117
Globular proteins
 apolar contribution variation, 332
 buried polar surface per hydrogen bond, 330–331
 calorimetric studies, 327–328
 folding/unfolding thermodynamics, calculations for, 337–338
 hydrogen bonding statistics, 330–331
 stability, general features, 339
Glutamate dehydrogenase
 ammonium sulfate-mediated chromatography, 8
 H. marismortui, affinity chromatography, 11
 large-scale purification, 10
Glutamyl residues, and Ca^{2+} binding to caseins, 92
Glyceraldehyde-3-phosphate dehydrogenase, solution structure, 42
Glyco-κ-casein, micellar distribution, 117–118
Golgi membranes, lactose synthase complex, 84
Group additivity
 and convergence temperatures, 333
 thermodynamics, for protein stability, 319–321
Guinier approximation, 34

H

Halobacteria
 cell envelope, 29–30
 DNA, 44
 enzyme purification methods
  affinity chromatography, 11
  ammonium sulfate-mediated chromatography, 7–10
  chromatography on hydroxylapatite gels, 11–12
  gel-permeation chromatography, 11
 enzymology, 14–15
 gene isolation, 49–50
 gene regulation, 51–53
 genomic organization, 43–46
 induced mutations, 46–49
 initiation of translation, 52–53
 metabolism, 12–14
 mutations, 44
 prototrophic, 46
 ribosomal subunits, 26–28
 RNA polymerases, 50–51
 salt concentration effects
  on catalytic properties, 20–24
  on protein stability, 15–19
 selectable markers for, 48
 solution studies, 31–35
 spheroplasts, DNA introduction, 47–48
 taxonomic relationships, 43–44
 transcript organization and structure, 50–53
Halobacterium cutirubrum
 slg gene, 49
 sod gene, 49
Halobacterium halobium
 gas vesicle protein genes, 49
 genetic analysis, complexity of, 46
 genetic instability, 45–46
 insertion sequences, 44–46
 mutations, 44
 purple membrane, 30–31
 surface glycoprotein gene, 49
Halobacterium marismortui
 ferredoxin from, 31
 ribosomal subunits, 26–28
Haloferax volcanii
 cosmid bank of, 46
 expression vector for, 48–49
 gene coding for dihydrofolate reductase, 49–50
 mutations, complementation with cosmids, 50
 natural genetic transfer system, 47
Halophilic bacteria, *see* Halobacteria
Halophilic enzymes
 affinity chromatography, 11
 ammonium sulfate-mediated chromatography, 7–10
 purification, 5–7
Halophilic proteins
 elongation factor Tu (*H. marismortui*), 41–42, 55
 negative charge distribution, 55
 stabilization, 39–41
 structure

elongation factor Tu, 41–42
ferredoxin, 31
glyceraldehyde-3-phosphate
   dehydrogenase, 42
heme-binding catalase-peroxidase,
   42
malate dehydrogenase, 36–39
purple membrane, 30–31
ribosomal subunits, 26–28
solution studies, 31–35
surface layers, 29–30
thermodynamics and, 35–36
Heat capacity, changes during protein
   folding/unfolding, 321–322
Heme-binding catalase-peroxidase,
   solution structure, 42
Heme–heme communication, through
   switch region, 291–292
Heme pockets, ^{1}H NMR
   in ligated form, 206–214
   in unligated form, 202–206
Hemoglobin A, ^{1}H NMR
   alkyl isocyanide binding, 229–234
   $\alpha_1\beta_2$ subunit interface
      contact area of, 159
      interactions in, 160, 294–295
   α chain
      helix and heme motion in, 163
      heme–heme communication, 291–295
      ligand binding to, 215–218
   β chain
      amino acid side chains around
         proximal histidyl residue, 163
      heme–heme communication, 291–295
      ligand binding to, 215–218
   cooperative oxygenation
      and ligand binding, 158–159
      molecular basis of, 158, 294–295
      molecular code for, 299
      nature of, 154–155
      nonconcerted structural changes
         during, 240–252
      partially ligated species and, 298–299
      role of ^{1}H NMR in, 174–175
      stereochemical mechanism for, 156–158
   deoxy form

conformational differences with
   ligated forms, 161
quaternary structural transition,
   156–158
oxy form, quaternary structural
   transition, 156–158
oxygen binding
   concerted mechanism for, 156
   2,3-diphosphoglycerate effects,
      164–165
   sequential mechanism for, 156,
      165, 300–302
   sigmoidal curve for, 155–156
proton resonances
   assignments, 192–200
   from subunit interfaces, 200–202
   table of, 195–199
quaternary structures, definition of,
   161
and structure–function studies, 165
Hemoglobin($\alpha\beta^{+CN}$)$_2$, ^{1}H NMR, 261–267
Hemoglobin($\alpha^{+CN}\beta$)$_2$, ^{1}H NMR, 261–267
Hemoglobin Chesapeake, ferrous hfs
   proton resonances, 243–244
Hemoglobin J Capetown, ferrous hfs
   proton resonances, 243–244
Hemoglobin Kempsey
   ferrous hfs proton resonances, 243–244
   IHP effects on ferrous hfs proton
      resonances, 242
Hemoglobin M Boston
   ferrous hfs proton resonances in
      D$_2$O, 204
   hfs proton resonances in H$_2$O, 205
Hemoglobin M Iwate, ^{1}H NMR, 260–261
Hemoglobin M Milwaukee, ^{1}H NMR
   ferrous hfs proton resonances in
      D$_2$O, 204
   hfs proton resonances in H$_2$O, 205
   structures and properties, 253–260
Hemoglobins, ^{1}H NMR
   asymmetric valency hybrids, 261–267
   equilibrium O$_2$ binding, 267–268
      tone cyanomet heme, 268–270
      two cyanomet hemes, 270–273
   ligation intermediates, 280–285

M-type, ^{1}H NMR, 253–260
mutant and modified
  nonconcerted structural changes
    during oxygenation, 240–252
  table of, 188–190
  partially ligated species
    structure of, 298–299
    x-ray crystallography, 285–291
  synthetic symmetric valency hybrids,
    ^{1}H NMR, 260–261
  tertiary and quarternary structure,
    effects of salt bridges, 273–280
Hemoglobin Yakima, ferrous hfs
  proton resonances, 243–244
Hemoglobin Ypsilanti, quaternary
  structure, 165–166
Hill coefficients
  for azide binding to met-Hb, 236
  binding of CO to Hb A, 218
  binding of O_2 to Hb A, 218
  for cyanide binding to met-Hb, 236
  for deoxy-Hb Kempsey, 243
  for mutant deoxyhemoglobins, 244
Hot spots, casein micelle surfaces, 141
Hydrogen bonding
  casein micelles, 131–132
  globular proteins, 330–331
  and partition function, 316
  and protein stability, 325–326
Hydrophobic bonding, casein micelles,
  131–132, 135
Hydrophobic effect
  in convergence temperature, 329
  and partition function, 316
  separation from other effects, 320–
    321
Hydrophobic hydration, 329
Hydrophobic interfaces, in
  cooperativity of protein folding/
  unfolding, 344–346
Hydroxylapatite gel chromatography,
  halophilic enzymes, 11–12
Hyperfine-shifted proton resonances
  chemical shifts, 177
  contact (Fermi) shifts, 177–178
  exchangeable, deoxy-Hb A
    in H_2O, 171
    $N_δH$ of proximal histidyl residues,
      166
  ferric, Hb M Milwaukee, deoxy, oxy,
    and carbon monoxy forms, 254–
      256
  ferrous
    deoxy-Hb A in D_2O, 171, 204
    sensitivity to Hb quaternary
      structural state, 240–244
  high-spin ferric, of Met-Hb A, 169
  low-spin ferric
    azidomet-Hb A, 170
    cyanomet-Hb A, 170
  in proton NMR, 177–178
  pseudocontact shifts, 178

I

Infrared spectroscopy
  Ca^{2+} binding to caseins, 93
  $α_{S1}$-casein, 86
Inositol hexaphosphate
  Adair constants for Hb A
    oxygenation, 248
  effects on
    deoxy-Hb Kempsey ferrous hfs
      proton resonances, 242
    O_2 binding to Hb A, 164
Insertion sequences, *H. halobium*, 44–46
Ion-exchange chromatography
  for casein composition, 71
  casein linkage to thioester-derived
    glass beads, 117
Isocitrate dehydrogenase, purification,
  6
Isoelectric precipitation, casein, 134,
  136

J

J-correlated spectroscopy, two-
  dimensional, 184, 209
Jump-and-return pulse sequence, 186

L

Lactose synthase complex, and calcium
  accumulation, 84
Liganded interfaces, in cooperativity of
  protein folding/unfolding, 347
Lipoamide dehydrogenase, *H. halobium*,
  13–14
Lysozyme denaturation, alcohol effects,
  332–333

M

Magnetic field strength, for ^{1}H NMR, 186–187
Malate dehydrogenase
  ammonium sulfate-mediated chromatography, 8
  H. marismortui
    affinity chromatography, 11
    composition of solution particles, 37–38
    dissociation, 18–19
    fluorescence spectra, 18
    gene coding for, 38
    inactivation–reactivation process, 17–18
    irreversible denaturation, 17–18
    sedimentation, 18–19
    solution structure
      methods, 31–35
      models for, 36–39
    UV absorption spectra, 18
    X-ray and neutron scattering curves, 37–38
  large-scale purification, 10
  purification, 6
  stability
    model for, 39–41
    salt concentration effects, 15–19
Mammary gland
  casein kinases, 80
  enzymes for glycosylation of κ-casein, 81
  Golgi membranes, 82–84
Messenger RNA
  caseins, 75–78
  polycistronic, of halobacteria, 50–52
Methemoglobin, ^{1}H NMR
  azide binding to, 236–240
  cyanide binding to, 236–240
  high-spin ferric hfs proton resonances, 169
Mevinolin resistance vectors, 48
Michaelis—Menten kinetics, rennetting reactions, 137–138
Milk, casein concentrations, 71–74
Molten globule intermediates, energetics of, 355–357
Multidimensional NMR spectroscopy, 185

Mutations, *H. halobium*, 44–49
Myoglobin
  configurational entropy change, 338
  cooperativity in folding/unfolding, 352–355
  free energy oif stabilization, 339
  polar enthalpy change, 338

N

NADH dehydrogenase, purification, 6
Neutron diffraction
  H. halobium purple membrane, 30–31
  ribosomal subunits of halophilic proteins, 26–28
Neutron scattering
  contrast variation experiments, 34, 38
  malate dehydrogenase dissociation, 18–19
  malate dehydrogenase (*H. marismortui*), 37–38
  small-angle, casein micelles, 109–110
  and ultracentrifugation, 36
NOESY, *see* Nuclear Overhauser effect and exchange spectroscopy
Novobiocin resistance vectors, 48
Nuclear magnetic resonance spectroscopy
  ^{1}H
    asymmetric valency hybrids, 261–267
    azidomethemoglobin, low-spin ferric hfs, 170
    carbon monoxyhemoglobin
      α and β chains, 209–211
      sample preparation, 187–191
    κ-casein, 91
    casein micelle hairy layer, 122–123
    cyanomethemoglobin, low-spin ferric hfs, 170
    deoxyhemoglobin, 167
      aliphatic proton resonances, 174
      aromatic proton resonances, 173
      2,3-DPG binding, 164
      ferrous hfs proton resonances in D_2O, 171
      hfs and exchangeable proton resonances in H_2O, 171
      hfs exchangeable, $N_\delta H$, of

proximal histidyl residues, 166
ring-current shifted and hfs proton resonances, 176
sample preparation, 187–191
dynamic range problem, 186
heme pockets in ligated form, 206–214
heme pockets in unligated form, 202–206
hemoglobin A
  alkyl isocyanide binding, 229–234
  partially oxygenated species, 240–252
hyperfine-shifted resonances, 177–178
jump-and-return pulse sequence, 186
methemoglobin
  azide binding to, 236
  cyanide binding to, 236
  high-spin ferric hfs, 169
methodology, 166–176
multidimensional, 185
nuclear Overhauser effect, 184–185
optimal magnetic field strength for, 186–187
oxyhemoglobin, 167
  aliphatic proton resonances, 174
  α and β chains, 209–211
  aromatic proton resonances, 173
  2,3-DPG binding, 164
  exchangeable proton resonances, 172
  ring-current shifted proton resonances, 175
  sample preparation, 187–191
proton chemical shift standard, 187
relaxation processes, 179–184
resonance characterization, 175–176
ring-current shifted resonances, 178–179
sample preparation, 187–191
spin–spin coupling, 184
symmetric valency hybrid hemoglobins, 252–253
  M-type hemoglobins, 253–260
  synthetic, 260–261
techniques, 185–187
^{2}H, β-caseins, 88
^{31}P
  $α_{S1}$-caseins, 89
  β-caseins, 88–89
2,3-DPG binding to oxy-Hb A and deoxy-Hb A, 164
Nuclear Overhauser effect, 184–185
Nuclear Overhauser effect and exchange spectroscopy (NOESY), 185

O

Oligonucleotide probes, synthetic, for halobacteria, 49–50
Optical rotatory dispersion
  $α_{S1}$-casein, 86
  $α_{S2}$-casein, 87
  β-casein backbone structure, 88
Ornithine carbamoyltransferase, purification, 7
Oxoacid dehydrogenase complexes, 13
2-Oxoacid:ferredoxin oxidoreductases, of halobacteria, 13
Oxygen, ^{1}H NMR
  binding to Hb A, 218–229
  as ligand for hemoglobin, 211–214
Oxyhemoglobin, ^{1}H NMR
  aliphatic proton resonances, 174
  α and β chains, 211–214
  aromatic proton resonances, 173
  exchangeable proton resonances, 172
  ligand binding, comparison with HbCO A, 218–219
  ring-current shifted proton resonances, 175
  sample preparation, 187–191
  spectra summary, 300- and 500 MHz, 167
  subunit motion from deoxy form to, 156–158

P

Particle aggregation theory, casein micelles and, 140–141
Partition function

energetics, forces required for, 316–318
enthalpic effects, 322–325
  apolar, 322–325
  polar, 325–326
entropic effects, 326–327
group additivity thermodynamics, 319–321
heat capacity changes, 321–322
hierarchical approach to, 340–341
model compound studies, 318–319
protein denaturation studies, 327–334
for protein folding/unfolding, 314–315
schematic of, 336
thermodynamic quantities necessary for, 335
Phosphoglycerate kinase (yeast), cooperativity in folding/unfolding, 348–351
Phosphophoryn, 92
Phosphorus, seryl phosphomonoesters in casein, 80–81
Phosphorylation, caseins, 80
Phosphoseryl residues, and Ca^{2+} binding to caseins, 92
Photoassimilation, of CO_2 by halobacteria, 14
Point mutations, $H.\ halobium$, 46–47
Polar contributions, to protein stability, 325–326
Polar enthalpy change, 317
  myoglobin, 338
Polar entropy change, 317
Promotor structures, consensus, 51–52
Protein denaturation
  alcohol effects, 332–333
  changes in polar/apolar buried surface area, 336–337
  cold, 339, 354
  convergence temperatures, 327–334
  heat, 339, 354
  hydration, 327–329
  irreversible, malate dehydrogenase ($H.\ marismortui$), 17
Protein folding/unfolding
  complementary regions in, 343–344
  cooperative folding units, 341
  cooperativity in

cooperative interactions, 342–344
folding units, 341–342
hydrophobic interfaces, 344–346
liganded interfaces, 347
myoglobin, 352–355
partition function, 314–315
phosphoglycerate kinase (yeast), 348–351
single-domain proteins, 351–352
thermodynamics
  calculation from protein structure, 335–340
  parameters for stability calculation, 335
two-domain proteins, 347–348
Protein stability
  calculation, equations for, 317–318
  and convergence temperatures, 329–334
  vs. cooperativity, 342
  estimation from crystal structure, thermodynamic parameters for, 335
  folding/unfolding partition function, 314–315
  globular proteins, 339
  hydrogen bonding and, 325–326
Protein structure, calculation of folding/unfolding thermodynamics from, 335–340
Proteolysis, and renneting reactions, 139–140
Proton NMR spectroscopy, see Nuclear magnetic resonance spectroscopy, ^1H
Proton resonances, hemoglobin A
  assignment, 192–194, 200
  subunit interfaces, 200–202
  table of, 195–199
Proximal histidyl $N_\delta H$ resonances, 206
Pseudocontact shifts, 178
Purple membrane, $H.\ halobium$, 29–30

Q

Quasi-elastic laser light scattering for, for translational diffusion coefficient, 33–34

R

Relaxation processes
  in proton NMR, 179–184
  total relaxation rate, 181
Renneting reactions, models of, 137–143
Ribosomal subunits, halophilic proteins, 26–28
Ribulose-bisphosphate carboxylase, in halobacteria, 14
Ring-current shifted proton resonances, 178–179
  deoxy-Hb A, 176
  oxy-Hb A, 175
  in proton NMR, 178–179
RNA polymerase, *H. halobium* and *H. morrhuae*, 50–51, 53–54
Runway hypothesis, of calcium binding, 92–93

S

Salt bridges, and Hb tertiary and quarternary structure, 273–280
Salt concentrations
  and catalytic properties of dihydrofolate reductase, 20–24
  and stability of malate dehydrogenase, 15–29
Scanning calorimetry, for protein folding/unfolding partition function, 315
Sedimentation
  experiments with halobacteria, 32–33
  malate dehydrogenase dissociation, 18–19
Selectable markers, for halobacteria, 48
Shine–Dalgarno sequences, halobacterial, 52–53
Sialyltransferase, mammary gland, 81
Signal peptides, caseins, 79
Site-specific mutagenesis, dihydrofolate reductase (*H. volcanii*), 22
Small-angle neutron scattering, casein micellar structure, 109–110
Small-angle scattering, theory of, 31–32
Smoluchowshi equation, for fully renneted micelles, 138–140
Solution studies
  halophilic proteins, 25–26, 31–35
  malate dehydrogenase (*H. marismortui*), 39–41
  thermodynamics of protein–salt–water interactions, 35–36
Spermidine, phosphorylation of casein, 80
Spermine, phosphorylation of casein, 80
Spin–lattice relaxation, 179–184
  paramagnetic contribution, 180–181
Spin–spin coupling, in proton NMR, 184
Spin–spin relaxation, 179–184
  apparent, 182
  paramagnetic contribution, 181
Stability, malate dehydrogenase *H. marismortui*, salt concentration effects, 15–19
Steric stabilization, casein micelles, 135–136
Structure–function studies, Hb A as model for, 165
Submicelles, casein, 64–65, 107–110
Superoxide dismutase, purification (*H. cutirubrum*), 10
Surface layers, halobacteria, 29–30
Surface structure, casein micelles, 119–123
Switch region, for heme–heme communication, 291–292
Synthetic oligonucleotide probes, for halobacteria, 49–50

T

Temperature dependence
  of casein coagulation, 135
  of free energy change, 316–317
  of halophilic protein stabilization, 40
  of hyperfine interactions, 178
Thermodynamics
  of multicomponent systems, 31–32
  protein folding/unfolding
    calculation from protein structure, 335–340
    protein stability and cooperative interactions, 340–347
  protein–salt–water interactions, 35–36

Thermus thermophilus, ribosomal subunits, 26–28
Transcription initiation factors, halobacteria, 52
Transfection, halobacterial phage DNA into spheroplasts (*H. halobium*), 47–48
Translational diffusion coefficient, quasi-elastic laser light scattering for, 33–34
Two-dimensional *J*-correlated spectroscopy, 184, 209
Two-domain proteins, cooperativity in, 347–348

U

UDP-*N*-acetyl-D-galactosamine:κ-casein polypeptide *N*-acetylgalactosaminyltransferase, 82
Ultracentrifugation, and forward scattering, unified approach to, 36
UV absorption spectra, malate dehydrogenase (*H. marismortui*), 18

W

Witten–Sander growth, 141

X

X-ray absorption spectroscopy, micellar calcium phosphate, 130
X-ray crystallography
 for halophilic proteins, 25–26
 Hb partially ligated species, 285–291
 heme pockets in HbO_2 A and HbCO A, 213–214, 218
 mutant hemoglobin Ypsilanti, 165–166
X-ray diffraction
 ferredoxin from *H. marismortui*, 31
 micellar calcium phosphate, 129
 ribosomal subunits of halophilic proteins, 26–28
X-ray irradiation, of *H. halobium*, 47
X-rays, contrast variation experiments, 34, 38
X-ray scattering, malate dehydrogenase (*H. marismortui*), 37–38

ISBN 0-12-034243-X

QD
431
.A3
v.43

PROPERTY OF
NOVO NORDISK
ENTOTECH, INC.